J. O. Bird
B.Sc.(H... ...Learn... ...g.(CEI), M.I...

A. J. C. May
B.A., C.Eng., M.I.M...

Technician mathematics
Levels 4 and 5

Longman London and New York

Longman Group Limited,
Longman House
Burnt Mill, Harlow, Essex, UK

Published in the United States of America
by Longman Inc., New York

© Longman Group Limited 1981

First published 1981

British Library Cataloguing in Publication Data

Bird, J O
    Technician mathematics, Levels 4 and 5.
    — (Longman technician series: mathematics and
    sciences).
    1. Shop mathematics
    I. Title   II. May, A J C
    510'.2 46   TJ1165

    ISBN 0–582–41762–7

Printed in Great Britain by
William Clowes (Beccles) Ltd
Beccles and London

General editor – Mathematics and Sciences

D. R. Browning, B.Sc., F.R.I.C., A.R.I.C.S.
Principal Lecturer and Head of Chemistry, Bristol Polytechnic

Contents

Preface

This textbook is one of a series which deal simply and carefully with the fundamental mathematics essential in the development of technicians.

The material for *Technician Mathematics Levels 4 and 5* has been selected from the Technician Education Council Mathematics bank of objectives U78/911. Twenty-nine general objectives have been selected and each general objective is covered in a separate chapter. The TEC objective bank reference is stated in both the contents list and with the title of each chapter (for example, Chapter 1 has a reference AB 16 and Chapter 17 has a reference CC 21, AB 16 and CC 21 being the general objective reference in the U78/911 bank of objectives).

Since it is recommended by TEC that a unit of design length 60 hours should consist of about 16 general objectives and a half unit consist of about 8 general objectives, it is anticipated that sufficient material is contained within the text to meet the demands of most college-devised units based on U78/911.

Each topic considered in the text is presented in a way that assumes in the reader only the knowledge attained at TEC level 3 Mathematics (syllabus U75/040) or selected objectives from U78/911 and the Calculus (1) level 3 half unit (syllabus TEC U77/403). This practical book contains about 50 illustrations, over 220 detailed worked problems, followed by some 700 further problems with answers. The text need not be confined to students studying on Higher TEC courses for it contains material suitable for study at H.N.C., H.N.D. or for degree courses. There is a companion textbook in this series, *Mathematics for Electrical Technicians Levels 4 and 5*.

The authors would like to thank Mr David Browning, Principal Lecturer, Bristol Polytechnic, for his continued valuable assistance in his capacity as General Editor of the Mathematics and Science Sector of the Longman Technician Series. They would also like to express their appreciation for the friendly co-operation and helpful advice given to them by the publishers.

Thanks are also due to Mrs Elaine Mayo for the excellent typing of the manuscript.

Finally, the authors would like to add a word of thanks to their wives, Elizabeth and Juliet, for their marvellous patience, help and encouragement during the preparation of this series of books.

J. O. BIRD
A. J. C. MAY

Highbury College of Technology
Portsmouth
1981

Acknowledgements

We are grateful to the following for permission to reproduce copyright material:

Biometrika Trustees for the table 'Percentage Points of the χ^2 Distribution' by Catherine M. Thompson from *Biometrika* Vol. 32, 1941; The Literary Executor of the late Sir Ronald A. Fisher, F.R.S., Dr. Frank Yates, F.R.S. and Longman Group Ltd., London, for an adaptation of Table III from *Statistical Tables for Biological, Agricultural and Medical Research* (6th Edition, 1974), previously published by Oliver & Boyd, Edinburgh; Macmillan Administration Ltd., London and Basingstoke, for the table 'Exponential and Hyperbolic Functions' from *Four Figure Mathematical Tables* by F. Castle.

Chapter 1

Solution of equations by iterative methods (AB 16)

1 Introduction

The solution of equations of the form $f(x) = 0$, where the function is linear, can be achieved by applying the basic rules of algebra or trigonometry. Other equations can be solved by applying a formula, as in the case of solving quadratic equations, where the equation $ax^2 + bx + c = 0$ can be solved by applying the formula $x = \dfrac{-b \pm \sqrt{(b^2 - 4ac)}}{2a}$. More difficult functions may be solved graphically, although this can be a long, tedious process if an accurate result is required. A more analytical approach is to use methods of successive approximation, in which a sequence of calculations is repeated as many times as necessary in order to give a result to the required degree of accuracy. For example, the equation $3x = 7$ can be solved as follows:

Try various integer values until the equation is approximately true; in this case, if $x = 2$, $3x = 6$ and if $x = 3$, $3x = 9$. This shows that the root lies between 2 and 3, and since $3x = 6$ is nearer to 7 than $3x = 9$, the root is nearer $x = 2$ than $x = 3$. Now try $x = 2.3$, giving $3x = 6.9$. When $x = 2.4$, $3x$ is equal to 7.2, thus the root is nearer 2.3 than 2.4, and so on. This process can be repeated as many times as it is necessary to give the required degree of accuracy and is called a **method of successive approximation**. Clearly this technique would not be used in the example given but only where straightforward algebraic methods fail. It is particularly suitable for using in cases where the solution is to be determined by a computer.

Several methods of successive approximation may be used to determine the value of the roots of an equation to a specified degree of accuracy, and two of these are introduced in this chapter.

2 An algebraic method of successive approximation

Methods of successive approximation may be used to solve equations of the form $a + bx + cx^2 + dx^3 + \ldots = 0$. Thus, they are suitable for solving not only polynomials of the form shown above, but also for the solution of other equations which can be expressed in polynomial form by means of, say, Maclaurin's series, introduced in Chapter 5. The method relies initially on an approximate value of a root being estimated, called the first approximation, and then a more accurate result being determined by the repetition of some procedure. The more accurately the first approximation is obtained, the less the number of repetitive cycles, called **iterations**, needed to obtain a given degree of accuracy. The first approximation can either be estimated graphically or by a method involving functional notation as shown below.

The approximate value of x at the point where the curve of $f(x) = 0$ crosses the x-axis is used as the first approximation. This occurs when the value of $f(x)$ changes from positive to negative or from negative to positive.

Consider the equation $x^3 - 2x - 5 = 0$.

When $x = 0, f(0) = (0)^3 - 2(0) - 5 = -5$,

when $x = 1, f(1) = (1)^3 - 2(1) - 5 = -6$,

when $x = 2, f(2) = (2)^3 - 2(2) - 5 = -1$, and

when $x = 3, f(3) = (3)^3 - 2(3) - 5 = 16$.

Since the sign of $f(x)$ changes from a negative value at $f(2)$ to a positive value at $f(3)$, then the first approximation is between $x = 2$ and $x = 3$. If a straight line is drawn between co-ordinates $(2, -1)$ and $(3, 16)$, it will cut the x-axis very near to 2. So a **first approximation** of x is taken as, say, 2.1.

The repetitive procedure used to determine the value of the root more accurately is shown below.

Second approximation

Let the true value of the root to be $x_1 = 2.1 + \delta_1$.

Substituting $2.1 + \delta_1$ for x in the original equation, $x^3 - 2x - 5 = 0$, gives:

$$(2.1 + \delta_1)^3 - 2(2.1 + \delta_1) - 5 = 0$$

If the first approximation is reasonably accurate, then δ_1 is small and δ_1^2 and δ_1^3 are very small. Using the binomial expansion to obtain the terms of $(2.1 + \delta_1)^3$ and neglecting terms containing δ_1^2 and δ_1^3 gives:

$$(2.1^3 + 3 \times 2.1^2 \times \delta_1 + \ldots) - 2(2.1 + \delta_1) - 5 \simeq 0$$

i.e. $9.261 + 13.23\delta_1 - 4.2 - 2\delta_1 - 5 \simeq 0$

$$\delta_1 \simeq - \frac{0.061}{11.23} \simeq -0.005$$

Thus a second (i.e. better) approximation to the root is $2.1 - 0.005$, that is, 2.095.

Third approximation

Let the true value of the root be $x_2 = 2.095 + \delta_2$.

Substituting $2.095 + \delta_2$ for x in the original equation, gives

$$(2.095 + \delta_2)^3 - 2(2.095 + \delta_2) - 5 = 0$$

Using the binomial series and neglecting terms containing δ_2^2 and δ_2^3 gives:

$(2.095^3 + 3 \times 2.095^2\,\delta_2 + \ldots) - 2\,(2.095 + \delta_2) - 5 \simeq 0$

i.e. $9.195 + 13.167\delta_2 - 4.19 - 2\delta_2 - 5 \simeq 0$

$$\delta_2 \simeq \frac{-0.005}{11.167} \simeq -0.000\,4$$

Thus, a third approximation to the root is $2.095 - 0.000\,4$, that is, $2.094\,6$.

This procedure can be repeated until the value of the required root on two consecutive iterations does not change when expressed to the stipulated degree of accuracy. The three approximations obtained of the root of the equation $x^3 - 2x - 5 = 0$ are: first approximation 2.1, second approximation 2.095 and third approximation $2.094\,6$. Since the value of the root, when expressed correct to four significant figures, does not change for the second and third approximations, then the value of the root is 2.095 correct to four significant figures.

A cubic equation can have one, two or three roots.

In this equation when x is larger than 3, the x^3 term predominates and $f(x)$ becomes large and positive. Hence there are no other roots when x is positive. When x is negative

$f(-1) = (-1)^3 - 2\,(-1) - 5 = -4$

$f(-2) = (-2)^3 - 2\,(-2) - 5 = -9$

As x becomes large and negative, the x^3 term again predominates and $f(x)$ becomes large and negative. Hence $x = 2.095$ correct to four significant figures is the only root of the equation $x^3 - 2x - 5 = 0$.

When solving a polynomial equation by an algebraic successive approximation method, the procedure can be summarised as follows:

(i)   Determine x, the approximate value of the root required, either graphically or by using a functional notation method.

(ii)   Let the true value of the root, x_1, be $(x + \delta_1)$.

(iii)   Estimate the approximate value of δ_1 by using $f(x + \delta_1) = 0$ and neglecting terms containing δ_1^2 and higher powers of δ_1.

(iv)   A better approximation of the root is $(x_1 + \delta_1)$. Repeat for $x_2, x_3, \ldots$ until the value of the root does not change on two consecutive iterations when expressed to the stipulated degree of accuracy.

Worked problems on an algebraic method of successive approximation

Problem 1. The equation $x^3 - 3x^2 + 4x - 7 = 0$ has only one positive root. Determine the value of this root, correct to four significant figures.

$$f(x) = x^3 - 3x^2 + 4x - 7.$$
When $x = 0$, $f(0) = (0)^3 - 3(0)^2 + 4(0) - 7 = -7$,
when $x = 1$, $f(1) = (1)^3 - 3(1)^2 + 4(1) - 7 = -5$,
when $x = 2$, $f(2) = (2)^3 - 3(2)^2 + 4(2) - 7 = -3$,
when $x = 3$, $f(3) = (3)^3 - 3(3)^2 + 4(3) - 7 = 5$

Thus, the value of the positive root is between $x = 2$ and $x = 3$ and dividing this interval in the ratio of 3 to 5 gives a value of the first approximation as $x = 2.4$.

Let the true value of the root x_1 be $(2.4 + \delta_1)$.
To estimate the approximate value of δ_1,

$$(2.4 + \delta_1)^3 - 3(2.4 + \delta_1)^2 + 4(2.4 + \delta_1) - 7 = 0$$

$$2.4^3 + 3 \times 2.4^2 \delta_1 - 3 \times 2.4^2 - 3 \times 2 \times 2.4 \times \delta_1 + 4 \times 2.4 + 4 \times \delta_1 - 7 \approx 0$$

$$13.824 + 17.28\delta_1 - 17.28 - 14.4\delta_1 + 9.6 + 4\delta_1 - 7 \approx 0$$

$$\delta_1 \approx \frac{0.856}{6.88} \approx 0.124$$

A second approximation to the root is $2.4 + 0.124 = 2.524$.

Let the true value of the root x_2 be $(2.524 + \delta_2)$

$$(2.524 + \delta_2)^3 - 3(2.524 + \delta_2)^2 + 4(2.524 + \delta_2) - 7 = 0$$

$$2.524^3 + 3 \times 2.524^2 \times \delta_2 - 3 \times 2.524^2 - 3 \times 2 \times 2.524\, \delta_2 + 4 \times 2.524$$
$$+ 4 \times \delta_2 - 7 \approx 0$$

$$16.079 + 19.112\delta_2 - 19.112 - 15.114\delta_2 + 10.096 + 4\delta_2 - 7 \approx 0$$

$$\delta_2 \approx \frac{-0.063}{7.998} \approx -0.008$$

A third approximation to the root is $2.524 - 0.008$, that is, 2.516, correct to four significant figures.

Let the true value of the root x_3 be $(2.516 + \delta_3)$

$$2.516^3 + 3 \times 2.516^2 \delta_3 - 3 \times 2.516^2 - 6 \times 2.516\delta_3 + 4 \times 2.516 + 4\delta_3$$
$$- 7 \approx 0$$

$$15.926\,9 + 18.990\,8\,\delta_3 - 18.990\,8 - 15.096\delta_3 + 10.064 + 4\delta_3 - 7 \approx 0$$

$$\delta_3 \approx \frac{0.000\,1}{7.894\,8} \approx -0.000\,01$$

A fourth approximation to the root is $2.516 - 0.000\,01$, that is, 2.516 correct to four significant figures.

Since the third approximation is equal to the fourth approximation, correct to the stipulated degree of accuracy, then the required root is **2.516, correct to four significant figures.**

Degree of accuracy of calculations
When determining the solution of equations iteratively, the accuracy of the calculations differs. In general, the accuracy to which the various calculations are done is arrived at intuitively. For example, there is little point in calculating δ_1 and δ_2 in the above problem to an accuracy of more than two or three decimal places. However, as the true value of the root is approached more closely, greater accuracy is needed and it can be seen that δ_3 is calculated

correct to five decimal places. If small errors are introduced, the worst result is that it may take, say, one extra iteration to achieve the stipulated degree of accuracy. Also, since terms containing δ^2 and higher powers of δ are neglected, the values obtained in calculations are only approximate values in any case.

Problem 2. Determine the value of the root of the equation

$x^4 - 3x^2 - 2x + 3 = 0$

in the interval $x = 0$ to $x = 1$, correct to three decimal places.

$f(0) = (0)^4 - 3(0)^2 - 2(0) + 3 = 3$

$f(1) = (1)^4 - 3(1)^2 - 2(1) + 3 = -1$

Let the first approximation be, say $x = 0.8$
Let the true value of the root be $x_1 = (0.8 + \delta_1)$
Substituting $x_1 = (0.8 + \delta_1)$ in the original equation gives

$(0.8 + \delta_1)^4 - 3(0.8 + \delta_1)^2 - 2(0.8 + \delta_1) + 3 = 0$

Using the binomial series and neglecting terms in δ_1^2, δ_1^3 and δ_1^4, gives

$0.8^4 + 4 \times 0.8^3 \delta_1 - 3 \times 0.8^2 - 6 \times 0.8\delta_1 - 2 \times 0.8 - 2\delta_1 + 3 \stackrel{\simeq}{=} 0$

$0.409\ 6 + 2.048\delta_1 - 1.92 - 4.8\delta_1 - 1.6 - 2\delta_1 + 3 \stackrel{\simeq}{=} 0$

i.e. $\delta_1 \stackrel{\simeq}{=} - \dfrac{0.110\ 4}{4.752} \stackrel{\simeq}{=} -0.023$

Hence, the second approximation is $0.8 - 0.023$, i.e. 0.777.
Let the true value of the root be $x_2 = (0.777 + \delta_2)$.
Substituting $x_2 = (0.777 + \delta_2)$ in the original equation, gives

$(0.777 + \delta_2)^4 - 3(0.777 + \delta_2)^2 - 2(0.777 + \delta_2) + 3 = 0$

Using the binomial series and neglecting terms in δ_2^2, δ_2^3 and δ_2^4, gives

$0.777^4 + 4 \times 0.777^3 \delta_2 - 3 \times 0.777^2 - 6 \times 0.777\delta_2 - 2 \times 0.777 - 2\delta_2 + 3 \stackrel{\simeq}{=} 0$

$0.364\ 5 + 1.876\ 4\delta_2 - 1.811\ 2 - 4.662\delta_2 - 1.554 - 2\delta_2 + 3 \stackrel{\simeq}{=} 0$

$\delta_2 \stackrel{\simeq}{=} - \dfrac{0.000\ 7}{4.785\ 6} \stackrel{\simeq}{=} -0.000\ 1$

Hence the third approximation is $0.777 - 0.000\ 1$, i.e. 0.777 correct to three decimal places. Since the second and third approximations are the same to the degree of accuracy required, the root is **0.777, correct to three decimal places.**

Further problems on an algebraic method of successive approximation may be found in Section 4 (Problems 1 to 7), page 10.

3 The Newton-Raphson method

Another iterative method of determining the values of the roots of an equation is the Newton-Raphson method, often referred to as 'Newton's method'. This states:

if r_1 is the approximate value for a real root of the equation $f(x) = 0$, then a closer approximation to the root, r_2, is given by:

$$r_2 = r_1 - \frac{f(r_1)}{f'(r_1)}$$

An explanation of this statement is given below.

Figure 1 shows part of a curve $y = f(x)$ in the region of the root, r, of the equation, shown as point A. Let the first approximation to the root be $(r_1, 0)$

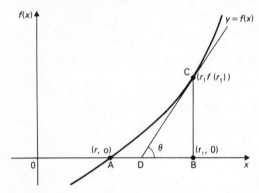

Figure 1

at point B. A vertical line BC is drawn through $(r_1, 0)$ to meet the curve at C, hence the co-ordinates of C are $(r_1, f(r_1))$. A tangent is drawn to the curve at C, cutting the x-axis at D. Let angle CDB be θ.

The slope of CD $= \dfrac{BC}{BD} = \tan\theta$,

thus $\qquad$ BD $= \dfrac{BC}{\tan\theta}$

But BC $= f(r_1)$ and the slope of the tangent drawn to the curve at C is the differential coefficient of $f(x)$ at point r_1, that is $f'(r_1)$.

Therefore, $\quad$ BD $= \dfrac{f(r_1)}{f'(r_1)}$

Now $\qquad$ OD $=$ OB $-$ BD

$$= r_1 - \frac{f(r_1)}{f'(r_1)}$$

It can be seen from Fig. 1 that D is nearer to the root at A than B and hence is a better approximation.

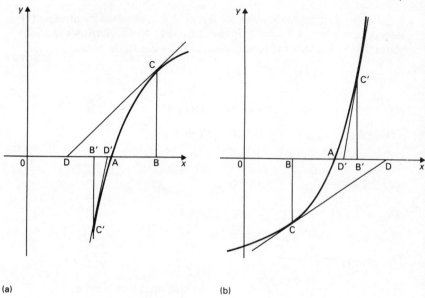

(a)   (b)

Figure 2

It will be found in practice that Newton's method sometimes does not give a closer approximation to the root. The reason for this is illustrated in Fig. 2. Occasions may occur where distance AB is less than AD, depending on the curvature, and the closeness of the first approximation to the true value of the root. In cases where successive approximations obtained by Newton's method are diverging from the root, a new value of r_1 should be chosen so that $f(r_1)$ has the same sign as $f''(r_1)$. With reference to Fig. 2(a), as x increases the slope of the curve, i.e. $f'(x)$ is decreasing, hence $f''(x)$ (the change of $f'(x)$ with respect to x) is negative. If $f(r_1)$ is negative then a first approximation is to the left of point A, say at B'; the tangent is then C' D' and the reasoning becomes similar to that applied to Fig. 1. In Fig. 2(b), as x increases, $f'(x)$ is increasing and $f''(x)$ is positive. If $f(r_1)$ is positive, the first approximation is to the right of point A at say B'; the tangent is then C' D' and again the reasoning becomes similar to that applied to Fig. 1.

Worked problems on the Newton–Raphson method

Problem 1. Use Newton's method to determine the root of the equation

$$x^4 - 11x - 5 = 0$$

in the range $x = 2$ to $x = 3$, correct to four significant figures.

$f(x) = x^4 - 11x - 5$
$f(2) = (2)^4 - 11(2) - 5 = -11$
$f(3) = (3)^4 - 11(3) - 5 = 43$

Let the first approximation be, say, $r_1 = 2.2$, obtained by dividing the interval $x = 2$ to $x = 3$ roughly in the ratio of 11 to 43. Newton's method states that r_2 is a closer approximation to the value of the root,

where $r_2 = r_1 - \dfrac{f(r_1)}{f'(r_1)}$

$f(x) = x^4 - 11x - 5, f(r_1) = (2.2)^4 - 11\,(2.2) - 5 \simeq -5.774$

$f'(x) = 4x^3 - 11, f'(r_1) = 4(2.2)^3 - 11 \simeq 31.59$

Hence, a second approximation,

$r_2 = 2.2 - \dfrac{(-5.774)}{31.59} \simeq 2.38$

$f(r_2) = (2.38)^4 - 11\,(2.38) - 5 \simeq 0.905\,4$

$f'(r_2) = 4(2.38)^3 - 11 \simeq 42.93$

The third approximation,

$r_3 = 2.38 - \dfrac{0.905\,4}{42.93} = 2.359$, correct to four significant figures.

$f(r_3) = (2.359)^4 - 11\,(2.359) - 5 \simeq 0.018\,9$

$f'(r_3) = 4(2.359)^3 - 11 \simeq 41.51$

The fourth approximation,

$r_4 = 2.359 - \dfrac{0.018\,9}{41.51}$

$\simeq 2.358\,5,$

i.e. 2.359 correct to four significant figures.

Since the third and fourth approximations are equal when expressed to the stipulated degree of accuracy, then the required root is **2.359, correct to four significant figures.**

Problem 2. Use Newton's method to determine the root of the equation

$3 \sin x + 4x - 5 = 0,$

in the range $x = 0$ to $x = 1$, correct to four decimal places.

$f(x) = 3 \sin x + 4x - 5$
$f(0) = 3 \sin (0) + 4(0) - 5 = -5$
$f(1) = 3 \sin 1 + 4(1) - 5 \simeq 1.52$
(Note, sin 1 means the sin of 1 radian)

Dividing the interval 0 to 1 roughly in the ratio of 5 to 1.5 gives a first approximation of $r_1 \simeq 0.75$.

$f(x) = 3 \sin x + 4x - 5, f(r_1) = 3 \sin 0.75 + 4(0.75) - 5 \simeq 0.044\,9$

$f'(x) = 3 \cos x + 4, f'(r_1) = 3 \cos 0.75 + 4 \simeq 6.195$

Hence, a second approximation,

$$r_2 = 0.75 - \frac{0.044\ 9}{6.195} = 0.742\ 8,$$

correct to four decimal places.

$$f(r_2) = 3 \sin 0.742\ 8 + 4\ (0.742\ 8) - 5 \simeq 0.000\ 259$$
$$f'(r_2) = 3 \cos 0.742\ 8 + 4 \simeq 6.210$$

Hence, a third approximation,

$$r_3 = 0.742\ 8 - \frac{0.000\ 259}{6.210}$$

$$= 0.742\ 8, \text{ correct to four decimal places.}$$

Since the second and third approximations are equal when expressed to the stipulated degree of accuracy, then the required root is **0.742 8, correct to four decimal places.**

The solution can be checked by using either a calculator or tables. When $x = 0.742\ 8$, $\sin x = \sin (0.742\ 8 \text{ radians}) = 0.676\ 4$.

Hence $3 \sin x + 4x - 5 = 2.029\ 1 + 2.971\ 2 - 5$

$$= 0.000\ 3$$

i.e. it is very near to zero, hence the answer stated is probably correct to the required accuracy.

Problem 3. The approximate value of the root of the equation

$$0.95\ e^x - 35 \ln x - 3 = 0$$

is $x = 4.0$. Determine the value of the root, correct to four significant figures.

Using functional notation to determine how close the root is to $x = 4.0$:

$$f(4.0) = 0.95e^{4.0} - 35 \ln 4.0 - 3 \simeq 0.348$$
$$f(4.1) = 0.95e^{4.1} - 35 \ln 4.1 - 3 \simeq 4.94$$

The results indicate that the root is slightly less than $x = 4.0$ and lies very near to it. Let the first approximation be, say, $r_1 = 3.99$

$$f(x) = 0.95e^x - 35 \ln x - 3, f(3.99) = 0.95e^{3.99} - 35 \ln 3.99 - 3 \simeq -0.08$$
$$f'(x) = 0.95e^x - \frac{35}{x}, f'(3.99) = 0.95e^{3.99} - \frac{35}{3.99}$$

$$= 42.58, \text{ correct to four significant figures.}$$

Hence, a second approximation is $r_2 = 3.99 - \left(\frac{-0.08}{42.58} \right)$

$$= 3.992$$

correct to four significant figures.

$$f(3.992) = 0.95e^{3.992} - 35 \ln 3.992 - 3 \approx 0.004\,72$$

$$f'(3.992) = 0.95e^{3.992} - \frac{35}{3.992} \approx 42.69$$

Hence, a third approximation, $r_3 = 3.992 - \dfrac{0.004\,72}{42.69}$

$$= 3.992, \text{ correct to four significant figures.}$$

Since the second and third approximations are the same when expressed to the stipulated degree of accuracy, the required root is **3.992, correct to four significant figures.**

Further problems on the Newton–Raphson method may be found in the following Section (4) (Problems 8 to 17).

4 Further problems

Algebraic method of successive approximation

Use an algebraic method of successive approximation to determine the value of the roots for the equations given in problems 1 to 7.

1. $x^3 + x = 1$ in the range $x = 0$ to $x = 1$, correct to four decimal places. [0.682 3]
2. $x^3 + 5x = 11$ in the range $x = 1$ to $x = 2$, correct to four significant figures. [1.511]
3. The two positive roots of the equation

   $$x^3 - 9x + 1 = 0,$$

   correct to four significant figures. [0.111 3, 2.943]
4. The negative root of the equation

   $$x^3 - 6x^2 + 12 = 0,$$

   correct to three decimal places. [−1.284]
5. The root of the equation

   $$x^4 - 3x^2 - 3x + 1 = 0$$

   in the interval $x = 2$ to $x = 3$, correct to two decimal places. [2.05]
6. The two roots of the equation

   $$x^4 - 12x = 5,$$

   correct to three significant figures. [−0.414, 2.41]
7. The four roots of the equation

   $$x^4 - 3x^3 - 4x^2 + 2x + 1 = 0,$$

   correct to three significant figures. [−1.19, −0.337, 0.644, 3.88]

The Newton–Raphson method

8. Determine the value of the root of the equation

$$2 \ln x + x = 2,$$

correct to two decimal places. [1.37]

9. Determine the two roots of the equation

$$x^2 = 4 \cos x$$

correct to three significant figures. [± 1.20]

10. Determine the roots of the equation

$$5 - \theta = 5e^{-\theta},$$

correct to three significant figures. [0, 4.97]

11. The value of the root of the equation

$$2 \tan x - 3 \ln 2x + 4e^{x/3} = 6.5$$

is roughly 1. Determine the value of this root, correct to three decimal places. [0.978]

12. The motion of a particle in an electrostatic field is described by the equation

$$y = x^3 + 3x^2 + 5x - 28$$

When $x = 2$, y is approximately equal to 0. Determine the value of x when $y = 0$, correct to two decimal places, using Newton's method. [1.93]

13. The velocity v mm s^{-1} of a point on an eccentric cam at a certain instant is given by $v = \frac{1}{3} x - \ln x$, where x is the displacement in mm. The velocity becomes zero when the displacement is between 4 mm and 5 mm. Determine this value of displacement for the velocity to be zero, correct to four significant figures. [4.536 mm]

14. The amplitude of the ripple, y, on part of a waveform displayed on a C.R.O. at time t μs is of the form

$$y = (t - \pi/3) + \sin t$$

Determine the time when the amplitude of the ripple is zero, correct to four decimal places. [0.536 3 μs]

15. The solution to a differential equation associated with the path taken by a projectile for which the resistance to motion is proportional to the velocity, is

$$y = 2.5 (e^x - e^{-x}) + x - 25$$

Determine the value of x, correct to two decimal places, for which the value of y is zero. [2.22]

16. An equation found in torsional oscillation problems is

$$\cot C = \frac{C}{2} - \frac{1}{2C}$$

Use Newton's method to determine the first two positive roots of the equation correct to four significant figures. [1.307, 3.673]

17. The critical speeds of oscillation (λ) of a loaded beam are given by the equation

$$\lambda^3 - 3.250\,\lambda^2 + \lambda - 0.063 = 0$$

Determine the value of λ which is approximately equal to 3.0, by Newton's method, correct to four decimal places. [2.9143]

Chapter 2

Partial fractions (BA 13)

1 Introduction

Consider the following addition of algebraic fractions:

$$\frac{1}{x+1} + \frac{1}{x+2} = \frac{(x+2)+(x+1)}{(x+1)(x+2)} = \frac{2x+3}{x^2+3x+2}$$

If we start with the expression $\dfrac{2x+3}{x^2+3x+2}$, and find the fractions whose

sum gives this result, the two fractions obtained (i.e. $\dfrac{1}{x+1}$ and $\dfrac{1}{x+2}$) are

called the **partial fractions** of $\dfrac{2x+3}{x^2+3x+2}$.

    This process of expressing a fraction in terms of simpler fractions, called resolving into partial fractions, is used as a preliminary to integrating certain functions (see Chapter 15, Section 5) and the techniques used are explained by example later in this chapter. However, before attempting to resolve an algebraic expression into partial fractions, the following points must be considered and appreciated:

(a) **The denominator of the algebraic expression** must **factorise.** In the above example the denominator $x^2 + 3x + 2$ factorises as $(x + 1)(x + 2)$.

(b) In the above example, the numerator, $2x + 3$, is said to be of degree one since the highest powered x term is x^1. The denominator, $x^2 + 3x + 2$, is said to be of degree two since the highest powered x term is x^2.
In order to resolve an algebraic expression into partial fractions, the

numerator must **be at least one degree less than the denominator**. When the degree of the numerator is equal to or higher than the degree of the denominator, the denominator must be divided into the numerator until the remainder is of lower degree than the denominator. For example, $\dfrac{x^2 + x - 5}{x^2 - 2x - 3}$ cannot be resolved into partial fractions as it stands, since the numerator and denominator are of the same degree. Dividing $x^2 + x - 5$ by $x^2 - 2x - 3$ gives:

$$
\begin{array}{r}
1 \\
x^2 - 2x - 3 \ \overline{)\ x^2 + x - 5} \\
x^2 - 2x - 3 \\
\hline
3x - 2
\end{array}
$$

Thus: $\dfrac{x^2 + x - 5}{x^2 - 2x - 3} = 1 + \dfrac{3x - 2}{x^2 - 2x - 3}$

Since $x^2 - 2x - 3$ factorises as $(x + 1)(x - 3)$ then $\dfrac{3x - 2}{x^2 - 2x - 3}$ may be resolved into partial fractions (see Type 1, Section 2).

(c) Given an identity such as:

$$5x^2 - 3x + 2 \equiv Ax^2 + Bx + C$$

(note: $\equiv$ means 'identically equal to'), then $A = 5, B = -3$ and $C = 2$, since the identity is true for all values of x and the coefficients of x^n (where $n = 0, 1, 2 \ldots$) on the left-hand side of the identity are equal to the coefficients of x^n on the right-hand side of the identity. Similarly, if $ax^3 + bx^2 - cx + d \equiv 2x^3 + 5x - 7$ then $a = 2, b = 0, c = -5$ and $d = -7$.

2 Type 1. Denominator containing linear factors

The corresponding partial fractions of an algebraic expression $\dfrac{f(x)}{(x - a)(x - b)}$ are of the form $\dfrac{A}{(x - a)} + \dfrac{B}{(x - b)}$, where $f(x)$ is a polynomial of degree less than 2. Similarly, the corresponding partial fractions of $\dfrac{f(x)}{(x + a)(x - b)(x - c)}$ are of the form $\dfrac{A}{(x + a)} + \dfrac{B}{(x - b)} + \dfrac{C}{(x - c)}$, where $f(x)$ is a polynomial of degree less than 3, and so on.

Problem 1. Resolve $\dfrac{x - 8}{x^2 - x - 2}$ into partial fractions.

The denominator factorises as $(x + 1)(x - 2)$ and the numerator is of less degree than the denominator. Thus $\dfrac{x-8}{x^2-x-2}$ may be resolved into partial fractions.

Let $\dfrac{x-8}{x^2-x-2} = \dfrac{x-8}{(x+1)(x-2)} \equiv \dfrac{A}{x+1} + \dfrac{B}{x-2}$ where A and B are constants to be determined.

Adding the two fractions on the right-hand side gives:

$$\dfrac{x-8}{(x+1)(x-2)} \equiv \dfrac{A(x-2)+B(x+1)}{(x+1)(x-2)}.$$

Since the denominators are the same on each side of the identity then the numerators must be equal to each other.

Hence $x-8 \equiv A(x-2)+B(x+1)$

There are two methods whereby A and B may be determined using the properties of identities introduced in Section 1.

Method 1.

Since an identity is true for all real values of x, substitute into the identity a value of x to reduce one of the unknown constants to zero.

Let $x=2$.　　Then　$2-8 = A(0)+B(3)$
　　　　　　　i.e.　　$-6 = 3B$
　　　　　　　　　　　$B = -2$

Let $x=-1$.　Then $-1-8 = A(-3)+B(0)$
　　　　　　　　　　　$-9 = -3A$
　　　　　　　　　　　$A = 3$

Method 2.

Since the coefficients of x^n ($n = 0, 1, 2 \ldots$) on the left-hand side of an identity equal the coefficients of x^n on the right-hand side, equate the respective coefficients on each side of the identity.

Since　$x-8 \equiv A(x-2)+B(x+1)$
then　　$x-8 \equiv Ax-2A+Bx+B = (A+B)x+(-2A+B)$
Thus　　$1 = A+B$ (by equating the coefficients of x)　　　　(1)
and　　　$-8 = -2A+B$ (by equating the constants)　　　　　(2)

Solving the two simultaneous equations gives $A=3$ and $B=-2$ as before.

Thus $\dfrac{x-8}{x^2-x-2} \equiv \dfrac{3}{x+1} + \dfrac{-2}{x-2} \equiv \dfrac{3}{x+1} - \dfrac{2}{x-2}$

　　　It is usually quicker and easier to adopt the first method as far as possible although with other types of partial fractions a combination of the two methods is necessary.

Problem 2. Express $\dfrac{6x^2+7x-25}{(x-1)(x+2)(x-3)}$ in partial fractions.

Let $\dfrac{6x^2 + 7x - 25}{(x-1)(x+2)(x-3)} \equiv \dfrac{A}{(x-1)} + \dfrac{B}{(x+2)} + \dfrac{C}{(x-3)}$

$$\equiv \dfrac{A(x+2)(x-3) + B(x-1)(x-3) + C(x-1)(x+2)}{(x-1)(x+2)(x-3)}$$

Equating the numerators gives:

$6x^2 + 7x - 25 \equiv A(x+2)(x-3) + B(x-1)(x-3) + C(x-1)(x+2)$

Let $x = 1$. Then $6 + 7 - 25 = A(3)(-2) + B(0)(-2) + C(0)(3)$

i.e. $\qquad\qquad\ -12 = -6A$

$\qquad\qquad\qquad A = 2$

Let $x = -2$. Then $6(-2)^2 + 7(-2) - 25 = A(0)(-5) + B(-3)(-5) + C(-3)(0)$

i.e. $\qquad\qquad\qquad -15 = 15B$

$\qquad\qquad\qquad\qquad B = -1$

Let $x = 3$. Then $6(3)^2 + 7(3) - 25 = A(5)(0) + B(2)(0) + C(2)(5)$

i.e $\qquad\qquad\qquad\quad 50 = 10C$

$\qquad\qquad\qquad\qquad C = 5$

Thus: $\dfrac{6x^2 + 7x - 25}{(x-1)(x+2)(x-3)} \equiv \dfrac{2}{(x-1)} - \dfrac{1}{(x+2)} + \dfrac{5}{(x-3)}$

Problem 3. Convert $\dfrac{x^3 - x^2 - 5x}{x^2 - 3x + 2}$ into partial fractions.

The numerator is of higher degree than the denominator, thus dividing gives:

$$
\begin{array}{r}
x + 2 \\
x^2 - 3x + 2 \ \overline{)\ x^3 - x^2 - 5x} \\
\underline{x^3 - 3x^2 + 2x} \\
2x^2 - 7x \\
\underline{2x^2 - 6x + 4} \\
-x - 4
\end{array}
$$

Thus $\dfrac{x^3 - x^2 - 5x}{x^2 - 3x + 2} = x + 2 + \dfrac{-x-4}{x^2 - 3x + 2} = x + 2 - \dfrac{x+4}{(x-1)(x-2)}$

Let $\dfrac{x+4}{(x-1)(x-2)} \equiv \dfrac{A}{x-1} + \dfrac{B}{x-2} \equiv \dfrac{A(x-2) + B(x-1)}{(x-1)(x-2)}$

Equating the numerators gives:

$$x + 4 \equiv A(x-2) + B(x-1)$$

Let $x = 1$. Then $\qquad\quad 5 = -A$

$\qquad\qquad\qquad\qquad A = -5$

Let $x = 2$. Then $\qquad\quad 6 = B$

Thus $\dfrac{x+4}{(x-1)(x-2)} \equiv \dfrac{-5}{(x-1)} + \dfrac{6}{(x-2)}$

Thus $\dfrac{x^3 - x^2 - 5x}{x^2 - 3x + 2} \equiv x + 2 - \dfrac{-5}{(x-1)} + \dfrac{6}{(x-2)}$

$\equiv x + 2 + \dfrac{5}{(x-1)} - \dfrac{6}{(x-2)}$

---

3 Type 2. Denominator containing repeated linear factors

When the denominator of an algebraic expression has a factor $(x - a)^n$ then the corresponding partial fractions are $\dfrac{A}{(x-a)} + \dfrac{B}{(x-a)^2} + \ldots + \dfrac{C}{(x-a)^n}$, since $(x - a)^n$ is assumed to hide the factors $(x - a)^{n-1}, (x - a)^{n-2} \ldots (x - a)$.

Problem 1. Express $\dfrac{x+5}{(x+3)^2}$ in partial fractions.

Let $\dfrac{x+5}{(x+3)^2} \equiv \dfrac{A}{(x+3)} + \dfrac{B}{(x+3)^2} \equiv \dfrac{A(x+3) + B}{(x+3)^2}$

Equating the numerators gives:

$$x + 5 \equiv A(x+3) + B$$

Let $x = -3$. Then $-3 + 5 = A(0) + B$

i.e. $B = 2$

Equating the coefficient of x gives $A = 1$.
(Check by equating constant terms gives: $5 = 3A + B$ which is true when $A = 1$ and $B = 2$.)

Thus $\dfrac{x+5}{(x+3)^2} \equiv \dfrac{1}{(x+3)} + \dfrac{2}{(x+3)^2}$

Problem 2. Resolve $\dfrac{5x^2 - 19x + 3}{(x-2)^2(x+1)}$ into partial fractions.

The given denominator is a combination of the linear factor type and repeated factor type.

Let $\dfrac{5x^2 - 19x + 3}{(x-2)^2(x+1)} \equiv \dfrac{A}{(x-2)} + \dfrac{B}{(x-2)^2} + \dfrac{C}{(x+1)}$

$\equiv \dfrac{A(x-2)(x+1) + B(x+1) + C(x-2)^2}{(x-2)^2(x+1)}$

Equating the numerators gives:

$$5x^2 - 19x + 3 \equiv A(x-2)(x+1) + B(x+1) + C(x-2)^2$$

Let $x = 2$. Then $5(2)^2 - 19(2) + 3 \quad = A(0)(3) + B(3) + C(0)^2$

i.e. $\qquad\qquad\qquad -15 \quad = 3B$

$\qquad\qquad\qquad\quad B \quad = -5$

Let $x = -1$. Then $5(-1)^2 - 19(-1) + 3 = A(-3)(0) + B(0) + C(-3)^2$

i.e. $\qquad\qquad\qquad\qquad 27 = 9C$

$\qquad\qquad\qquad\qquad\quad C = 3$

$$\begin{aligned}
5x^2 - 19x + 3 &\equiv A(x-2)(x+1) + B(x+1) + C(x-2)^2 \\
&\equiv A(x^2 - x - 2) + B(x+1) + C(x^2 - 4x + 4) \\
&\equiv (A+C)x^2 + (-A+B-4C)x + (-2A+B+4C)
\end{aligned}$$

Equating the coefficients of x^2 gives:

$$5 = A + C$$

Since $C = 3$ then $A = 2$

[Check: Equating the coefficients of x gives: $\qquad -19 \quad = -A + B - 4C$.

If $A = 2, B = -5$ and $C = 3$ then $-A + B - 4C \quad = -19 = $ L.H.S.

Equating the constant terms gives: $\qquad 3 \quad = -2A + B + 4C$.

If $A = 2, B = -5$ and $C = 3$ then $-2A + B + 4C \quad = 3 = $ L.H.S.]

Thus $\dfrac{5x^2 - 19x + 3}{(x-2)^2(x+1)} \equiv \dfrac{2}{(x-2)} - \dfrac{5}{(x-2)^2} + \dfrac{3}{(x+1)}$

Problem 3. Convert $\dfrac{2x^2 - 13x + 13}{(x-4)^3}$ into partial fractions.

Let $\dfrac{2x^2 - 13x + 13}{(x-4)^3} \equiv \dfrac{A}{(x-4)} + \dfrac{B}{(x-4)^2} + \dfrac{C}{(x-4)^3}$

$\qquad\qquad\qquad\qquad \equiv \dfrac{A(x-4)^2 + B(x-4) + C}{(x-4)^3}$

Equating the numerators gives:

$$2x^2 - 13x + 13 \equiv A(x-4)^2 + B(x-4) + C$$

Let $x = 4$. Then $2(4)^2 - 13(4) + 13 = A(0)^2 + B(0) + C$

i.e. $\qquad\qquad\qquad\qquad -7 = C$

Also $2x^2 - 13x + 13 \equiv A(x^2 - 8x + 16) + B(x-4) + C$

$\qquad\qquad\qquad\quad \equiv Ax^2 + (-8A + B)x + (16A - 4B + C)$

Equating the coefficients of x^2 gives: $\quad 2 = A$

Equating the coefficients of x gives: $\quad -13 = -8A + B$, from which $B = 3$.

(Check: Equating the constant terms gives: $\quad 13 = 16A - 4B + C$.

If $A = 2, B = 3$ and $C = -7$

then R.H.S. $= 16(2) - 4(3) - 7 = 13 = $ L.H.S.)

Thus: $\dfrac{2x^2 - 13x + 13}{(x-4)^3} \equiv \dfrac{2}{(x-4)} + \dfrac{3}{(x-4)^2} - \dfrac{7}{(x-4)^3}$

---

4 Type 3. Denominator containing a quadratic factor

When the denominator contains a quadratic factor of the form $px^2 + qx + r$ (where p, q and r are constants), which does not factorise without containing surds or imaginary values, then the corresponding partial fraction is of the form $\dfrac{Ax + B}{px^2 + qx + r}$, i.e. the numerator is assumed to be a polynomial of degree one less than the denominator. Hence the corresponding partial fractions of an algebraic expression $\dfrac{f(x)}{(px^2 + qx + r)(x - a)}$ are of the form

$$\dfrac{Ax + B}{px^2 + qx + r} + \dfrac{C}{x - a}.$$

Problem 1. Resolve $\dfrac{8x^2 - 3x + 19}{(x^2 + 3)(x - 1)}$ into partial fractions.

Let $\dfrac{8x^2 - 3x + 19}{(x^2 + 3)(x - 1)} \equiv \dfrac{Ax + B}{(x^2 + 3)} + \dfrac{C}{(x - 1)}$

$$\equiv \dfrac{(Ax + B)(x - 1) + C(x^2 + 3)}{(x^2 + 3)(x - 1)}$$

Equating the numerators gives:

$8x^2 - 3x + 19 \equiv (Ax + B)(x - 1) + C(x^2 + 3)$

Let $x = 1$. Then $8(1)^2 - 3(1) + 19 = (A + B)(0) + C(4)$

i.e. $\qquad\qquad 24 = 4C$

$\qquad\qquad\qquad C = 6$

$8x^2 - 3x + 19 \equiv (Ax + B)(x - 1) + C(x^2 + 3)$

$\equiv Ax^2 - Ax + Bx - B + Cx^2 + 3C$

$\equiv (A + C)x^2 + (-A + B)x + (-B + 3C)$

Equating the coefficients of the x^2 terms gives:

$\qquad\qquad 8 = A + C$

Since $C = 6$, $\qquad A = 2$

Equating the coefficients of the x terms gives:

$\qquad\qquad -3 = -A + B$

Since $A = 2$, $\qquad B = -1$

(Check: Equating the constant terms gives: $19 = -B + 3C$
R.H.S. $= -B + 3C = -(-1) + 3(6) = 19 = $ L.H.S.)

Hence: $\dfrac{8x^2 - 3x + 19}{(x^2 + 3)(x - 1)} \equiv \dfrac{2x - 1}{(x^2 + 3)} + \dfrac{6}{(x - 1)}$

Problem 2. Resolve $\dfrac{2 + x + 6x^2 - 2x^3}{x^2\,(x^2 + 1)}$ into partial fractions.

Terms such as x^2 may be treated as $(x + 0)^2$, i.e. it is a repeated linear factor type.

Let $\dfrac{2 + x + 6x^2 - 2x^3}{x^2\,(x^2 + 1)} \equiv \dfrac{A}{x} + \dfrac{B}{x^2} + \dfrac{Cx + D}{(x^2 + 1)}$

$$\equiv \dfrac{Ax\,(x^2 + 1) + B(x^2 + 1) + (Cx + D)\,x^2}{x^2\,(x^2 + 1)}$$

Equating the numerators gives:

$2 + x + 6x^2 - 2x^3 \quad \equiv \quad Ax(x^2 + 1) + B(x^2 + 1) + (Cx + D)\,x^2$

$$\equiv \quad (A + C)x^3 + (B + D)x^2 + Ax + B$$

Let $x = 0$. Then $2 = B$

Equating the coefficients of x^3 gives:

$-2 = A + C$ (1)

Equating the coefficients of x^2 gives:

$$6 = B + D$$

Since $B = 2$, $D = 4$

Equating the coefficients of x gives:

$$1 = A$$

From equation (1), $C = -3$

(Check: Equating the constant terms: $2 = B$ as before.)

Hence $\dfrac{2 + x + 6x^2 - 2x^3}{x^2\,(x^2 + 1)} \equiv \dfrac{1}{x} + \dfrac{2}{x^2} + \dfrac{4 - 3x}{x^2 + 1}$

5 Summary

Type. Denominator containing	Expression	Form of partial fractions
1. Linear factors	$\dfrac{f(x)}{(x + a)(x + b)(x + c)}$	$\dfrac{A}{(x + a)} + \dfrac{B}{(x + b)} + \dfrac{C}{(x + c)}$
2. Repeated linear factors	$\dfrac{f(x)}{(x - a)^3}$	$\dfrac{A}{(x - a)} + \dfrac{B}{(x - a)^2} + \dfrac{C}{(x - a)^3}$

Type. Denominator containing	Expression	Form of partial fractions
3. Quadratic factors	$\dfrac{f(x)}{(ax^2 + bx + c)\,(x - d)}$	$\dfrac{Ax + B}{(ax^2 + bx + c)} + \dfrac{C}{(x - d)}$
General Example:	$\dfrac{f(x)}{(x^2 + a)(x + b)^2(x + c)}$	$\dfrac{Ax + B}{(x^2 + a)} + \dfrac{C}{(x + b)} + \dfrac{D}{(x + b)^2} + \dfrac{E}{(x + c)}$

In each of the above cases $f(x)$ must be of less degree than the relevant denominator. If it is not, then the denominator must be divided into the numerator. For every possible factor of the denominator there is a corresponding partial fraction.

6 Further Problems

Resolve the following into partial fractions:

1. $\dfrac{8}{x^2 - 4}$ $\left[\dfrac{2}{(x - 2)} - \dfrac{2}{(x + 2)}\right]$

2. $\dfrac{3x + 5}{x^2 + 2x - 3}$ $\left[\dfrac{2}{(x - 1)} + \dfrac{1}{(x + 3)}\right]$

3. $\dfrac{y - 13}{y^2 - y - 6}$ $\left[\dfrac{3}{(y + 2)} + \dfrac{2}{(y - 3)}\right]$

4. $\dfrac{17x^2 - 21x - 6}{x(x + 1)(x - 3)}$ $\left[\dfrac{2}{x} + \dfrac{8}{(x + 1)} + \dfrac{7}{(x - 3)}\right]$

5. $\dfrac{6x^2 + 7x - 49}{(x - 4)(x + 1)(2x - 3)}$ $\left[\dfrac{3}{(x - 4)} - \dfrac{2}{(x + 1)} + \dfrac{4}{(2x - 3)}\right]$

6. $\dfrac{x^2 + 2}{(x + 4)(x - 2)}$ $\left[1 - \dfrac{3}{(x + 4)} + \dfrac{1}{(x - 2)}\right]$

7. $\dfrac{2x^2 + 4x + 19}{2(x - 3)(x + 4)}$ $\left[1 + \dfrac{7}{2(x - 3)} - \dfrac{5}{2(x + 4)}\right]$

8. $\dfrac{2x^3 + 7x^2 - 2x - 27}{(x - 1)(x + 4)}$ $\left[2x + 1 - \dfrac{4}{(x - 1)} + \dfrac{7}{(x + 4)}\right]$

9. $\dfrac{2t - 1}{(t + 1)^2}$ $\left[\dfrac{2}{(t + 1)} - \dfrac{3}{(t + 1)^2}\right]$

10. $\dfrac{8x^2 + 12x - 3}{(x + 2)^3}$ $\left[\dfrac{8}{(x + 2)} - \dfrac{20}{(x + 2)^2} + \dfrac{5}{(x + 2)^3}\right]$

22

11. $\dfrac{6x+1}{(2x+1)^2}$ $\left[\dfrac{3}{(2x+1)} - \dfrac{2}{(2x+1)^2}\right]$

12. $\dfrac{1}{x^2(x+2)}$ $\left[\dfrac{1}{2x^2} - \dfrac{1}{4x} + \dfrac{1}{4(x+2)}\right]$

13. $\dfrac{9x^2 - 73x + 150}{(x-7)(x-3)^2}$ $\left[\dfrac{5}{(x-7)} + \dfrac{4}{(x-3)} - \dfrac{3}{(x-3)^2}\right]$

14. $\dfrac{-(9x^2 + 4x + 4)}{x^2(x^2 - 4)}$ $\left[\dfrac{1}{x} + \dfrac{1}{x^2} + \dfrac{2}{(x+2)} - \dfrac{3}{(x-2)}\right]$

15. $\dfrac{-(a^2 + 5a + 13)}{(a^2 + 5)(a - 2)}$ $\left[\dfrac{2a-1}{(a^2 + 5)} - \dfrac{3}{(a-2)}\right]$

16. $\dfrac{3-x}{(x^2 + 3)(x + 3)}$ $\left[\dfrac{1-x}{2(x^2 + 3)} + \dfrac{1}{2(x+3)}\right]$

17. $\dfrac{12 - 2x - 5x^2}{(x^2 + x + 1)(3 - x)}$ $\left[\dfrac{2x+5}{(x^2 + x + 1)} - \dfrac{3}{(3-x)}\right]$

18. $\dfrac{x^3 + 7x^2 + 8x + 10}{x(x^2 + 2x + 5)}$ $\left[1 + \dfrac{2}{x} + \dfrac{3x-1}{(x^2 + 2x + 5)}\right]$

19. $\dfrac{5x^3 - 3x^2 + 41x - 64}{(x^2 + 6)(x - 1)^2}$ $\left[\dfrac{2-3x}{(x^2 + 6)} + \dfrac{8}{(x-1)} - \dfrac{3}{(x-1)^2}\right]$

20. $\dfrac{6x^3 + 5x^2 + 4x + 3}{(x^2 + x + 1)(x^2 - 1)}$ $\left[\dfrac{2x-1}{(x^2 + x + 1)} + \dfrac{3}{(x-1)} + \dfrac{1}{(x+1)}\right]$

Chapter 3

Maclaurin's and Taylor's Series (BD 16)

1 Maclaurin's series

Conditions

Certain mathematical functions may be represented as a power series, containing terms in ascending powers of the variable. This enables mixed functions containing, say, algebraic, trigonometric and exponential functions to be expressed solely as algebraic functions. In this form, subsequent operations such as differentiation or integration often can be more readily performed. Maclaurin's theorem may be used to express any function, say $f(x)$, as a power series, provided that, at $x = 0$, the three conditions given below are met:

(i) The function is finite, i.e. $f(0) \neq \infty$. For the two functions $f(x) = \sin x$ and $g(x) = \ln x$, $f(0) = \sin 0 = 0$ and $g(0) = \ln 0 = -\infty$, hence $f(x) = \sin x$ meets this condition but $g(x) = \ln x$ does not.

(ii) The derivatives of the function are finite, i.e. $f'(0), f''(0), f'''(0) \ldots \neq \infty$. For the function $f(x) = \sin x$, $f'(0) = \cos 0 = 1$, $f''(0) = -\sin 0 = 0$, $f'''(0) = -\cos(0) = -1$ and so on. For the function $g(x) = \ln x$, $g'(x) = \dfrac{1}{x}$ and $g'(0) = \dfrac{1}{0} = \infty$. Hence the function $f(x) = \sin x$ meets this condition, but again $g(x) = \ln x$ does not meet this condition, and therefore cannot be expressed as a power series by using Maclaurin's theorem.

(iii) The power series arising from the use of Maclaurin's theorem must be convergent, that is, in general, the values of the terms (or groups of terms) must get progressively smaller and the sum of the term reach a limiting value. For example, the series $1 + \dfrac{1}{2} + \dfrac{1}{4} + \dfrac{1}{8} + \ldots$ is convergent, since the values of the terms are getting smaller and the sum of the terms is approaching a limiting value of 2. However, for the series $1 + \frac{1}{2} + \frac{1}{3} + \frac{1}{4} + \frac{1}{5} + \ldots$, groups of terms are getting successively larger, i.e. $(\frac{1}{3} + \frac{1}{4}) > \frac{1}{2}$, $(\frac{1}{5} + \frac{1}{6} + \frac{1}{7} + \frac{1}{8}) > \frac{1}{2}$ and so on, so although the values of the terms are getting smaller, groups of terms are not and this series diverges to an infinite number. A full study of the conditions for a series to be convergent is a topic in its own right but one test which applies to most series is as follows. If two consecutive terms in a series are u_n and u_{n+1}, then for the series to be convergent, the limiting value of $\frac{u_{n+1}}{u_n}$ as $n \to \infty$ must be less than 1, i.e.

$$\lim_{n \to \infty} \left| \frac{u_{n+1}}{u_n} \right| < 1, \text{ for the series to converge,}$$

where the vertical lines indicate 'the modulus of' or 'the positive value' of the quantity. For example, in the series $1 + \frac{1}{2!} + \frac{1}{3!} + \frac{1}{4!} + \ldots$, the third term is $\frac{1}{3!}$, the fourth term is $\frac{1}{4!}$ and the nth term is $\frac{1}{n!}$, i.e. $u_n = \frac{1}{n!}$. The term after the nth term is the $(n + 1)^{th}$ term, hence,

$$u_{n+1} = \frac{1}{(n+1)!}. \text{ Thus } \frac{u_{n+1}}{u_n} = \frac{\frac{1}{(n+1)!}}{\frac{1}{n!}} = \frac{n!}{(n+1)!} = \frac{1}{n+1}.$$

As $n \to \infty$, $\frac{1}{n+1} \to 0$, hence this series is convergent.

Derivation of Maclaurin's theorem

Let the power series for $f(x)$ be:

$$f(x) = a_0 + a_1 x + a_2 x^2 + a_3 x^3 + a_4 x^4 + a_5 x^5 + \ldots, \tag{1}$$

where $a_0, a_1, a_2, \ldots$ are constants.
When $x = 0, f(0) = a_0$, since all terms having a factor of x^n in them become zero. Thus, $a_0 = f(0)$.

Differentiating equation (1) with respect to x, gives:

$$f'(x) = a_1 + 2a_2 x + 3a_3 x^2 + 4a_4 x^3 + 5a_5 x^4 + \ldots \tag{2}$$

When $x = 0, f'(0) = a_1$, i.e. $a_1 = f'(0)$.

Differentiating equation (2) with respect to x, gives:

$$f''(x) = 2a_2 + 3 \cdot 2a_3 x + 4 \cdot 3a_4 x^2 + 5 \cdot 4a_5 x^3 + \ldots \tag{3}$$

When $x = 0, f''(0) = 2a_2 = 2!a_2$, i.e. $a_2 = \frac{f''(0)}{2!}$

Differentiating equation (3) with respect to x, gives:

$$f'''(x) = 3 \cdot 2a_3 + 4 \cdot 3 \cdot 2a_4 x + 5 \cdot 4 \cdot 3a_5 x^2 + \ldots \tag{4}$$

When $x = 0, f'''(0) = 3 \cdot 2a_3 = 3!a_3$, i.e. $a_3 = \dfrac{f'''(0)}{3!}$

The same procedure may be repeated to find $f^{iv}(x), f^{v}(x), \ldots$ giving $f^{iv}(0) = 4!a_4, f^{v}(0) = 5!a_5$ and so on. Thus

$$a_4 = \frac{f^{iv}(0)}{4!}, a_5 = \frac{f^{v}(0)}{5!} \ldots$$

Substituting for $a_0, a_1, a_2, a_3, \ldots$ in equation (1), gives

$$f(x) = f(0) + f'(0)x + \frac{f''(0)}{2!} x^2 + \frac{f'''(0)}{3!} x^3 + \ldots$$

i.e. $f(x) = f(0) + xf'(0) + \dfrac{x^2}{2!} f''(0) + \dfrac{x^3}{3!} f'''(0) + \ldots$ \hfill (5)

This mathematical statement is called **Maclaurin's theorem** or **Maclaurin's series**. The uses of Maclaurin's series include expressing mixed functions, containing some or all of trigonometric, exponential, algebraic and logarithmic terms, in a form where such techniques as Newton's method (see Chapter 1), or approximate integration (see Section 3 of this chapter), can be employed. At a more advanced level, such theories as Webb's buckling of struts are based on an understanding of Maclaurin's series.

Worked problems on Maclaurin's series

Problem 1. Determine the first four terms of the power series for $\sin x$. Hence determine the value of $\sin 0.5$ radians correct to three decimal places. What will be the power series of $\sin 2x$?

The values of $f(0), f'(0), f''(0) \ldots$ in Maclaurin's series are determined as follows:

$$
\begin{array}{llll}
f(x) & = \sin x & f(0) & = \sin 0 = 0 \\
f'(x) & = \cos x & f'(0) & = \cos 0 = 1 \\
f''(x) & = -\sin x & f''(0) & = -\sin 0 = 0 \\
f'''(x) & = -\cos x & f'''(0) & = -\cos 0 = -1 \\
f^{iv}(x) & = \sin x & f^{iv}(0) & = \sin 0 = 0
\end{array}
$$

Since $f^{iv}(x) = f(x)$, then $f^{v}(x) = f'(x)$, $f^{vi}(x) = f''(x)$ and so on. Thus the coefficients are $0, 1, 0, -1, 0, 1, 0, -1, \ldots$. Substituting the values of $f(0)$, $f'(0), f''(0), \ldots$ in equation (5) gives:

$$\sin x = 0 + (1)x + 0 + (-1)\frac{x^3}{3!} + 0 + (1)\frac{x^5}{5!} + 0 + (-1)\frac{x^7}{7!} + \ldots$$

i.e. $\sin x = x - \dfrac{x^3}{3!} + \dfrac{x^5}{5!} - \dfrac{x^7}{7!} + \ldots$

When $x = 0.5$ radians

Sin 0.5 $= 0.5 - \dfrac{0.5^3}{6} + \dfrac{0.5^5}{120} - \dfrac{0.5^7}{5040} + \ldots$

$= 0.500\,00 - 0.020\,83 + 0.000\,26 - 0.000\,00$

$= \textbf{0.479 correct to three decimal places.}$

Since $\sin x = x - \dfrac{x^3}{3!} + \dfrac{x^5}{5!} - \dfrac{x^7}{7!} + \ldots$

then $\sin u = u - \dfrac{u^3}{3!} + \dfrac{u^5}{5!} - \dfrac{u^7}{7!} + \ldots$

Replacing u by $2x$, gives

$\sin 2x \quad = 2x - \dfrac{(2x)^3}{3!} + \dfrac{(2x)^5}{5!} - \dfrac{(2x)^7}{7!} + \ldots$

i.e. $\sin 2x = 2x - \dfrac{4x^3}{3} + \dfrac{4x^5}{15} - \dfrac{8x^7}{315} + \ldots$

Problem 2. Determine the first five terms of the series for $\ln(1 + x)$.

The values of $f(0), f'(0), f''(0), \ldots$ in Maclaurin's series are defined as follows:

$f(x) \quad = \ln(1 + x) \qquad f(0) \quad = \ln(1 + 0) = 0$

$f'(x) \quad = \dfrac{1}{1 + x} \qquad f'(0) \quad = \dfrac{1}{1 + 0} = 1$

$f''(x) \quad = -\dfrac{1}{(1 + x)^2} \quad f''(0) \quad = -\dfrac{1}{(1 + 0)^2} = -1$

$f'''(x) \quad = \dfrac{2}{(1 + x)^3} \qquad f'''(0) \quad = \dfrac{2}{(1 + 0)^3} = 2!$

$f^{iv}(x) \quad = -\dfrac{3 \cdot 2}{(1 + x)^4} \quad f^{iv}(0) \quad = -\dfrac{3!}{(1 + 0)^4} = (-1)(3!)$

$f^{v}(x) \quad = \dfrac{4 \cdot 3 \cdot 2}{(1 + x)^5} \qquad f^{v}(0) \quad = \dfrac{4!}{(1 + 0)^5} = 4!$

Substituting the values of $f(0), f'(0), f''(0), \ldots$ in equation (5), gives:

$\ln(1 + x) \quad = 0 + (1)x + (-1)\dfrac{x^2}{2!} + \dfrac{(2!)x^3}{3!} + \dfrac{(-1)(3!)x^4}{4!} + \dfrac{4!x^5}{5!} + \ldots$

i.e. $\ln(1 + x) = x - \dfrac{x^2}{2} + \dfrac{x^3}{3} - \dfrac{x^4}{4} + \dfrac{x^5}{5} - \ldots$

In this series, $|U_n| = \dfrac{x^n}{n}$, i.e. when n is, say, 4 the term in x^4 is $\dfrac{x^4}{4}$, and so on.

Also $|U_{n+1}| = \dfrac{x^{n+1}}{n+1}$.

Testing for convergence (see page 24)

$$\lim_{n \to \infty} \left| \frac{U_{n+1}}{U_n} \right| = \lim_{n \to \infty} \left| \frac{x^{n+1}}{n+1} \times \frac{n}{x^n} \right|$$

$$= \lim_{n \to \infty} \left| \frac{nx}{n+1} \right|$$

As $n \to \infty$, $n \to n+1$ and $\dfrac{n}{n+1} \to 1$

Hence, $\displaystyle\lim_{n \to \infty} \left| \frac{nx}{n+1} \right| = |x|$

The modulus of x must be less than 1 for the series to converge, hence this series only converges for values of x such that $-1 < x < 1$ and hence is only valid for these values of x.

Problem 3. Use the power series for $\ln \left(\dfrac{1+x}{1-x} \right)$ to determine the value of $\ln 3$ correct to three decimal places.

The power series for $\ln (1 + x)$ is derived in Problem 2 above, and is

$$\ln (1 + x) = x - \frac{x^2}{2} + \frac{x^3}{3} - \frac{x^4}{4} + \frac{x^5}{5} - \cdots$$

Replacing x by $(-x)$ in this power series, gives:

$$\ln (1 + (-x)) = (-x) - \frac{(-x)^2}{2} + \frac{(-x)^3}{3} - \frac{(-x)^4}{4} + \frac{(-x)^5}{5} - \cdots$$

i.e. $\ln (1-x) = -x - \dfrac{x^2}{2} - \dfrac{x^3}{3} - \dfrac{x^4}{4} - \dfrac{x^5}{5} - \cdots$

A law of logarithms states that $\ln \left(\dfrac{A}{B} \right) = \ln A - \ln B$.

Applying this law to $\ln \left(\dfrac{1+x}{1-x} \right)$ gives:

$$\ln\left(\frac{1+x}{1-x}\right) = \ln (1 + x) - \ln (1 - x) = \left(x - \frac{x^2}{2} + \frac{x^3}{3} - \frac{x^4}{4} + \frac{x^5}{5} - \cdots \right)$$

$$- \left(-x - \frac{x^2}{2} - \frac{x^3}{3} - \frac{x^4}{4} - \frac{x^5}{5} - \cdots \right)$$

$$= 2x + \frac{2x^3}{3} + \frac{2x^5}{5} + \frac{2x^7}{7} + \cdots$$

i.e. $\ln \left(\dfrac{1+x}{1-x} \right) = 2 \left(x + \dfrac{x^3}{3} + \dfrac{x^5}{5} + \dfrac{x^7}{7} + \cdots \right)$

The series for $\ln\left(\dfrac{1+x}{1-x}\right)$ is convergent provided $-1 < x < 1$, and is far more suitable than the power series for $\ln(1+x)$ or $\ln(1-x)$ when determining the values of Naperian logarithms of numbers. This is because it converges far more rapidly since its terms contain only odd powers of x and it reaches the term containing, say, x^7 in four terms instead of the seven terms in the expansion of $\ln(1 \pm x)$. Hence less terms have to be calculated to give the value of a Naperian logarithm to a required degree of accuracy; also it may be used to determine the values of Naperian logarithms for numbers equal to or greater than 2.

Since the series for $\ln\left(\dfrac{1+x}{1-x_|}\right)$ is being used to find the value of $\ln 3$, then

$$\frac{1+x}{1-x} = 3, \therefore 1+x = 3-3x$$

i.e. $x = \dfrac{1}{2}$

Thus $\ln\left(\dfrac{1+\frac{1}{2}}{1-\frac{1}{2}}\right) = \ln 3 = 2\left(\dfrac{1}{2} + \dfrac{(\frac{1}{2})^3}{3} + \dfrac{(\frac{1}{2})^5}{5} + \dfrac{(\frac{1}{2})^7}{7} + \dfrac{(\frac{1}{2})^9}{9} + \dots\right)$

$$= 2\,(0.500\,0 + 0.041\,7 + 0.006\,3 + 0.001\,1$$
$$+ 0.000\,2 + \dots)$$

$$= 2\,(0.549\,3) = 1.098\,6$$

i.e. $\qquad\qquad$ **$\ln 3 = 1.099$ correct to three decimal places.**

Problem 4. Use Maclaurin's series to find the expansion of $(3+x)^4$.

The values of $f(0), f'(0), f''(0), \dots$ in Maclaurin's series are determined as follows:

$$\begin{aligned}
f(x) &= (3+x)^4 & f(0) &= 3^4 \\
f'(x) &= 4\,(3+x)^3 & f'(0) &= 4\cdot3^3 \\
f''(x) &= 4\cdot3\,(3+x)^2 & f''(0) &= 4\cdot3\cdot3^2 \\
f'''(x) &= 4\cdot3\cdot2\,(3+x) & f'''(0) &= 4\cdot3\cdot2\cdot3^1 \\
f^{iv}(x) &= 4\cdot3\cdot2\cdot1\,(3+x)^0 & f^{iv}(0) &= 4\cdot3\cdot2\cdot1\cdot3^0
\end{aligned}$$

Substituting the values of $f(0), f'(0), f''(0), \dots$ in equation (5) gives:

$$(3+x)^4 = 3^4 + 4\cdot3^3 x + \frac{4\cdot3\cdot3^2\cdot x^2}{2!} + \frac{4\cdot3\cdot2\cdot3^1\cdot x^3}{3!} + \frac{4\cdot3\cdot2\cdot1\cdot x^4}{4!}$$

i.e. the expansion which would have been obtained by applying the binomial series. Thus

$$(3+x)^4 = 81 + 108x + 54x^2 + 12x^3 + x^4$$

Problem 5. Use Maclaurin's series to determine the terms of the power series for $\ln(1+e^x)$ as far as the term in x^3.

$$f(x) = \ln(1 + e^x) \qquad\qquad f(0) = \ln(1 + 1) = \ln 2$$

$$f'(x) = \frac{e^x}{1 + e^x} \qquad\qquad f'(0) = \frac{1}{1+1} = \frac{1}{2}$$

$$f''(x) = \frac{(1 + e^x) e^x - e^x (e^x)}{(1 + e^x)^2}$$

$$= \frac{e^x (1 + e^x - e^x)}{(1 + e^x)^2} = \frac{e^x}{(1 + e^x)^2} \cdot \qquad f''(0) = \frac{1}{2^2} = \frac{1}{4}$$

$$f'''(x) = \frac{(1 + e^x)^2 e^x - e^x \cdot 2 (1 + e^x) e^x}{(1 + e^x)^4} \cdot \qquad f'''(0) = \frac{2^2 - 2(2)}{2^4} = 0$$

Substituting the values of $f(0), f'(0), f''(0), \ldots$ in equation (5) gives:

$$\ln(1 + e^x) = \ln 2 + \frac{x}{2} + \frac{x^2}{4(2!)} + 0 \cdot x^3$$

i.e. $\ln(1 + e^x) = \ln 2 + \dfrac{x}{2} + \dfrac{x^2}{8}$, as far as the term in x^3

Power series

Many mathematical functions may be determined along similar lines to those shown in the worked problems above. Some of the results which may be obtained by applying Maclaurin's theorem to various functions are listed below.

$$\sin x = x - \frac{x^3}{3!} + \frac{x^5}{5!} - \frac{x^7}{7!} + \ldots$$

$$\cos x = 1 - \frac{x^2}{2!} + \frac{x^4}{4!} - \frac{x^6}{6!} + \ldots$$

$$e^x = 1 + x + \frac{x^2}{2!} + \frac{x^3}{3!} + \ldots$$

$$\sinh x = x + \frac{x^3}{3!} + \frac{x^5}{5!} + \frac{x^7}{7!} + \ldots$$

$$\cosh x = 1 + \frac{x^2}{2!} + \frac{x^4}{4!} + \frac{x^6}{6!} + \ldots$$

$$\ln(1 + x) = x - \frac{x^2}{2!} + \frac{x^3}{3!} - \frac{x^4}{4!} + \ldots \text{(Only valid for } -1 < x < 1)$$

$$(a + x)^n = a^n + na^{n-1}x + \frac{n(n - 1)}{2!} a^{n-2}x^2 + \frac{n(n - 1)(n - 2)}{3!} a^{n-3}x^3 + \ldots$$

(This last power series is called the **binomial series.**)

These results can be used to determine the power series for more complicated functions. For example, since the series for e^x is $1 + x + \dfrac{x^2}{2!} + \dfrac{x^3}{3!} + \ldots$

the series for e^{2x} is $1 + (2x) + \dfrac{(2x)^2}{2!} + \dfrac{(2x)^3}{3!} + \ldots$

A power series for the product of two functions can be determined by multiplying the corresponding power series. For example, the power series for $e^x \cos x$ is $\left(1 + x + \dfrac{x^2}{2!}\right)\left(1 - \dfrac{x^2}{2!}\right)$, that is, $1 - \dfrac{x^2}{2!} + x - \dfrac{x^3}{2!} + \dfrac{x^2}{2!}$,

i.e. $1 + x - \dfrac{x^3}{2!}$, as far as the term in x^3. Functions involving quotients may be determined in a similar way.

Further problems on Maclaurin's series may be found in Section 4 (Problems 1 to 10), page 36.

2 Taylor's series

Maclaurin's series, introduced in section 1, expresses the height of a curve at any point x in terms of its height at the origin. With reference to Fig. 1(a),

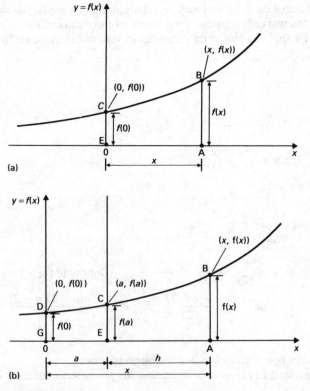

(a)

(b)

Figure 1

the height of the curve at B is AB, that is, $f(x) =$ AB and the height at the origin is EC, that is, $f(0) =$ EC. Maclaurin's series, introduced in equation (5), expresses $f(x)$ in terms of $f(0)$ and is

$$f(x) = f(0) + x f'(0) + \frac{x^2}{2!} f''(0) + \dots$$

Taylor's series expresses the height of a curve at distance $(a + h)$ from the origin in terms of its height at distance a from the origin. With reference to Fig. 1(b), the origin of the vertical axis is moved until it lies along GD, that is, it is moved distance a to the left. Since the shape of ECBA in Fig. 1(b) is the same as in Fig. 1(a), then Maclaurin's series must still relate the height of AB to the height of EC. However, since the origin has been moved, EC is now of height $f(a)$ and distance x in Fig. 1(a) now becomes $(x - a)$ with reference to the new axis. Writing $f(a)$ for $f(0)$ and $(x - a)$ for x in Maclaurin's series gives the height of the curve at AB as:

$$f(x) = f(a) + (x-a)f'(a) + \frac{(x-a)^2}{2!} f''(a) + \frac{(x-a)^3}{3!} f'''(a) + \dots \quad (6)$$

This equation is called **Taylor's theorem** or **Taylor's series**.

If a is made zero in Taylor's series given in equation (6), then:

$$f(x) = f(0) + x f'(0) + \frac{x^2}{2!} f''(0) + \dots,$$

i.e. Maclaurin's series is a special case of Taylor's series when a is equal to 0. Some of the applications of Taylor's series are to such techniques as numerical differentiation, limits, small errors and the numerical solution of certain differential equations.

An alternative way of writing Taylor's series is obtained by writing the series in terms of h rather than x. Since, from Fig. 1(b), $x = (a + h)$, and $(x - a) = h$, then substituting for x in equation (6) gives:

$$f(a + h) = f(a) + h f'(a) + \frac{h^2}{2!} f''(a) + \frac{h^3}{3!} f'''(a) + \dots \quad (7)$$

Taylor's series may be used for the following two applications:

(i) To determine the approximate values of functions which are close to known values of that function. For example, the value of cos 60° is known (i.e. cos 60° = 0.5). Taylor's series may be used to determine the value of, say, cos 61°. In the power series for $f(a + h)$, given in equation (7), a is taken as 60°, i.e. $\pi/3$ radians and h is taken as 1°, i.e. $\frac{1 \times \pi}{180}$ radians. These values are then substituted in the power series for cos $(a + h)$.

(ii) Newton's method of finding the approximate value of the roots of an equation is introduced in Chapter 1, and may be derived as follows. Let the equation $f(x) = 0$ have a root which is approximately equal to a. Also, let the true value of the root be $(a + a \text{ small quantity})$, say, $(a + h)$. Then $f(a + h) = 0$. By Taylor's series:

$$f(a + h) = f(a) + h f'(a) + \frac{h^2}{2!} \ f''(c) + \frac{h^3}{3!} \ f'''(a) + \ldots$$

Since $f(a + h) = 0$, then

$$f(a) + h f'(a) + \frac{h^2}{2!} \ f''(a) + \ldots = 0$$

Because h is small, terms containing h^2 and higher powers of h are very small and may be neglected, hence

$$f(a) + h f'(a) \ \simeq \ 0$$

i.e. $\quad h \ \simeq \ \dfrac{-f(a)}{f'(a)}$

Since the true value of the root is at $(a + h)$ and the value obtained for h is only an approximate value, then

$$a + h \ \simeq \ a - \frac{f(a)}{f'(a)}$$

Let this latter approximation to the value of the root be a_1 then

$$a_1 \ = \ a - \frac{f(a)}{f'(a)} \tag{8}$$

This is the basis of Newton's method of determining the approximate value of roots of equations.

Worked problems on Taylor's series

Problem 1. Determine the power series for $\sin (a + h)$ and hence determine the value of $\sin 30° \ 30'$ correct to five decimal places.

Taylor's series for $f(a + h)$ is given in equation (7) and is:

$$f(a + h) = f(a) + h f'(a) + \frac{h^2}{2!} \ f''(a) + \frac{h^3}{3!} \ f'''(a) + \ldots$$

The function f is the sine function in this problem, hence

$f(a + h) = \sin (a + h)$ and $f(a) = \sin a$.

By repeated differentiation, $f'(a) = \cos a$, $f''(a) = -\sin a$, $f'''(a) = -\cos a$, $f^{iv}(a) = \sin a$, and so on.

Substituting for $f(a), f'(a), f''(a), \ldots$ in Taylor's series, gives:

$$\sin (a + h) = \sin a + h \cos a - \frac{h^2}{2!} \ \sin a - \frac{h^3}{3!} \ \cos a + \ldots$$

To evaluate $\sin 30° \ 30'$, radian measure must be used and a is taken as $30°$, i.e. $\pi/6$ radians, and h as $30'$ or $0.5°$, i.e. $\dfrac{0.5 \times \pi}{180}$ radians.

Thus, $\sin 30° 30' = \sin \dfrac{\pi}{6} + \dfrac{0.5\pi}{180} \cos \dfrac{\pi}{6} - \dfrac{1}{2!} \left(\dfrac{0.5\pi}{180} \right)^2 \sin \dfrac{\pi}{6}$

$$- \dfrac{1}{3!} \left(\dfrac{0.5\pi}{180} \right)^3 \cos \pi/6 + \ldots$$

$$= 0.500\,000 + 0.008\,727 \cdot \dfrac{\sqrt{3}}{2} - \dfrac{1}{2} \cdot 0.000\,076 \cdot \dfrac{1}{2}$$

$$- \dfrac{1}{6} \cdot 0.000\,0007 \dfrac{\sqrt{3}}{2}$$

$$= 0.500\,000 + 0.007\,558 - 0.000\,019 - 0.000\,0001,$$

i.e. $\sin 30° 30' = \mathbf{0.507\,54}$, **correct to five decimal places.**

Problem 2. Given that $\cos 60° = 0.5$, determine the value of $\cos 70°$ by using Taylor's series, correct to four decimal places.

Taylor's series for $f(x)$ is given in equation (6) and is:

$$f(x) = f(a) + (x-a)f'(a) + \dfrac{(x-a)^2}{2!}f''(a) + \dfrac{(x-a)^3}{3!}f'''(a) + \ldots$$

With reference to Fig. 1(b), x corresponds to $70°$, i.e. $\dfrac{7\pi}{18}$ radians, a corresponds to $60°$, i.e. $\dfrac{\pi}{3}$ radians, and $(x-a)$ corresponds to $10°$, i.e. $\dfrac{\pi}{18}$ radians. Also, $f(a) = \cos \dfrac{\pi}{3}$, and by repeated differentiation,

$f'(a) = - \sin \dfrac{\pi}{3}, f''(a) = - \cos \dfrac{\pi}{3}, f'''(a) = \sin \dfrac{\pi}{3}$ and so on.

Thus, substituting these values in Taylor's series, gives:

$$f(x) = \cos 70° = \cos \dfrac{\pi}{3} - \left(\dfrac{\pi}{18} \right) \sin \dfrac{\pi}{3} - \dfrac{1}{2!} \left(\dfrac{\pi}{18} \right)^2 \cos \dfrac{\pi}{3}$$

$$+ \dfrac{1}{3!} \left(\dfrac{\pi}{18} \right)^3 \sin \dfrac{\pi}{3} + \dfrac{1}{4!} \left(\dfrac{\pi}{18} \right)^4 \cos \dfrac{\pi}{3} - \ldots$$

$$= 0.500\,00 - 0.151\,15 - 0.007\,62 + 0.000\,77 + 0.000\,02$$

$$= 0.342\,02$$

i.e. $\cos 70° = \mathbf{0.342\,0}$, **correct to four decimal places.**

Problem 3. Determine the value of the positive root of the equation $x^3 - 2x - 5 = 0$, correct to three decimal places, using Newton's method.

The first approximation to the value of the root is found by determining the approximate point where the graph cuts the x-axis. Using functional notation:

$$f(x) = x^3 - 2x - 5$$

$$f(0) = (0)^3 - 2(0) - 5 = -5$$
$$f(1) = (1)^3 - 2(1) - 5 = -6$$
$$f(2) = (2)^3 - 2(2) - 5 = -1$$
$$f(3) = (3)^3 - 2(3) - 5 = +16$$

This shows that $x^3 - 2x - 5$ is very nearly equal to 0 at $x = 2$, i.e. the first approximation is $a = 2$.

Since $f(x) = x^3 - 2x - 5$, $f'(x) = 3x^2 - 2$.

From equation (8), a second approximation is given by

$$a_1 = a - \frac{f(a)}{f'(a)}$$

When $a = 2$, $a_1 = 2 - \frac{(-1)}{3(2)^2 - 2} = 2.1$

The third approximation is given by:

$$a_2 = a_1 - \frac{f(a_1)}{f'(a_1)}$$

When $a_1 = 2.1$, $a_2 = 2.1 - \left(\frac{(2.1)^3 - 2(2.1) - 5}{3(2.1)^2 - 2} \right)$

$$= 2.1 - \frac{0.06}{11.23} = 2.095$$

The fourth approximation is given by:

$$a_3 = a_2 - \frac{f(a_2)}{f'(a_2)}$$

When $a_2 = 2.095$, $a_3 = 2.095 - \left(\frac{(2.095)^3 - 2(2.095) - 5}{3(2.095)^2 - 2} \right)$

$$= 2.095 - \frac{0.005}{11.17}$$

$$= 2.095 - 0.000\,4 = 2.094\,6$$

Since the values of the third and fourth approximations are the same, when expressed to the required degree of accuracy (i.e. three decimal places), then

the positive root of $x^3 - 2x - 5 = 0$ is 2.095, correct to three decimal places.

Further problems on Taylor's series may be found in Section 4 (Problems 11 to 17), page 37.

3 Approximate values of definite integrals by using series expansions

The value of many definite integrals cannot be determined by the various analytical methods introduced at Level 3 and in Chapters 18 and 19. Several methods exist of finding the approximate value of definite integrals. One of these is to express the function as a power series and then to integrate the terms of the power series. This method is particularly suitable for use with functions whose power series converge fairly rapidly, where evaluation of a few terms gives the required degree of accuracy. This method is shown in the worked problems following.

Worked problems on approximate values of definite integrals by using series expansions

---

Problem 1. Determine the value of $\int_{0.1}^{0.5} e^{\sin x} dx$, correct to three significant figures.

The power series for $e^{\sin x}$ is obtained by applying Maclaurin's theorem given in equation (5), i.e.

$$f(x) \;=\; f(0) + f'(0)\,x + f''(0)\frac{x^2}{2!} + f'''(0)\,\frac{x^3}{3!} + \dots$$

$$f(x) \;=\; e^{\sin x}. \text{ Hence, } f(0) = e^{\sin 0} = 1$$

$$f'(x) \;=\; \cos x\,e^{\sin x}. \text{ Hence, } f'(0) = 1 \cdot e^0 = 1$$

$$f''(x) \;=\; \cos^2 x\,e^{\sin x} + (-\sin x)\,e^{\sin x}$$
$$\;=\; e^{\sin x}(\cos^2 x - \sin x). \text{ Hence, } f''(0) = 1(1 - 0) = 1$$

$$f'''(x) \;=\; e^{\sin x}(2\cos x \sin x - \cos x) + \cos x\,e^{\sin x}(\cos^2 x - \sin x)$$
$$\;=\; \cos x\,e^{\sin x}(\sin x - 1 + \cos^2 x). \text{ Hence, } f'''(0) = 1(0 - 1 + 1) = 0$$

Thus, the power series for $e^{\sin x}$ is $1 + x + \dfrac{x^2}{2} + 0 + \dots$

It follows that

$$\int_{0.1}^{0.5} e^{\sin x}\,dx \;\simeq\; \int_{0.1}^{0.5}\left(1 + x + \frac{x^2}{2}\right)\,dx$$

$$\simeq \left[x + \frac{x^2}{2} + \frac{x^3}{6} \right]_{0.1}^{0.5}$$

$$\simeq\; 0.645\,8 - 0.105\,2$$
$$\simeq\; 0.540\,6$$

i.e. $\int_{0.1}^{0.5} e^{\sin x}\,dx = 0.541$, **correct to three significant figures.**

Problem 2. Determine the value of $\int_{0}^{1} \dfrac{\cos 2x}{x^{\frac{1}{3}}}\,dx$, correct to two decimal places.

Maclaurin's theorem is used to express $\cos 2x$ as a power series.

$f(x) = \cos 2x$	$f(0) = \cos 0 = 1$
$f'(x) = -2 \sin 2x$	$f'(0) = -2(0) = 0$
$f''(x) = -4 \cos 2x$	$f''(0) = -4(1) = -4$
$f'''(x) = 8 \sin 2x$	$f'''(0) = 8(0) = 0$
$f^{iv}(x) = 16 \cos 2x$	$f^{iv}(0) = 16(1) = 16$

$f^{v}(x) = -32 \sin 2x$ $\qquad$ $f^{v}(0) = -32\,(0) = 0$

$f^{vi}(x) = -64 \cos 2x$ $\qquad$ $f^{vi}(0) = -64\,(1) = -64$

Hence, $\cos 2x = 1 - \dfrac{4x^2}{2!} + \dfrac{16x^4}{4!} - \dfrac{64x^6}{6!} + \cdots$

$$= 1 - 2x^2 + \frac{2x^4}{3} - \frac{4x^6}{45} + \cdots$$

Thus, $\displaystyle\int_0^1 \frac{\cos 2x}{x^{\frac{1}{3}}}\,dx = \int_0^1 x^{-\frac{1}{3}}(\cos 2x)\,dx \doteq \int_0^1 \left\{ x^{-\frac{1}{3}}\left(1 - 2x^2 + \frac{2x^4}{3} - \frac{4x^6}{45}\right)\right\}dx$

$$\doteq \int_0^1 \left(x^{-\frac{1}{3}} - 2x^{\frac{5}{3}} + \frac{2x^{\frac{11}{3}}}{3} - \frac{4x^{\frac{17}{3}}}{45} \right)dx$$

$$\doteq \left[\frac{3x^{\frac{2}{3}}}{2} - \frac{(2)3x^{\frac{8}{3}}}{8} + \frac{(2)3x^{\frac{14}{3}}}{(3)(14)} - \frac{(4)3x^{\frac{20}{3}}}{(45)(20)} \right]_0^1$$

$$\doteq \left[\frac{3x^{\frac{2}{3}}}{2} - \frac{3x^{\frac{8}{3}}}{4} + \frac{x^{\frac{14}{3}}}{7} - \frac{x^{\frac{20}{3}}}{75} \right]_0^1$$

$$\doteq \left(\frac{3}{2} - \frac{3}{4} + \frac{1}{7} - \frac{1}{75} \right)$$

$$\doteq 0.879\,5$$

i.e. $\displaystyle\int_0^1 x^{-\frac{1}{3}}(\cos 2x)\,dx = 0.88$, **correct to two decimal places**

Further problems on approximate values of definite integrals by series expansions may be found in the following Section (4) (Problems 18 to 27).

4 Further problems

Maclaurin's series

In Problems 1 to 9, use Maclaurin's theorem or the series given on page 29, to determine the power series stated.

1. $\cos x$, as far as the term in x^6. Hence determine the value of $\cos 0.38$ radians correct to six decimal places. $\left[1 - \dfrac{x^2}{2!} + \dfrac{x^4}{4!} - \dfrac{x^6}{6!} \; ; 0.928\,665 \right]$

2. $e^{\frac{x}{2}}$, as far as the term in x^5. $\left[1 + \dfrac{x}{2} + \dfrac{x^2}{8} + \dfrac{x^3}{48} + \dfrac{x^4}{384} + \dfrac{x^5}{3\,840} \right]$

3. $\tan ax$, as far as the term in x^5. $\left[ax + \dfrac{a^3x^3}{3} + \dfrac{2a^5x^5}{15} \right]$

4. sec $\dfrac{x}{2}$, as far as the term in x^4. $\left[1 + \dfrac{x^2}{8} + \dfrac{5x^4}{384}\right]$

5. $\cos^2 2x$, as far as the term in x^6. $\left[1 - 4x^2 + \dfrac{16x^4}{3} - \dfrac{128x^6}{45}\right]$

6. $(1-x)^{-3}$, as far as the term in x^4. $[1 + 3x + 6x^2 + 10x^3 + 15x^4]$

7. $e^{2x} \cos 3x$, as far as the term in x^2. $\left[1 + 2x - \dfrac{5}{2} x^2\right]$

8. $\dfrac{1+x}{(1-x)^3}$, as far as the term in x^3. $[1 + 4x + 9x^2 + 16x^3]$

9. In $(1 + 3x) \sin \dfrac{3x}{2}$, as far as the term in x^4. $\left[\dfrac{9x^2}{2} - \dfrac{27x^3}{4} + \dfrac{189x^4}{16}\right]$

10. By expressing $\ln\left(\dfrac{1+x}{1-x}\right)$ as a power series, determine the value of $\ln 0.8$, correct to five decimal places. $[-0.223\,14]$

Taylor's series

11. Determine the power series for $\cos (a + h)$ and hence determine the value of $\cos 31°$, correct to five decimal places.

$[\cos (a + h) = \cos a - h \sin a - \dfrac{h^2}{2!} \cos a + \dfrac{h^3}{3!} \sin a + \ldots; 0.857\,17]$

12. Find the value of $\tan 31°$ correct to five decimal places by determining the power series for $\tan (a + h)$.

$[\tan (a + h) = \tan a + h \sec^2 a + h^2 \sec^2 a \tan a + \dfrac{h}{3} \sec^2 a$

$(\sec^2 a + 2 \tan^2 a) + \ldots; 0.600\,86]$

13. Given that $\ln 10 = 2.302\,585$, determine the value of $\ln 12$, correct to five decimal places. $[2.484\,91]$

14. Use Taylor's series to determine the value of $\tan 50°$, correct to six significant figures, given that $\tan 40° = 0.839\,100$. $[1.191\,75]$

15. Determine the three roots of $x^3 = 7x + 1$, correct to four significant figures, by Newton's method. $[2.714, -0.143\,3, -2.571]$

16. Find the value of the real root of the equation

$9 - 9\left(\dfrac{7}{6}x - 1\right)^{\frac{8}{7}} - 10x = 0$, correct to three significant figures, by

using Newton's method. $[0.883]$

17. Determine the value of the root of the equation

$1 - \dfrac{x}{6} + \dfrac{x^2}{180} - \dfrac{x^3}{12\,960} = 0,$

which lies near to 8, correct to four significant figures. $[7.815]$

Approximate values of definite integrals by using series expansions

In Problems 18 to 22, determine the values of the definite integrals given, correct to two significant figures.

18. $\int_0^1 x^{\frac{1}{2}} \cos x \, dx$ [0.53]

19. $\int_0^{\frac{1}{2}} \sqrt{(x)} \ln(x+1) \, dx$ [0.061]

20. $\int_0^{\pi/3} (\sec x)^{\frac{1}{2}} \, dx$ [1.2]

21. $\int_1^2 \frac{1}{2} \cos 3\theta \, d\theta$ [−0.070]

22. $\int_{\frac{1}{3}}^1 \sec^2(3x-1) \, dx$ [−0.73]

23. Determine the value of $\int_0^{\frac{1}{2}} x^2 \ln(1 + \sin x) \, dx$, correct to two significant figures.   [0.013]

24. By determining the power series for $\sec \theta$ as far as the term in θ^4, determine the approximate value of

$$\int_0^{\pi/6} \sec \theta \, d\theta$$

$$\left[\int_0^{\pi/6} \left(1 + \frac{\theta^2}{2} + \frac{5}{24} \theta^4 \right) d\theta \simeq 0.667 \text{ correct to three decimal places.} \right]$$

25. Use Maclaurin's theorem to determine the power series for e^{-x^2} as far as the term in x^4 and hence determine the value of $\int_0^1 e^{-x^2} \, dx$, correct to three significant figures.   [0.767]

26. Determine the value of $\int_0^{0.3} \frac{(1-y^2)^{\frac{1}{2}}}{4\sqrt{y}} \, dy$, correct to three significant figures.   [0.534]

27. Use the power series for $\sin \phi$ to determine the value of $\int_0^{\pi/6} \phi^{\frac{1}{2}} \sin \phi \, d\phi$, correct to three decimal places.   [0.077]

Chapter 4

Hyperbolic functions (BA 17)

1 Definitions of hyperbolic functions

Trigonmetric functions are called '**circular functions**' since they arise naturally in connection with the geometry of the circle. There are other functions which are associated with the geometry of the conic section called a **hyperbola** (see page 155) and are thus classified as **hyperbolic functions**. Such functions have several applications, in particular with transmission line theory and with catenary problems, a catenary being a curve formed by a chain or rope of uniform density hanging freely from two fixed points not in the same vertical line.

Six hyperbolic functions are defined below, each of them being closely connected with the six trigonometrical ratios.

By definition:

(a) Hyperbolic sine of x, $\sinh x = \dfrac{e^x - e^{-x}}{2}$ (1)

'$\sinh x$' is often abbreviated to 'sh x' and is pronounced as 'shine x'.

(b) Hyperbolic cosine of x, $\cosh x = \dfrac{e^x + e^{-x}}{2}$ (2)

'$\cosh x$' is often abbreviated to 'ch x' and is pronounced as 'kosh x'.

(c) Hyperbolic tangent of x, $\tanh x = \dfrac{\sinh x}{\cosh x} = \dfrac{e^x - e^{-x}}{e^x + e^{-x}}$ (3)

'$\tanh x$' is often abbreviated to 'th x' and is pronounced as 'than x'.

(d) Hyperbolic cosecant of x, **cosech** $x = \dfrac{1}{\sinh x} = \dfrac{2}{e^x - e^{-x}}$ (4)

'cosech x' is pronounced as 'coshec x'.

(e) Hyperbolic secant of x, **sech** $x = \dfrac{1}{\cosh x} = \dfrac{2}{e^x + e^{-x}}$ (5)

'sech x' is pronounced as 'shec x'.

(f) Hyperbolic contangent of x, **coth** $x = \dfrac{1}{\tanh x} = \dfrac{e^x + e^{-x}}{e^x - e^{-x}}$ (6)

'coth x' is pronounced as 'koth x'.

2 Some properties of hyperbolic functions

(a) If for a function of x, $f(-x) = f(x)$, i.e. $f(x)$ is unchanged when x is replaced by $-x$, then $f(x)$ is called an **even function** of x. Cos $(-x) = \cos x$, thus the cosine function is an even function. Replacing x by $-x$ in the formula for cosh x gives:

$$\cosh(-x) = \frac{e^{-x} + e^{-(-x)}}{2} = \frac{e^{-x} + e^x}{2} = \cosh x$$

Thus the hyperbolic cosine, **cosh** x, **is an even function**. The graph of an even function is symmetrical about the y-axis (see section 4). Since cosh x is an even function then sech $x = \left(\dfrac{1}{\cosh x} \right)$ is an even function also.

(b) If for a function of x, $f(-x) = -f(x)$ then $f(x)$ is called an **odd function** of x. Sin $(-x) = -\sin x$, thus the sine function is an odd function. Replacing x by $-x$ in the formula for sinh x gives:

$$\sinh(-x) = \frac{e^{-x} - e^{-(-x)}}{2} = \frac{e^{-x} - e^x}{2} = -\left(\frac{e^x - e^{-x}}{2} \right) = -\sinh x$$

Thus the hyperbolic sine, **sinh** x, **is an odd function**. The graph of an odd function is symmetrical about the origin (see Section 4).
Since sinh x is an odd function then cosech $x = \left(\dfrac{1}{\sinh x} \right)$ is an odd function also.

(c) Replacing x by $-x$ in the formula for tanh x gives:

$$\tanh(-x) = \frac{e^{-x} - e^{-(-x)}}{e^{-x} + e^{-(-x)}} = \frac{e^{-x} - e^x}{e^{-x} + e^x} = -\left(\frac{e^x - e^{-x}}{e^x + e^{-x}} \right) = -\tanh x.$$

Thus the hyperbolic tangent, **tanh** x, **is an odd function**. It follows that coth $x = \left(\dfrac{1}{\tanh x} \right)$ is also an odd function.

(d) replacing x by 0 in the formula for sinh x gives:

$$\sinh 0 = \frac{e^0 - e^{-0}}{2} = \frac{1-1}{2} = 0$$

(e) Replacing x by 0 in the formula for cosh gives:

$$\cosh 0 = \frac{e^0 + e^{-0}}{2} = \frac{1+1}{2} = 1$$

3 Evaluation of hyperbolic functions

Tables of exponential and hyperbolic functions are readily available and one such table is shown in Table 1, where it is seen that values of sinh x and cosh x may be read directly from an argument of $x = 0$ to $x = 6.0$. (Note that 'argument' means 'the value on which calculation of another quantity depends'.) For example,

sh 0.32 = $0.325\ 5$,
ch 0.80 = $1.337\ 4$,
sh 4.6 = 49.737 and
ch 5.9 = 182.52

Values of hyperbolic functions for values of x greater than 6.0 and for values of x of greater accuracy than two significant figures are evaluated as shown in the following worked problems.

Worked problems on evaluating hyperbolic functions

Problem 1. Evaluate, using tables, sinh 6.3, correct to four significant figures.

$$\sinh 6.3 = \frac{1}{2}(e^{6.3} - e^{-6.3}) = \frac{1}{2}\left[(e^{6.0})(e^{0.3}) - (e^{-6.0})(e^{-0.3})\right]$$

$$= \frac{1}{2}\left[(403.43)(1.349\ 9) - (0.002\ 48)(0.740\ 8)\right]$$

$$= \frac{1}{2}\left[544.59 - 0.001\ 837\right]$$

Hence sinh 6.3 = 272.3 correct to four significant figures.

Problem 2. Using tables, find the value of cosh 2.54 correct to four significant figures.

$$\cosh 2.54 = \frac{1}{2}(e^{2.54} + e^{-2.54}) = \frac{1}{2}\left[(e^{2.5})(e^{0.04}) + (e^{-2.5})(e^{-0.04})\right]$$

$$= \frac{1}{2}\left[(12.182)(1.040\ 8) + (0.082\ 1)(0.960\ 8)\right]$$

Table 1 Exponential and hyperbolic functions

x	e^x	e^{-x}	$\sinh x$	$\cosh x$	x	e^x	e^{-x}	$\sinh x$	$\cosh x$
0.02	1.0202	0.9802	0.0200	1.0002	1.0	2.7183	0.3679	1.1752	1.5431
0.04	1.0408	0.9608	0.0400	1.0008	1.1	3.0042	0.3329	1.3356	1.6685
0.06	1.0618	0.9418	0.0600	1.0018	1.2	3.3201	0.3012	1.5095	1.8107
0.08	1.0833	0.9231	0.0801	1.0032	1.3	3.6693	0.2725	1.6984	1.9709
0.10	1.1052	0.9048	0.1002	1.0050	1.4	4.0552	0.2466	1.9043	2.1509
0.11	1.1163	0.8958	0.1102	1.0061	1.5	4.4817	0.2231	2.1293	2.3524
0.12	1.1275	0.8869	0.1203	1.0072	1.6	4.9530	0.2109	2.3756	2.5775
0.13	1.1388	0.8781	0.1304	1.0085	1.7	5.4739	0.1827	2.6456	2.8283
0.14	1.1503	0.8694	0.1405	1.0098	1.8	6.0497	0.1653	2.9422	3.1075
0.15	1.1618	0.8607	0.1506	1.0113	1.9	6.6859	0.1496	3.2682	3.5177
0.16	1.1735	0.8521	0.1607	1.0128	2.0	7.3891	0.1353	3.6269	3.7622
0.17	1.1853	0.8437	0.1708	1.0145	2.1	8.1662	0.1225	4.0219	4.1443
0.18	1.1972	0.8353	0.1810	1.0162	2.2	9.0250	0.1108	4.4571	4.5679
0.19	1.2092	0.8270	0.1911	1.0181	2.3	9.9742	0.1003	4.9370	5.0372
0.20	1.2214	0.8187	0.2013	1.0201	2.4	11.023	0.0907	5.4662	5.5569
0.21	1.2337	0.8106	0.2115	1.0221	2.5	12.182	0.0821	6.0502	6.1323
0.22	1.2461	0.8025	0.2218	1.0243	2.6	13.464	0.0743	6.6947	6.7690
0.23	1.2586	0.7945	0.2320	1.0266	2.7	14.880	0.0672	7.4063	7.4735
0.24	1.2712	0.7866	0.2423	1.0289	2.8	16.445	0.0608	8.1919	8.2527
0.25	1.2840	0.7788	0.2526	1.0314	3.0	20.085	0.0498	10.018	10.068
0.26	1.2969	0.7711	0.2629	1.0340	3.0	20.085	0.0498	10.018	10.068
0.27	1.3100	0.7634	0.2733	1.0367	3.1	22.198	0.0450	11.076	11.121
0.28	1.3231	0.7558	0.2837	1.0395	3.2	24.532	0.0408	12.246	12.287
0.29	1.3364	0.7483	0.2941	1.0423	3.3	27.113	0.0369	13.538	13.575
0.30	1.3499	0.7408	0.3045	1.0453	3.4	29.964	0.0334	14.965	14.999
0.31	1.3634	0.7335	0.3150	1.0484	3.5	33.115	0.0302	16.543	16.573
0.32	1.3771	0.7261	0.3255	1.0516	3.6	36.598	0.0273	18.285	18.313
0.33	1.3910	0.7189	0.3360	1.0550	3.7	40.447	0.0247	20.211	20.236
0.34	1.4050	0.7118	0.3466	1.0584	3.8	44.701	0.0224	22.339	22.362
0.35	1.4191	0.7047	0.3572	1.0619	3.9	49.402	0.0202	24.691	24.711
0.36	1.4333	0.6977	0.3678	1.0655	4.0	54.598	0.0183	27.290	27.308
0.37	1.4477	0.6907	0.3785	1.0692	4.1	60.340	0.0166	30.162	30.178
0.38	1.4623	0.6839	0.3892	1.0731	4.2	66.686	0.0150	33.336	33.351
0.39	1.4770	0.6771	0.4000	1.0770	4.3	73.700	0.0136	36.843	36.857
0.40	1.4918	0.6703	0.4107	1.0811	4.4	81.451	0.0123	40.719	40.732
0.41	1.5068	0.6636	0.4216	1.0852	4.5	90.017	0.0111	45.003	45.014
0.42	1.5220	0.6570	0.4325	1.0895	4.6	99.484	0.0100	49.737	49.747
0.43	1.5373	0.6505	0.4434	1.0939	4.7	109.95	0.00910	54.969	54.978
0.44	1.5527	0.6440	0.4543	1.0984	4.8	121.51	0.00823	60.751	60.759
0.45	1.5683	0.6376	0.4653	1.1033	4.9	134.29	0.00745	67.141	67.149
0.46	1.5841	0.6313	0.4764	1.1077	5.0	148.41	0.00674	74.203	74.210
0.47	1.6000	0.6250	0.4875	1.1125	5.1	164.02	0.00610	82.008	82.014
0.48	1.6161	0.6188	0.4986	1.1174	5.2	181.27	0.00552	90.633	90.639
0.49	1.6323	0.6126	0.5098	1.1225	5.3	200.34	0.00499	100.17	100.17
0.50	1.6487	0.6065	0.5211	1.1276	5.4	221.41	0.00452	110.70	110.71
0.6	1.8221	0.5488	0.6367	1.1855	5.5	244.69	0.00409	122.34	122.35
0.7	2.0138	0.4966	0.7586	1.2552	5.6	270.43	0.00370	135.21	135.21
0.8	2.2255	0.4493	0.8881	1.3374	5.7	298.87	0.00335	149.43	149.43
0.9	2.4596	0.4066	1.0265	1.4331	5.8	330.30	0.00303	165.15	165.15
					5.9	365.04	0.00274	182.52	182.52
					6.0	403.43	0.00248	201.71	201.72

$$= \frac{1}{2}\left[12.679 + 0.078\,88\right]$$

Hence, cosh 2.54   **= 6.379 correct to four significant figures.**

Problem 3. Evaluate sinh 1.487 2 correct to four significant figures.

$$\sinh 1.487\,2 = \frac{1}{2}\,(e^{1.4872} - e^{-1.4872})$$

$e^{1.4872}$ may be evaluated using a calculator or by the following method (which was explained in *Technician Mathematics level 3* by J.O. Bird and A.J.C. May, Chapter 3):

Let $e^{1.4872} = y$

Taking logarithms to base e of both sides gives:

$$\ln e^{1.4872} = \ln y$$
$$1.487\,2 \ln e = \ln y$$
i.e. $1.487\,2 = \ln y$ (since $\ln e = 1$)

Taking antilogarithms, using four figure tables of hyperbolic or Naperian logarithms,

$$y = 4.425$$
Hence $e^{1.4872} = 4.425$

Similarly, let $e^{-1.4872} = y$
then $-1.487\,2 = \ln y$
$$-1.487\,2 = -2 + 0.512\,8$$
Hence $\bar{2}.512\,8 = \ln y$
$$\bar{2}.512\,8 = \bar{3}.697\,4 + (\bar{2}.512\,8 - \bar{3}.697\,4)$$
Therefore $\ln y = \bar{3}.697\,4 + 0.815\,4$
$$y = (10^{-1})\,(2.26)$$
$$y = 0.226$$

Hence $e^{-1.4872} = 0.226$

$$\sinh 1.487\,2 = \frac{1}{2}\,(4.425 - 0.226) = 2.099\,5$$

= 2.100 correct to four significant figures.

Problem 4. Evaluate correct to four significant figures:

(a) th 0.62   (b) cosech 1.6   (c) sech 0.96   (d) coth 0.44

(a) $\tanh 0.62 = \dfrac{e^{0.62} - e^{-0.62}}{e^{0.62} + e^{-0.62}}$

44

$$e^{0.62} = (e^{0.6})(e^{0.02}) = (1.822\ 1)(1.020\ 2) = 1.858\ 9$$
$$e^{-0.62} = (e^{-0.6})(e^{-0.02}) = (0.548\ 8)(0.980\ 2) = 0.537\ 9$$

Hence $\tanh 0.62 = \dfrac{1.858\ 9 - 0.537\ 9}{1.858\ 9 + 0.537\ 9} = \dfrac{1.321\ 0}{2.396\ 8} = \mathbf{0.551\ 2}$

(b) $\operatorname{cosech} 1.6 = \dfrac{1}{\sinh 1.6} = \dfrac{1}{2.375\ 6}$ (from Table 1) $= \mathbf{0.420\ 9}$

(c) $\operatorname{sech} 0.96 = \dfrac{1}{\cosh 0.96} = \dfrac{2}{e^{0.96} + e^{-0.96}}$

$$e^{0.96} \quad = (e^{0.9})(e^{0.06}) = (2.459\ 6)(1.061\ 8) = 2.611\ 6$$
$$e^{-0.96} \quad = (e^{-0.9})(e^{-0.06}) = (0.406\ 6)(0.941\ 8) = 0.382\ 9$$

Hence $\operatorname{sech} 0.96 = \dfrac{2}{2.611\ 6 + 0.382\ 9} = \dfrac{2}{2.994\ 5} = \mathbf{0.667\ 9}$

(d) $\coth 0.44 = \dfrac{1}{\operatorname{th} 0.44} = \dfrac{\operatorname{ch} 0.44}{\operatorname{sh} 0.44} = \dfrac{1.098\ 4}{0.454\ 3}$ (from Table 1) $= \mathbf{2.418}$

Further problems on evaluating hyperbolic functions may be found in Section 9 (Problems 1 to 14), page 58.

4 Graphs of hyperbolic functions

(a) $y = \cosh x$

A graph of $y = \cosh x$ may be plotted using the values in Table 1, or by taking the average of the graphs of $y = e^x$ and $y = e^{-x}$. Either method will produce the curve shown in Fig. 1.

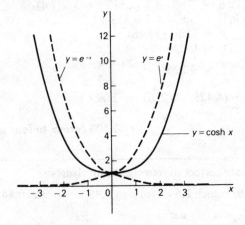

Figure 1 Graphs of $y = e^x$, $y = e^{-x}$ and $y = \cosh x$

It is noted that: (i) ch 0 = 1

                (ii) The graph is symmetrical about the y-axis, thus, cosh x is an even function.

               (iii) As $x \to \infty$, ch $x \to \frac{1}{2} e^x$ (since $e^{-x} \to 0$).

                    and as $x \to -\infty$, ch $x \to \frac{1}{2} e^{-x}$ (since $e^x \to 0$)

The shape of the curve $y = \cosh x$ is that of a heavy rope or chain hanging freely under gravity and is called a **catenary**. Examples include a telegraph wire, a fisherman's line or a transmission line.

(b) $y = \sinh x$

A graph of $y = \sinh x$ may be plotted using the values in Table 1 or by taking half the difference of the graphs $y = e^x$ and $y = e^{-x}$. Either of these methods will produce the curve shown in Fig. 2.

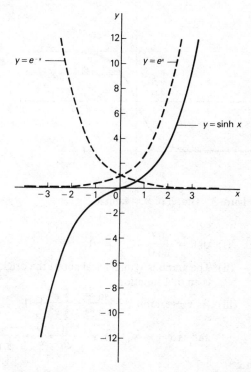

Figure 2   Graphs of $y = e^x$, $y = e^{-x}$ and $y = \sinh x$

It is noted that: (i) sh 0 = 0

                (ii) The graph is symmetrical about the origin, thus sinh x is an odd function.

               (iii) As $x \to \infty$, sh $x \to \frac{1}{2} e^x$ (since $e^{-x} \to 0$)

                    and as $x \to -\infty$, sh $x \to -\frac{1}{2} e^{-x}$ (since $e^x \to 0$)

(c) $y = \tanh x$

Since $\tanh x = \dfrac{\operatorname{sh} x}{\operatorname{ch} x}$, Table 1 may be used to draw up the table of values shown below.

x		-4	-3	-2	-1	0	1	2	3	4
$\operatorname{sh} x$		-27.29	-10.02	-3.63	-1.18	0	1.18	3.63	10.02	27.29
$\operatorname{ch} x$		27.31	10.07	3.76	1.54	1	1.54	3.76	10.07	27.31
$y = \operatorname{th} x = \dfrac{\operatorname{sh} x}{\operatorname{ch} x}$		-0.999	-0.995	-0.97	-0.77	0	0.77	0.97	0.995	0.999

The graph $y = \tanh x$ is shown in Fig. 3.

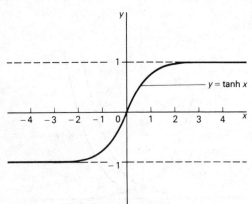

Figure 3 Graph of $y = \tanh x$

It is noted that:

(i) $\operatorname{th} 0 = \dfrac{\operatorname{sh} 0}{\operatorname{ch} 0} = \dfrac{0}{1} = 0$

(ii) The graph is symmetrical about the origin, thus $\tanh x$ is an odd function.

(iii) As $x \to \infty$, $\tanh x \to \dfrac{\operatorname{sh} \infty}{\operatorname{ch} \infty} \to \dfrac{\frac{1}{2} e^x}{\frac{1}{2} e^x} \to 1$,

and as $x \to -\infty$, $\tanh x \to \dfrac{\operatorname{sh} -\infty}{\operatorname{ch} -\infty} \to \dfrac{-\frac{1}{2} e^{-x}}{\frac{1}{2} e^{-x}} \to -1$

(d) $y = \operatorname{cosech} x$

Since $\operatorname{cosech} x = \dfrac{1}{\sinh x}$, Table 1 may be used to draw up the table of values shown below

x		-4	-3	-2	-1	0	1	2	3	4
$\operatorname{sh} x$		-27.29	-10.02	-3.63	-1.18	0	1.18	3.63	10.02	27.29
$\operatorname{cosech} x = \dfrac{1}{\operatorname{sh} x}$		-0.04	-0.10	-0.28	-0.85	$\pm\infty$	0.85	0.28	0.10	0.04

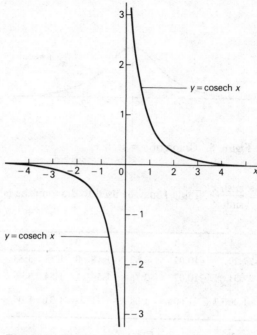

Figure 4 Graph of y = cosech x

The graph $y = $ cosech x is shown in Fig. 4.

It is noted that: (i) cosech $0 = \pm\infty$

                 (ii) The graph is symmetrical about the origin, thus cosech x is an odd function.

                (iii) As $x \to \infty$, cosech $x \to 0$ and as $x \to -\infty$, cosech $x \to 0$.

(e) $y = $ sech x

Since sech $x = \dfrac{1}{\cosh x}$, Table 1 may be used to draw up the table of values shown below.

x	−4	−3	−2	−1	0	1	2	3	4
ch x	27.31	10.07	3.76	1.54	1	1.54	3.76	10.07	27.31
sech $x = \dfrac{1}{\text{ch } x}$	0.04	0.10	0.27	0.65	1	0.65	0.27	0.10	0.04

The graph $y = $ sech x is shown in Fig. 5.

It is noted that: (i) sech $0 = 1$.

                (ii) The graph is symmetrical about the y-axis, thus sech x is an even function.

                (iii) As $x \to \infty$, sech $x \to 0$ and as $x \to -\infty$, sech $x \to 0$.

48

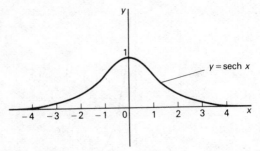

Figure 5 Graph of $y = \text{sech } x$

(f) $y = \coth x$.

Since $\coth x = \dfrac{\cosh x}{\sinh x}$, Table 1 may be used to draw up the table of values shown below.

x	-4	-3	-2	-1	0	1	2	3	4
sh x	-27.29	-10.02	-3.63	-1.18	0	1.18	3.63	10.02	27.29
ch x	27.31	-10.07	3.76	1.54	1	1.54	3.76	10.07	27.31
$\coth x = \dfrac{\text{ch } x}{\text{sh } x}$	-1.000 7	-1.005	-1.04	-1.31	$\pm\infty$	1.31	1.04	1.005	1.000 7

The graph $y = \coth x$ is shown in Fig. 6.

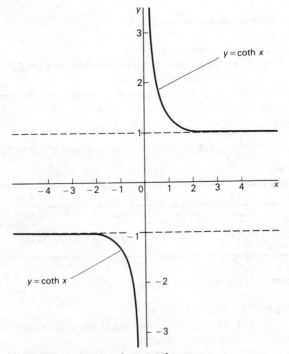

Figure 6 Graph of $y = \coth x$

It is noted that: (i) $\coth 0 = \pm \infty$

               (ii) The graph is symmetrical about the origin, thus $\coth x$ is an odd function.

               (iii) As $x \to \infty$, $\coth x \to 1$ and as $x \to -\infty$, $\coth x \to -1$.

5 Hyperbolic identities – Osborne's rule

$$\cosh x + \sinh x = \left(\frac{e^x + e^{-x}}{2}\right) + \left(\frac{e^x - e^{-x}}{2}\right) = e^x$$

$$\cosh x - \sinh x = \left(\frac{e^x + e^{-x}}{2}\right) - \left(\frac{e^x - e^{-x}}{2}\right) = e^{-x}$$

$$(\text{ch } x + \text{sh } x)(\text{ch } x - \text{sh } x) = (e^x)(e^{-x}) = e^0 = 1$$

i.e. $\text{ch}^2 x - \text{sh}^2 x = 1$ (1)

Dividing each term of equation (1) by $\text{ch}^2 x$ gives:

$$\frac{\text{ch}^2 x}{\text{ch}^2 x} - \frac{\text{sh}^2 x}{\text{ch}^2 x} = \frac{1}{\text{ch}^2 x}$$

i.e. $1 - \text{th}^2 x = \text{sech}^2 x$ (2)

Dividing each term of equation (1) by $\text{sh}^2 x$ gives:

$$\frac{\text{ch}^2 x}{\text{sh}^2 x} - \frac{\text{sh}^2 x}{\text{sh}^2 x} = \frac{1}{\text{sh}^2 x}$$

i.e. $\text{coth}^2 x - 1 = \text{cosech}^2 x$ (3)

Osborne's rule *states that the six trigonometrical ratios used in trigonometrical identities relating general angles may be replaced by their corresponding hyperbolic functions, but the sign of any direct or implied product of two sines must be changed.*

Hence, since $\cos^2 x + \sin^2 x = 1$, by Osborne's rule $\text{ch}^2 x - \text{sh}^2 x = 1$, i.e. the trigonometrical functions have been changed to their corresponding hyperbolic functions and since $\sin^2 x$ is a direct product of two sines the sign is changed from $+$ to $-$.

Similarly, since $1 + \tan^2 x = \sec^2 x$, then $1 - \text{th}^2 x = \text{sech}^2 x$.

In this case $\tan^2 x = \left(\frac{\sin^2 x}{\cos^2 x}\right)$ is considered as an implied product of two sines, hence the sign change.

Also, since $\cot^2 x + 1 = \text{cosec}^2 x$, then, from Osborne's rule,

                       $-\text{coth}^2 x + 1 = -\text{cosech}^2 x$

               i.e. $\text{coth}^2 x - 1 = \text{cosech}^2 x$.

In this case, $\cot^2 x = \frac{\cos^2 x}{\sin^2 x}$ and $\text{cosec}^2 x = \frac{1}{\sin^2 x}$ are both implied products of two sines, hence the sign changes.

Below are listed some trigonometrical identities, with their corresponding hyperbolic identities which may be proved either by replacing sh A by $\left(\dfrac{e^A - e^{-A}}{2}\right)$ and ch A by $\left(\dfrac{e^A + e^{-A}}{2}\right)$ or by using Osborne's rule. Those identities marked with an asterisk involve sign changes due to a product of two sines.

Trigonometrical identities	Corresponding hyperbolic identities	
$\cos^2 A + \sin^2 A = 1$	$\mathrm{ch}^2 A - \mathrm{sh}^2 A = 1$	*
$1 + \tan^2 A = \sec^2 A$	$1 - \tanh^2 A = \mathrm{sech}^2 A$	*
$\cot^2 A + 1 = \mathrm{cosec}^2 A$	$\coth^2 A - 1 = \mathrm{cosech}^2 A$	*

Compound angle addition and subtraction formulae

$\sin (A + B)$ $= \sin A \cos B + \cos A \sin B$	$\mathrm{sh} (A + B)$ $= \mathrm{sh} A \, \mathrm{ch} B + \mathrm{ch} A \, \mathrm{sh} B$	
$\sin (A - B)$ $= \sin A \cos B - \cos A \sin B$	$\mathrm{sh} (A - B)$ $= \mathrm{sh} A \, \mathrm{ch} B - \mathrm{ch} A \, \mathrm{sh} B$	
$\cos (A + B)$ $= \cos A \cos B - \sin A \sin B$	$\mathrm{ch} (A + B)$ $= \mathrm{ch} A \, \mathrm{ch} B + \mathrm{sh} A \, \mathrm{sh} B$	*
$\cos (A - B)$ $= \cos A \cos B + \sin A \sin B$	$\mathrm{ch} (A - B)$ $= \mathrm{ch} A \, \mathrm{ch} B - \mathrm{sh} A \, \mathrm{sh} B$	*

Dougle angles

$\sin 2A = 2 \sin A \cos A$	$\mathrm{sh} \, 2A = 2 \, \mathrm{sh} A \, \mathrm{ch} A$	
$\cos 2A = \cos^2 A - \sin^2 A$	$\mathrm{ch} \, 2A = \mathrm{ch}^2 A + \mathrm{sh}^2 A$	*
$\qquad = 2 \cos^2 A - 1$	$\qquad = 2 \, \mathrm{ch}^2 A - 1$	
$\qquad = 1 - 2 \sin^2 A$	$\qquad = 1 + 2 \, \mathrm{sh}^2 A$	*

Worked problems on hyperbolic identities

Problem 1. Prove that (a) $2 \, \mathrm{ch}^2 \, \theta - 1 \equiv \mathrm{ch} \, 2\theta$.

and (b) $\dfrac{\mathrm{sech} \, x - \mathrm{cosech} \, x}{\mathrm{th} \, x - \coth x} \equiv \mathrm{ch} \, x - \mathrm{sh} \, x$

(a) Left hand side (L.H.S.) $\equiv 2 \, \mathrm{ch}^2 \, \theta - 1$

$\equiv 2 \left(\dfrac{e^\theta + e^{-\theta}}{2} \right)^2 - 1$

$\equiv 2 \left(\dfrac{e^{2\theta} + 2 e^\theta e^{-\theta} + e^{-2\theta}}{4} \right) - 1$

$\equiv \dfrac{e^{2\theta} + 2 + e^{-2\theta}}{2} - 1$

$\equiv \dfrac{e^{2\theta} + e^{-2\theta}}{2} + \dfrac{2}{2} - 1$

$\equiv \dfrac{e^{2\theta} + e^{-2\theta}}{2} \equiv \mathrm{ch} \, 2\theta = \text{R.H.S.}$

(b) L.H.S. $\equiv \dfrac{\text{sech } x - \text{cosech } x}{\text{th } x - \text{coth } x}$

$\equiv \dfrac{\dfrac{1}{\text{ch } x} - \dfrac{1}{\text{sh } x}}{\dfrac{\text{sh } x}{\text{ch } x} - \dfrac{\text{ch } x}{\text{sh } x}} \equiv \dfrac{\dfrac{\text{sh } x - \text{ch } x}{\text{ch } x \text{ sh } x}}{\dfrac{\text{sh}^2 x - \text{ch}^2 x}{\text{ch } x \text{ sh } x}}$

$\equiv \dfrac{\text{sh } x - \text{ch } x}{\text{sh}^2 x - \text{ch}^2 x} \equiv \dfrac{\text{sh } x - \text{ch } x}{-1}$

$\left(\begin{array}{l} \text{since ch}^2 x - \text{sh}^2 x = 1 \\ \text{then sh}^2 x - \text{ch}^2 x = -1 \end{array} \right)$

$\equiv \text{ch } x - \text{sh } x = \text{R.H.S.}$

Problem 2. Prove that $\text{ch } A \text{ ch } B - \text{sh } A \text{ sh } B \equiv \text{ch } (A - B)$

L.H.S. $\equiv \text{ch } A \text{ ch } B - \text{sh } A \text{ sh } B$

$\equiv \left(\dfrac{e^A + e^{-A}}{2} \right) \left(\dfrac{e^B + e^{-B}}{2} \right) - \left(\dfrac{e^A - e^{-A}}{2} \right) \left(\dfrac{e^B - e^{-B}}{2} \right)$

$\equiv \tfrac{1}{4} \left(e^A e^B + e^A e^{-B} + e^{-A} e^B + e^{-A} e^{-B} \right)$

$\qquad\qquad - \tfrac{1}{4} \left(e^A e^B - e^A e^{-B} - e^{-A} e^B + e^{-A} e^{-B} \right)$

$\equiv \tfrac{1}{4} \left(e^{A+B} + e^{A-B} + e^{-A+B} + e^{-A-B} \right.$

$\qquad\qquad \left. - e^{A+B} + e^{A-B} + e^{-A+B} - e^{-A-B} \right)$

$\equiv \tfrac{1}{2} e^{A-B} + \tfrac{1}{2} e^{-A+B}$

$\equiv \tfrac{1}{2} \left(e^{A-B} + e^{-(A-B)} \right) \equiv \text{ch } (A - B) = \text{R.H.S.}$

Problem 3. (a) If $P e^x + Q e^{-x} \equiv 3 \text{ ch } x - 4 \text{ sh } x$ find the values of P and Q.

(b) If $5 e^x - 2 e^{-x} \equiv A \text{ sh } x + B \text{ ch } x$ find the values of A and B.

(a) $P e^x + Q e^{-x} \equiv 3 \text{ ch } x - 4 \text{ sh } x$

$\equiv 3 \left(\dfrac{e^x + e^{-x}}{2} \right) - 4 \left(\dfrac{e^x - e^{-x}}{2} \right)$

$P e^x + Q e^{-x} \equiv \tfrac{3}{2} e^x + \tfrac{3}{2} e^{-x} - 2 e^x + 2 e^{-x}$

$P e^x + Q e^{-x} \equiv -\tfrac{1}{2} e^x + \tfrac{7}{2} e^{-x}$

Equating coefficients gives $P = -\tfrac{1}{2}$ and $Q = 3\tfrac{1}{2}$

(b) $5 e^x - 2 e^{-x} \equiv A \text{ sh } x + B \text{ ch } x$

$\equiv A \left(\dfrac{e^x - e^{-x}}{2} \right) + B \left(\dfrac{e^x + e^{-x}}{2} \right)$

$\equiv \dfrac{A}{2} e^x - \dfrac{A}{2} e^{-x} + \dfrac{B}{2} e^x + \dfrac{B}{2} e^{-x}$

$$5e^x - 2e^{-x} = \left(\frac{A+B}{2}\right)e^x + \left(\frac{B-A}{2}\right)e^{-x}$$

Equating coefficients gives: $5 = \dfrac{A+B}{2}$ and $-2 = \dfrac{B-A}{2}$

i.e. $A + B = 10$ (1)

and $-A + B = -4$ (2)

Adding equations (1) and (2) gives: $2B = 6$, i.e. $B = 3$

Substituting in equation (1) gives: $A = 7$

Further problems on hyperbolic identities may be found in Section 9 (Problems 15 to 21), page 59.

6. Differentiation of hyperbolic functions

(a) $\dfrac{d}{dx} (\sinh x) = \dfrac{d}{dx} \left(\dfrac{e^x - e^{-x}}{2}\right) = \frac{1}{2}[e^x - (-e^{-x})]$

$$= \frac{1}{2} (e^x + e^{-x}) = \cosh x$$

More generally, if $y = \sinh ax$ then $\dfrac{dy}{dx} = a \cosh ax$

(b) $\dfrac{d}{dx} (\cosh x) = \dfrac{d}{dx} \left(\dfrac{e^x + e^{-x}}{2}\right) = \frac{1}{2}[e^x + (-e^{-x})]$

$$= \frac{1}{2} (e^x - e^{-x}) = \sinh x$$

More generally, if $y = \cosh ax$ then $\dfrac{dy}{dx} = a \sinh ax$

(c) Since $\tanh x = \dfrac{\sinh x}{\cosh x}$ then by the quotient rule:

$$\dfrac{d}{dx} (\tanh x) = \dfrac{d}{dx} \left(\dfrac{\sinh x}{\cosh x}\right) = \dfrac{(\cosh x)(\cosh x) - (\sinh x)(\sinh x)}{\cosh^2 x}$$

$$= \dfrac{\cosh^2 x - \sinh^2 x}{\cosh^2 x} = \dfrac{1}{\cosh^2 x} = \text{sech}^2 x$$

More generally, if $y = \tanh ax$ then $\dfrac{dy}{dx} = a \,\text{sech}^2\, ax$

(d) $\dfrac{d}{dx} (\text{sech}\, x) = \dfrac{d}{dx} \left(\dfrac{1}{\cosh x}\right) = \dfrac{(\cosh x)(0) - (1)(\sinh x)}{\cosh^2 x} = \dfrac{-\sinh x}{\cosh^2 x}$

$$= -\left(\dfrac{1}{\cosh x}\right)\left(\dfrac{\sinh x}{\cosh x}\right) = -\,\text{sech}\, x \tanh x$$

More generally, if $y = \text{sech } ax$ then $\dfrac{dy}{dx} = -a \text{ sech } ax \text{ tanh } ax$

(e) $\dfrac{d}{dx} (\text{cosech } x) = \dfrac{d}{dx}\left(\dfrac{1}{\sinh x}\right) = \dfrac{(\sinh x)(0) - (1)(\cosh x)}{\sinh^2 x} = -\dfrac{\cosh x}{\sinh^2 x}$

$= -\left(\dfrac{1}{\sinh x}\right)\left(\dfrac{\cosh x}{\sinh x}\right) = -\text{ cosech } x \text{ coth } x$

More generally, if $y = \text{cosech } ax$ then $\dfrac{dy}{dx} = -a \text{ cosech } ax \text{ coth } ax$

(f) $\dfrac{d}{dx} (\text{coth } x) = \dfrac{d}{dx}\left(\dfrac{\cosh x}{\sinh x}\right) = \dfrac{(\sinh x)(\sinh x) - (\cosh x)(\cosh x)}{\sinh^2 x}$

$= \dfrac{\sinh^2 x - \cosh^2 x}{\sinh^2 x} = -\dfrac{(\cosh^2 x - \sinh^2 x)}{\sinh^2 x}$

$= \dfrac{-1}{\sinh^2 x} = -\text{ cosech}^2 x$

More generally, if $y = \text{coth } ax$ then $\dfrac{dy}{dx} = -a \text{ cosech}^2 ax$

Summary

y or $f(x)$	$\dfrac{dy}{dx}$ or $f'(x)$
sinh ax	a cosh ax
cosh ax	a sinh ax
tanh ax	a sech^2 ax
sech ax	$-a$ sech ax tanh ax
cosech ax	$-a$ cosech ax coth ax
coth ax	$-a$ cosech^2 ax

Worked problems on differentiation of hyperbolic functions

Problem 1. Differentiate the following with respect to x:

(a) $y = 3 \text{ sh } 4x + \frac{2}{3} \text{ ch } 2x$ (b) $y = 5 \text{ th } 3x - 4 \text{ coth } \dfrac{x}{2}$

(c) $y = \frac{1}{2} (\text{sech } 6x - 3 \text{ cosech } 5x)$

(a) $y = 3 \text{ sh } 4x + \frac{2}{3} \text{ ch } 2x$

$\dfrac{dy}{dx} = (3)(4) \text{ ch } 4x + (\frac{2}{3})(2) \text{ sh } 2x = 12 \text{ ch } 4x + \frac{4}{3} \text{ sh } 2x$

(b) $y = 5 \text{ th } 3x - 4 \text{ coth } \dfrac{x}{2}$

$$\frac{dy}{dx} = (5)(3) \text{ sech}^2 \, 3x - (4)(-\tfrac{1}{2}) \text{ cosech}^2 \, \frac{x}{2}$$

$$= 15 \text{ sech}^2 \, 3x + 2 \text{ cosech}^2 \, \frac{x}{2}$$

(c) $y = \tfrac{1}{2}$ (sech $6x$ − 3 cosech $5x$)

$$\frac{dy}{dx} = \tfrac{1}{2}[-6 \text{ sech } 6x \text{ th } 6x - (3)(-5) \text{ cosech } 5x \text{ coth } 5x]$$

$$= \tfrac{3}{2}[5 \text{ cosech } 5x \text{ coth } 5x - 2 \text{ sech } 6x \text{ th } 6x]$$

Problem 2. Find the differential coefficients of the following with respect to the variable:

(a) ln (sh $2x$) + 3 ch^2 $4x$ (b) 3 sin $2t$ ch $5t$ (c) $\dfrac{2\theta^3}{\text{ch } 3\theta}$

(a) $f(x)$ = ln (sh $2x$) + 3 ch^2 $4x$ (i.e. 'a function of a function')

$$f'(x) = \left(\frac{1}{\text{sh } 2x}\right)(2 \text{ ch } 2x) + (3)(2 \text{ ch } 4x)(4 \text{ sh } 4x)$$

$$= 2 (\coth 2x + 12 \text{ sh } 4x \text{ ch } 4x)$$

(b) $f(t)$ = 3 sin $2t$ ch $5t$ (i.e. a product)
$f'(t)$ = (3 sin $2t$) (5 sh $5t$) + (ch $5t$) (6 cos $2t$)
$= 3 (5 \sin 2t \text{ sh } 5t + 2 \cos 2t \text{ ch } 5t)$

(c) $f(\theta)$ = $\dfrac{2\theta^3}{\text{ch } 3\theta}$ (i.e. a quotient)

$$f'(\theta) = \frac{(\text{ch } 3\theta)(6\theta^2) - (2\theta^3)(3 \text{ sh } 3\theta)}{(\text{ch } 3\theta)^2} = 6\theta^2\left[\frac{\text{ch } 3\theta}{\text{ch}^2 3\theta} - \frac{\theta \text{ sh } 3\theta}{\text{ch}^2 \, 3\theta}\right]$$

$$= 6\theta^2\left[\frac{1}{\text{ch } 3\theta} - \frac{\theta \text{ sh } 3\theta}{\text{ch } 3\theta}\left(\frac{1}{\text{ch } 3\theta}\right)\right]$$

$$= 6\theta^2 \text{ sech } 3\theta \, (1 - \theta \text{ th } 3\theta)$$

Further problems on differentiation of hyperbolic functions may be found in Section 9 (Problems 22 to 27), page 60.

7 Solution of equations of the form $a \cosh x + b \sinh x = c$

Equations of the form a ch $x + b$ sh $x = c$, where a, b and c are constants, may be solved either by plotting graphs of $y = a$ ch $x + b$ sh x and $y = c$ and noting the points of intersection or, more accurately, as follows:

(i) Change ch x to $\left(\dfrac{e^x + e^{-x}}{2}\right)$ and sh x to $\left(\dfrac{e^x - e^{-x}}{2}\right)$.

(ii) Rearrange the equation into the form $m\,e^x + n\,e^{-x} + p = 0$, where m, n and p are constants.

(iii) Multiply each term by e^x. This produces an equation of the form: $m(e^x)^2 + p\,e^x + n = 0$, since $(e^{-x})(e^x) = 1$.

(iv) Solve the quadratic equation $m(e^x)^2 + p\,e^x + n = 0$ for e^x either by factorising or by using the quadratic formula.

(v) Given $e^x = k$, take logarithms to base e of both sides to give $x = \ln k$.

Worked problems on solving equations of the form a ch $x + b$ sh $x = c$

Problem 1. Solve the equation sh $x = 1$ correct to four significant figures.

Following the above procedure:

(i) sh $x = \dfrac{e^x - e^{-x}}{2} = 1$

(ii) $e^x - e^{-x} = 2$

  i.e. $e^x - e^{-x} - 2 = 0$

(iii) $(e^x)^2 - (e^{-x})e^x - 2e^x = 0$
   $(e^x)^2 - 2e^x - 1 = 0$

(iv) $e^x = \dfrac{-(-2) \pm \sqrt{[-2]^2 - 4(1)(-1)]}}{2(1)} = \dfrac{2 \pm \sqrt{(4+4)}}{2}$

   $= \dfrac{2 \pm \sqrt{8}}{2} = \dfrac{2 \pm 2.828\,4}{2} = \dfrac{4.828\,4}{2}$ or $\dfrac{-0.828\,4}{2}$

Hence $e^x = 2.414\,2$ or $-0.414\,2$

(v) $x = \ln 2.414\,2$ or $x = \ln(-0.414\,2)$ which has no real solution.

Hence $x = 0.881\,4$ correct to four significant figures.

Problem 2. Solve the equation 3 ch$\theta - 5 = 0$ correct to four decimal places.

(i) 3 ch $\theta - 5 = 0$

  $3\left(\dfrac{e^\theta + e^{-\theta}}{2}\right) - 5 = 0$

(ii) $\tfrac{3}{2}e^\theta + \tfrac{3}{2}e^{-\theta} - 5 = 0$
   i.e. $3e^\theta + 3e^{-\theta} - 10 = 0$

(iii) $3(e^\theta)^2 + 3 - 10\,e^\theta = 0$
   $3(e^\theta)^2 - 10\,e^\theta + 3 = 0$

(iv) $(3\,e^\theta - 1)(e^\theta - 3) = 0$
   $e^\theta = \tfrac{1}{3}$ or $e^\theta = 3$

(v) $\theta = \ln\left(\tfrac{1}{3}\right)$ or $\theta = \ln 3$

Hence $\theta = -1.098\,6$ or $1.098\,6$ correct to four decimal places.

Problem 3. Solve the equation $3.0 \operatorname{ch} x + 2.0 \operatorname{sh} x = 14.31$, correct to four decimal places.

(i) $3.0 \operatorname{ch} x + 2.0 \operatorname{sh} x = 14.31$

$$3.0 \; \frac{e^x + e^{-x}}{2} + 2.0 \; \frac{e^x - e^{-x}}{2} = 14.31$$

(ii) $\begin{aligned} 1.5 \, e^x + 1.5 \, e^{-x} + e^x - e^{-x} &= 14.31 \\ 2.5 \, e^x + 0.5 \, e^{-x} - 14.31 &= 0 \end{aligned}$

(iii) $2.5 \, (e^x)^2 - 14.31 \, e^x + 0.5 = 0$

(iv) $e^x = \dfrac{-(-14.31) \pm \sqrt{[(-14.31)^2 - 4\,(2.5)\,(0.5)]}}{2\,(2.5)}$

$$= \frac{14.31 \pm 14.134\,22}{5} = \frac{28.444\,22}{5} \quad \text{or} \quad \frac{0.175\,78}{5}$$

Hence $e^x = 5.688\,844$ or $0.035\,156$

(v) $x = \ln 5.688\,844$ or $x = \ln 0.035\,156$

Hence $x = \mathbf{1.738\,5}$ or $\mathbf{-3.348\,0}$ **correct to four decimal places.**

Further problems on solving equations of the form $a \operatorname{ch} x + b \operatorname{sh} x = c$ may be found in Section 9 (Problems 28 to 38), page 60.

8 Series expansions for $\cosh x$ and $\sinh x$

e^x is defined by the power series:

$$e^x = 1 + x + \frac{x^2}{2!} + \frac{x^3}{3!} + \frac{x^4}{4!} + \frac{x^5}{5!} + \dots$$

If x is replaced by $-x$ then $e^{-x} = 1 - x + \frac{x^2}{2!} - \frac{x^3}{3!} + \frac{x^4}{4!} - \frac{x^5}{5!} + \dots$

$$\cosh x = \tfrac{1}{2} \, (e^x + e^{-x}) = \tfrac{1}{2}\left[\left(1 + x + \frac{x^2}{2!} + \frac{x^3}{3!} + \frac{x^4}{4!} + \frac{x^5}{5!} + \dots\right)\right.$$
$$\left.+ \left(1 - x + \frac{x^2}{2!} - \frac{x^3}{3!} + \frac{x^4}{4!} - \frac{x^5}{5!} + \dots\right)\right]$$
$$= \tfrac{1}{2}\left[2 + \frac{2x^2}{2!} + \frac{2x^4}{4!} \dots \right]$$

i.e. $\cosh x = 1 + \dfrac{x^2}{2!} + \dfrac{x^4}{4!} + \dots$

$\cosh x$ is an even function and contains only even powers of x

Similarly, $\sinh x = \tfrac{1}{2} \, (e^x - e^{-x}) = \tfrac{1}{2}\left[\left(1 + x + \frac{x^2}{2!} + \frac{x^3}{3!} + \frac{x^4}{4!} + \frac{x^5}{5!} + \dots\right)\right.$

$$-\left(1 - x + \frac{x^2}{2!} - \frac{x^3}{3!} + \frac{x^4}{4!} - \frac{x^5}{5!} + \dots\right)\Bigg]$$

$$= \tfrac{1}{2}\left[2x + \frac{2x^3}{3!} + \frac{2x^5}{5!} + \dots\right]$$

i.e. **sinh** x $\quad = x + \dfrac{x^3}{3!} + \dfrac{x^5}{5!} + \dots$

sinh x is an odd function and contains only odd powers of x.
The series expansions for cosh x and sinh x are true for all values of x

Worked problems on series expansion for cosh x and sinh x

Problem 1. Using the series expansion for ch x evaluate ch 1 correct to four decimal places.

$$\cosh x = 1 + \frac{x^2}{2!} + \frac{x^4}{4!} + \dots$$

Let $x = 1$ then cosh $1 = 1 + \dfrac{1}{2\times1} + \dfrac{1}{4\times3\times2\times1} + \dfrac{1}{6\times5\times4\times3\times2\times1}$

$$= 1 + 0.5 + 0.041\,67 + 0.001\,389$$

i.e. **ch 1** $\quad = $ **1.543 1 correct to four decimal places,**

which is the same as the value of ch 1 given in Table 1.

Problem 2. Using the series expansion for sh x evaluate sh 2 correct to four decimal places.

$$\sinh x = x + \frac{x^3}{3!} + \frac{x^5}{5!} + \dots$$

Let $x = 2$ then sh $2 = 2 + \dfrac{2^3}{3\times2\times1} + \dfrac{2^5}{5\times4\times3\times2\times1}$

$$+ \dfrac{2^7}{7\times6\times5\times4\times3\times2\times1} + \dfrac{2^9}{9\times8\times7\times6\times5\times4\times3\times2\times1}$$

$$+ \dfrac{2^{11}}{11\times10\times9\times8\times7\times6\times5\times4\times3\times2\times1}$$

$$= 2 + 1.333\,33 + 0.266\,67 + 0.025\,40$$
$$+ 0.001\,41 + 0.000\,05$$

i.e. **sh 2** $\quad = $ **3.626 9 correct to four decimal places,**

which is the same as the value of sh 2 given in Table 1.

Problem 3. Find the series expansion for ch 2θ − sh 2θ as far as the term in θ^5.

In the series expansions for sh x and ch x let $x = 2\theta$ then:

$$\text{ch } 2\theta \ = 1 \ + \ \frac{(2\theta)^2}{2!} + \frac{(2\theta)^4}{4!} + \ \dots \ = 1 \ + \ 2\theta^2 + \frac{2\theta^4}{3} + \ \dots$$

$$\text{sh } 2\theta \ = 2\theta + \frac{(2\theta)^3}{3!} + \frac{(2\theta)^5}{5!} + \ \dots \ = 2\theta + \tfrac{4}{3}\theta^3 + \tfrac{4}{15}\theta^5 + \dots$$

Hence ch 2θ − sh 2θ $= (1 + 2\theta^2 + \tfrac{2}{3}\theta^4 + \dots) - (2\theta + \tfrac{4}{3}\theta^3 + \tfrac{4}{15}\theta^5 + \dots)$

$$= 1 - 2\theta + 2\theta^2 - \tfrac{4}{3}\theta^3 + \tfrac{2}{3}\theta^4 - \tfrac{4}{15}\theta^5 + \dots$$

as far as the term in θ^5.

Further problems on series expansions for cosh x and sinh x may be found in the following Section (9) (Problems 39 to 43), page 61.

9 Further problems

Evaluation of hyperbolic functions

In problems 1 to 10, evaluate correct to four significant figures.

1. (a) sh 0.52   (b) sh 2.24   (c) sh 5.36.
(a)[0.543 8]   (b)[4.643]   (c)[106.4]

2. (a) sh 0.212   (b) sh 7.7   (c) sh 8.6.
(a)[0.213 6]   (b)[1 104]   (c)[2 716]

3. (a) ch 0.67   (b) ch 1.84   (c) ch 4.78.
(a)[1.233]   (b)[3.228]   (c)[59.56]

4. (a) ch 0.346   (b) ch 6.5   (c) ch 7.4.
(a)[1.060]   (b)[332.6]   (c)[818.0]

5. (a) sh 1.527 3   (b) sh 2.689 1   (c) sh 5.326 4.
(a)[2.194]   (b)[7.325]   (c)[102.8]

6. (a) ch 0.478 8   (b) ch 1.723 1   (c) ch 4.681 5.
(a)[1.117]   (b)[2.890]   (c)[53.97]

7. (a) th 0.43   (b) th 0.76   (c) th 1.54.
(a)[0.405 3]   (b)[0.641 1]   (c)[0.912 1]

8. (a) cosech 0.26   (b) cosech 0.624   (c) cosech 2.45.
(a)[3.804]   (b)[1.503]   (c)[0.173 9]

9. (a) sech 0.15   (b) sech 0.82   (c) sech 2.324.
(a)[0.988 8]   (b)[0.737 8]   (c)[0.193 9]

10. (a) coth 0.33   (b) coth 0.746   (c) coth 1.168.
(a)[3.140]   (b)[1.580]   (c)[1.214]

11. A telegraph wire hangs so that its shape is described by

$y \ = \ 60 \text{ ch } \dfrac{x}{60}$. Evaluate, correct to four significant figures,

y when x is 30.   [67.66]

12. The length, l of a heavy cable hanging under gravity is given by:

$$l = 2c \sinh \frac{L}{2c}$$

Find l when $c = 50$ and $L = 40$.   [41.07]

13. The speed, V, of waves over the bottom of shallow water is given by the

formula: $V^2 = 0.55 L \tanh\left(\frac{6.3 d}{L}\right)$ where d is the depth and L is the

wavelength. If $d = 10$ and $L = 100$, calculate the value of V.   [5.540]

14. The increase in resistance of strip conductors due to eddy currents at

power frequencies is given by: $\lambda = \dfrac{\alpha t}{2}\left[\dfrac{\sinh \alpha t + \sin \alpha t}{\cosh \alpha t - \cos \alpha t}\right]$. Calculate λ

correct to five significant figures when $\alpha = 1.08$ and $t = 1$.   [1.007 5]

Hyperbolic identities

In Problems 15 to 20 prove the given identities.

15. (a) $\text{ch}(A + B) \equiv \text{ch}\,A\,\text{ch}\,B + \text{sh}\,A\,\text{sh}\,B$.
    (b) $\text{ch}\,2\theta \equiv \text{ch}^2\,\theta + \text{sh}^2\,\theta$
    (c) $1 + 2\,\text{sh}^2 y \equiv \text{ch}\,2y$.

16. (a) $1 + \text{sh}^2 A \equiv \text{ch}^2 A$.
    (b) $\text{sech}^2 B + \text{th}^2 B \equiv 1$
    (c) $\text{cosech}^2 C + 1 \equiv \text{coth}^2 C$.

17. (a) $\text{th}\,\alpha\,(1 + \text{sh}^2\alpha) \equiv \tfrac{1}{2}\,\text{sh}\,2\alpha$.
    (b) $1 + \text{ch}\,2\beta \equiv \text{sh}\,2\beta\,\text{coth}\,\beta$
    (c) $\text{coth}\,t - 2\,\text{cosech}\,2t \equiv \text{th}\,t$

18. (a) $\text{th}\,2\theta \equiv \dfrac{2\,\text{th}\,\theta}{1 + \text{th}^2\,\theta}$

    (b) $\text{ch}\,2\theta + \text{sh}\,2\theta \equiv \dfrac{1 + \text{th}\,\theta}{1 - \text{th}\,\theta}$

    (c) $\text{sh}\,2\theta \equiv 2\,\text{sh}\,\theta\,\text{ch}\,\theta$.

19. (a) $\text{sh}\,A\,\text{ch}\,B + \text{ch}\,A\,\text{sh}\,B \equiv \text{sh}(A + B)$

    (b) $\text{th}(A + B) \equiv \dfrac{\text{th}\,A + \text{th}\,B}{1 + \text{th}\,A\,\text{th}\,B}$

    (c) $\text{sh}^2 x \equiv \tfrac{1}{2}(\text{ch}\,2x - 1)$.

20. (a) $\text{ch}(A + B) + \text{ch}(A - B) \equiv 2\,\text{ch}\,A\,\text{ch}\,B$
    (b) $\text{ch}(A + B) - \text{ch}(A - B) \equiv 2\,\text{sh}\,A\,\text{sh}\,B$
    (c) $\text{sh}\,3A - 3\,\text{sh}\,A \equiv 4\,\text{sh}^3 A$.

21. (a) If $4e^x + 3e^{-x} \equiv P \text{ sh } x + Q \text{ ch } x$, find P and Q.　　　$[P = 1, Q = 7]$
　　(b) If $A \text{ ch } x - B \text{ sh } x \equiv 3e^x - 4e^{-x}$, find A and B.　$[A = -1, B = -7]$
　　(c) If $\alpha e^x - \beta e^{-x} \equiv 5 \text{ ch } x - 2 \text{ sh } x$, find α and β.　　$[\alpha = 1\tfrac{1}{2}, \beta = -3\tfrac{1}{2}]$

Differentiation of hyperbolic functions

In Problems 22 to 27, differentiate the given functions with respect to the variable.

22. (a) $4 \text{ sh } 3x$　　　　　(b) $2 \text{ ch } 4t$　　　　　(c) $5 \text{ th } 7x$.
　　(a)$[12 \text{ ch } 3x]$　　　(b)$[8 \text{ sh } 4t]$　　　(c)$[35 \text{ sech}^2 7x]$
23. (a) $\tfrac{3}{4} \text{ cosech } \tfrac{3}{2} \theta$　　　(b) $\tfrac{5}{8} \text{ sech } 4x$　　　(c) $3 \text{ coth } 5y$.

　　(a)$\left[-\tfrac{9}{8} \text{ cosech } \dfrac{3\theta}{2} \text{ coth } \dfrac{3\theta}{2}\right]$　　　(b)$\left[-\tfrac{5}{2} \text{ sech } 4x \tanh 4x\right]$

　　(c)$[-15 \text{ cosech}^2 5y]$

24. (a) $3 \ln (\text{ch } x)$　　　(b) $2 \ln \text{ th} \dfrac{3x}{2}$　　　(c) $\text{cosech}^3 2t$.

　　(a)$[3 \text{ th } x]$　　　(b)$[3 \text{ sech } \tfrac{3}{2} x \text{ cosech } \tfrac{3}{2} x]$
　　(c)$[-6 \text{ cosech}^3 2t \text{ coth } 2t]$

25. (a) $7 \text{ sh}^2 x$　　　　(b) $2 \text{ coth}^4 3p$　　　(c) $\text{ch}^3 4x + 3 \text{ th}^2 2x$.
　　(a)$[14 \text{ sh } x \text{ ch } x \text{ or } 7 \text{ sh } 2x]$　　(b)$[-24 \text{ coth}^3 3p \text{ cosech}^2 3p]$
　　(c)$[12 \text{ ch}^2 4x \text{ sh } 4x + \text{ th } 2x \text{ sech}^2 2x]$

26. (a) $\text{sh } 3x \text{ ch } 3x$　　　(b) $2 \cos 3t \text{ ch } 3t$　　(c) $e^{2x} \text{ ch } 3x$.
　　(a)$[3 (\text{sh}^2 3x + \text{ch}^2 3x) \text{ or } 3 \text{ ch } 6x]$　　(b) $[6(\cos 3t \text{ sh } 3t - \text{ch } 3t \sin 3t)]$
　　(c)$[e^{2x} (3 \text{ sh } 3x + 2 \text{ ch } 3x)]$

27. (a) $\dfrac{\text{sh } 2\theta}{\sin 2\theta}$　　　(b) $\dfrac{2 \text{ ch } 5t}{3t^4}$　　　(c) $x^{\text{ch} x}$

　　(a) $\left[\dfrac{2 (\sin 2\theta \text{ ch } 2\theta - \text{sh } 2\theta \cos 2\theta)}{\sin^2 2\theta} \right]$

　　(b)$\left[\dfrac{2}{3t^5} (5t \text{ sh } 5t - 4 \text{ ch } 5t) \right]$　　(c)$\left[x^{\text{ch } x} \left(\dfrac{\text{ch } x}{x} + \text{sh } x \ln x \right) \right]$

Solution of equations of the form $a \text{ ch } x + b \text{ sh } x = c$

In problems 28 to 37 solve the equations correct to four decimal places.

28. $\text{sh } x = 2$.　　　　　　　　　　　　　　　$[1.443\ 6]$
29. $152 \text{ ch } x = 6.70$　　　　　　　　　　　$[\pm 2.163\ 4]$
30. $3.0 \text{ sh } x + 2.0 \text{ ch } x = 0$.　　　　　　$[-0.804\ 7]$
31. $\text{ch } x + 2 \text{ sh } x = 4$.　　　　　　　　　$[1.024\ 7]$
32. $6.80 \text{ sh } \theta - 3.72 \text{ ch } \theta = 1.68$.　　　$[0.905\ 2]$
33. $2.97 - 3.16 \text{ sh } x - 4 \text{ ch } x = 0$.　　$[-0.432\ 7 \text{ or } -1.710\ 2]$
34. $3.94 \text{ ch } x + 10.82 \text{ sh } x = 4.77$.　　　$[0.075\ 6]$
35. $20.2 \text{ sh } y - 16.4 \text{ ch } y = 11.7$　　　　$[2.008\ 3]$

36. th $x = $ sh x [0]
37. 7 th $x - 3 = 0$.   [0.458 1]

38. A chain hangs in the form $y = 50$ ch $\dfrac{x}{50}$. Find, correct to four

significant figures, (i) the value of y when $x = 30$, and (ii) the
value of x when $y = 64.73$.

               (i) [59.28]   (ii) [37.49]

Series expansions for cosh x and sinh x
39. Use the series expansion for cosh x to evaluate correct to four
decimal places: (a) ch 2   (b) ch 0.5   (c) ch 2.5.
(a) [3.762 2]   (b) [1.127 6]   (c) [6.132 3]
40. Use the series expansion for sinh x to evaluate correct to four
decimal places: (a) sh 0.4   (b) sh 1.5   (c) sh 2.4
(a) [0.410 8]   (b) [2.129 3]   (c) [5.466 2]
41. Expand the following as a power series as far as the term in x^5:
(a) sh $2x$ (b) ch $4x$ (c) sh $3x$.

(a) $\left[2x + \dfrac{4}{3}x^3 + \dfrac{4}{15}x^5 \right]$ (b) $\left[1 + 8x^2 + \dfrac{32}{3}x^4 \right]$

                        (c) $\left[3x + \dfrac{9}{2}x^3 + \dfrac{81}{40}x^5 \right]$

42. Prove the following identities, the series being taken as far as the
term in x^5 only.

(a) ch $2x -$ ch $x \equiv \dfrac{x^2}{2}\left(3 + \dfrac{5}{4}x^2 \right)$

(b) sh $x +$ sh $2x \equiv 3x\left(1 + \dfrac{x^2}{2} + \dfrac{11}{120}x^4 \right)$

(c) ch $3x +$ ch $2x +$ ch $x \equiv 3 + 7x^2 + \dfrac{49}{12}x^4$

43. The value of C_v for an Einstein crystal is given by

$$C_v = \frac{3No\,\epsilon^2 k}{(kT)^2} \cdot \frac{e^{\epsilon/kT}}{(e^{\epsilon/kT}-1)^2}$$

Express C_v in terms of sinh $\dfrac{\epsilon}{2kT}$, and by using Taylor's series

show that when $\dfrac{\epsilon}{2kT}$ is very small $C_v = 3Nok$.

$$\left[C_v = \frac{3Nok}{4} \cdot \frac{(\epsilon/kT)^2}{\sinh^2(\epsilon/2kT)} \right]$$

Chapter 5

Scalar and vector products (DE 14)

1 Introduction

Many physical quantities such as time, volume, density, mass, energy and temperature are defined entirely by a numerical value together with the appropriate units. Such quantities are called **scalar quantities** or just **scalars**, and obey the fundamental laws of algebra. Other physical quantities such as displacement, velocity, acceleration, force, moment and momentum are defined in terms of both a numerical value and a direction in space. These quantities are called **vector quantities** or just **vectors**.

A vector quantity representing, say, a force, has two properties:

(i) the magnitude or length of the vector is directly proportional to the magnitude of the force, and
(ii) the direction of the vector is in the same direction as the line of action of the force.

Vector quantities are indicated by using small letters and **bold** print and line *oa* shown in Fig. 1 is *oa*, that is vector *oa*, where the magnitude of line *oa* is directly proportional to 9 newtons and its direction is from *o* to *a*, i.e. *oa* is 9 newtons at an angle of 45° to the horizontal.

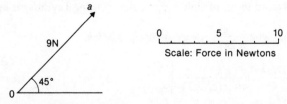

Fig. 1   Vector *oa* = 9N at 45° to the horizontal

Vectors can be added or subtracted by means of triangle or parallelogram laws. For example, Fig. 2 shows a force of F_2 newtons inclined at an angle of $\theta°$ to a force of F_1 newtons. These forces can be added vectorially by the 'nose-to-tail' method. The force F_1 is represented by line oa and is drawn F_1 units long in the direction of the F_1 newton force. From its nose (arrowhead), ar is drawn inclined at an angle of $\theta°$ to oa and F_2 units long. The resultant vector is or.

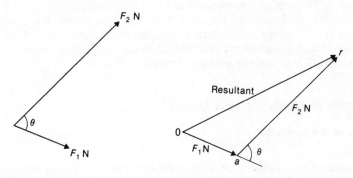

Fig. 2 Nose-to-tail method of adding vectors $oa + ar = or$

The vector equation is $oa + ar = or$.

The vector $(-ar)$ is the vector represented by line ra, that is, a minus sign in front of a vector has the effect of rotating it through $180°$ with respect to its positive direction, i.e. reversing its direction. Thus, $oa + (-ar) = oa + ra$.

A force of F_2 newtons is shown inclined at an angle of $\theta°$ to a force of F_1 newtons in Fig. 3. The vector representing the force of F_2 newtons can be subtracted from the vector representing the force of F_1 newtons by using the nose-to-tail method. oa is drawn F_1 units long in the direction of the F_1 newton force. ra is drawn in the direction of the F_2 newton force with its nose coinciding with a. The resultant is vector or. The vector equation is:

$oa - ar = oa + ra = or$

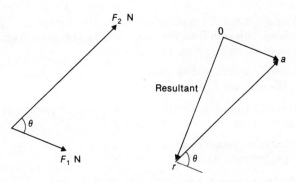

Fig. 3 Vector subtraction $oa - ar = oa + ra = or$

The magnitude or numerical value of a vector is shown by using modulus lines. Thus the magnitude of *oa* is indicated by $|oa|$ and $|oa| = oa$. When vector *oa* is multiplied by a scalar quantity, say k, the resultant is the vector $k\,oa$ which has a magnitude of $k\,|oa|$ and acts in the direction of *oa*.

A vectorial approach is used quite extensively in solving problems in mechanics. In this chapter, scalar and vector products of two vectors are introduced, followed by scalar and vector products of three vectors in Chapter 6. The applications of scalar and vector products to mechanics at this stage are very limited and the one or two obvious applications are included in the worked problems in the various sections. It will not be until differentiation and integration of vector quantities has been mastered at a higher level that their real potential will become apparent.

2 Scalar products and unit vectors

Scalar or dot products

Definition

The **scalar** or **dot** product of two vectors is found by multiplying the product of their magnitudes by the cosine of the angle between them. Thus, with reference to Fig. 4, if *oa* represents a force F_1 and *ob* represents a force F_2, the scalar product is written as $oa \cdot ob$. From the definition given above:

$$
\begin{aligned}
oa \cdot ob &= |oa|\,|ob| \cos \theta \\
&= (oa)\,(ob) \cos \theta \\
&= F_1 F_2 \cos \theta.
\end{aligned}
\tag{1}
$$

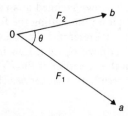

Fig. 4   Scalar product: $oa \cdot ob = (oa)(ob) \cos \theta = F_1 F_2 \cos \theta$.

In algebra $a \times b = b \times a$, and this law is also true for vectors. This can be shown as follows. It is seen from Fig. 5(a) that $F_2 \cos \theta$ is the magnitude of the projection of F_2 on F_1, thus

$$
\begin{aligned}
oa \cdot ob &= (oa)\,(\text{projection of } ob \text{ on } oa) \\
&= (F_1)\,(F_2 \cos \theta) = F_1 F_2 \cos \theta.
\end{aligned}
$$

Similarly, it is seen from Fig. 5(b) that $F_1 \cos \theta$ is the magnitude of the projection of F_1 on F_2, thus

$$
\begin{aligned}
ob \cdot oa &= (ob)\,(\text{projection of } oa \text{ on } ob) \\
&= (F_2)\,(F_1 \cos \theta) = F_1 F_2 \cos \theta.
\end{aligned}
$$

It follows that $oa \cdot ob = ob \cdot oa$.

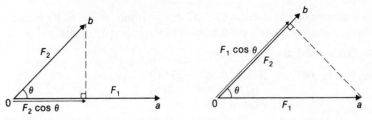

Fig. 5 (a) $oa \cdot ob = F_1F_2 \cos \theta$. (b) $ob \cdot oa = F_1F_2 \cos \theta$.

Unit vectors

A unit vector is a vector whose magnitude is unity. For any vector oa, the unit vector is $\dfrac{oa}{oa}$ in the direction of oa. Unit vectors are frequently used to specify directions in space. To completely specify the direction of a vector with reference to some point, three unit vectors, mutually at right angles to each other and associated with the x-, y- and z- axes of a Cartesian reference system, are used. In this context, the three unit vectors are called the **unit triad**. In

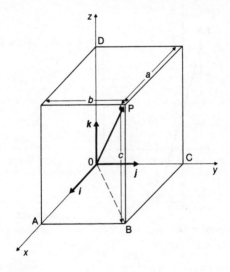

Fig. 6

Fig. 6, Ox, Oy and Oz represent the three axes mutually at right angles. Oi, Oj and Ok represent the three unit vectors. The system usually adopted is a 'right-handed system' and the vectors are labelled in such a way that the head of a right-threaded screw being screwed along Oz in the direction of Ok turns from Ox to Oy.

An alternative notation to the geometrical 'oa' is usually used in analytical work on vectors, using a single letter only. In this notation oa is written as a and ao as $-a$. The unit triad vectors shown in Fig. 6 are i, j and k in this notation. This notation will be largely adopted for the remainder of this text. Any point P can be specified by its distance from planes Oyz, Oxz and Oxy, these distances being shown as a, b and c respectively in Fig. 6. By vector addition, using the nose-to-tail method and the geometry of Fig. 6:

OP = OA + AB + BP

From Fig. 6, $|\textbf{OP}|$ = OP = $\sqrt{(OB^2 + BP^2)}$
= $\sqrt{(OA^2 + AB^2 + BP^2)}$
= $\sqrt{(a^2 + b^2 + c^2)}$

That is, the magnitude of **OP** is $\sqrt{(a^2 + b^2 + c^2)}$.
OP is called the **position vector** of P with respect to O. Using the unit triad notation and single letters for vectors, if **OP** = r, then, position

vector, $r = ai + bj + ck$
and $|r| = \sqrt{(a^2 + b^2 + c^2)}$ (2)

For example, the vector $r = 3i + 4j - 4k$ is a position vector and is shown in Fig. 7. From equation (2) above, the magnitude of r, i.e. $|r|$, is given by

$|r| = r = \sqrt{(3^2 + 4^2 + (-4)^2)}$
$= \sqrt{(9 + 16 + 16)} = \sqrt{41}.$

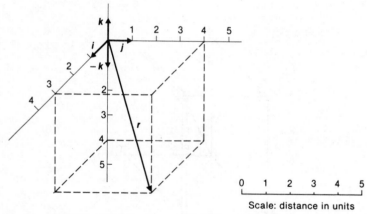

Fig. 7 $r = 3i + 4j - 4k$

Scale: distance in units

To determine the angle between two vectors

One of the ways of expressing the scalar product of two vectors, say a and b, is given in equation (1), and is

$a \cdot b = a\, b \cos \theta,$

where a and b are the magnitudes of a and b respectively and θ is the acute angle between the positive directions of a and b. Thus

$\cos \theta = \dfrac{a \cdot b}{a\, b}$

Let the position vectors a and b be $a_1 i + a_2 j + a_3 k$ and $b_1 i + b_2 j + b_3 k$ respectively. Then

$$a \cdot b = (a_1 i + a_2 j + a_3 k) \cdot (b_1 i + b_2 j + b_3 k)$$
$$= a_1 i \cdot b_1 i + a_1 i \cdot b_2 j + a_1 i \cdot b_3 k + a_2 j \cdot b_1 i + a_2 j \cdot b_2 j + \ldots$$

It is shown in Worked Problem 1(b) following that these terms can be written as:

$$a \cdot b = a_1 b_1 \, i \cdot i + a_1 b_2 \, i \cdot j + a_1 b_3 \, i \cdot k + a_2 b_1 \, i \cdot j + a_2 b_2 \, j \cdot j + \ldots$$

By the definition of a scalar product, $(a \cdot b = ab \cos \theta)$, for the unit vectors i, j and k having a magnitude of unity and being mutually at right angles:

$$i \cdot i = (1)(1) \cos 0 = 1, \text{ since } \theta = 0°$$
$$i \cdot j = (1)(1) \cos 90° = 0, \text{ since } \theta = 90°$$
and $$i \cdot k = (1)(1) \cos 90° = 0.$$

Thus $a \cdot b = (a_1 b_1)(1) + (a_1 b_2)(0) + (a_1 b_3)(0) + (a_2 b_1)(0)$
$$+ (a_2 b_2)(1) + \ldots$$

i.e. $\quad a \cdot b = a_1 b_1 + a_2 b_2 + a_3 b_3 \qquad (3)$

Substituting for $a \cdot b$ in the equation for $\cos \theta$, gives:

$$\cos \theta = \frac{a_1 b_1 + a_2 b_2 + a_3 b_3}{ab}$$

But from equation (2), $a = |a| = \sqrt{(a_1^2 + a_2^2 + a_3^2)}$
and $b = |b| = \sqrt{(b_1^2 + b_2^2 + b_3^2)}$

Thus, $\cos \theta = \dfrac{a \cdot b}{|a| \, |b|} = \dfrac{a_1 b_1 + a_2 b_2 + a_3 b_3}{\sqrt{(a_1^2 + a_2^2 + a_3^2)} \sqrt{(b_1^2 + b_2^2 + b_3^2)}} \qquad (4)$

To determine the angle between vectors a and b, where, for example, $a = 2i - j + 3k$ and $b = -4i + 6j + 5k$, the constants in equation (4) are $a_1 = 2, a_2 = -1, a_3 = 3, b_1 = -4, b_2 = 6$ and $b_3 = 5$, giving

$$\cos \theta = \frac{(2)(-4) + (-1)(6) + (3)(5)}{\sqrt{(2^2 + (-1)^2 + 3^2)} \sqrt{((-4)^2 + 6^2 + 5^2)}}$$

$$= \frac{1}{\sqrt{14} \sqrt{77}} = \frac{1}{32.83} = 0.030\,46$$

Hence $\theta = 88.25°$ or $88° \, 15'$

Work done

A typical application of scalar products is that of determining the work done by a force when moving a body. The amount of work done is the product of the applied force and the distance moved in the direction of the applied force. With reference to Fig. 8, if force F applied at A moves the point of application to A', through vector displacement d, then the work done by F is given by $F \cdot d$. Thus, if a constant force of $F = i + 2j - 5k$ newtons acts on a body at a point and the point is displaced by $d = 2i + 3j - k$ metres, then from equation (3):

68

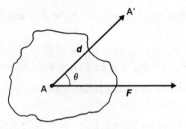

Fig. 8   Work done $= F \cdot d$

work done, $F \cdot d = (1)(2) + (2)(3) + (-5)(-1)$
$\qquad\qquad\quad = 13 \text{ N m}.$

Worked problems on scalar products and unit vectors

Problem 1. Show that (a) $p \cdot (q + r) = p \cdot q + p \cdot r$
$\qquad\qquad\qquad$ (b) $k(p \cdot q) = (kp) \cdot q = p \cdot (kq)$

(a) By the definition of a scalar product:

$p \cdot q = p(q \cos \theta)$, where θ is the angle between p and q
$\qquad = p$ (the magnitude of the projection of q on p)

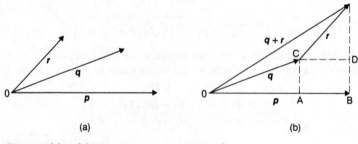

$\qquad\qquad$ (a) $\qquad\qquad\qquad\qquad\qquad$ (b)

Fig. 9   (a)   (b) $p \cdot q + p \cdot r = p \cdot (q + r)$.

$\qquad$ Let p, q and r be any vectors as shown in Fig. 9(a). Using the nose-to-tail method, $(q + r)$ is as shown in Fig. 9(b). Then

$p \cdot q + p \cdot r = p(OA) + p(CD)$
$\qquad\qquad\quad = p(OA) + p(AB) = p(OA + AB)$
$\qquad\qquad\quad = p(OB)$
$\qquad\qquad\quad = p$ (projection of $(q + r)$ on p)
i.e. $p \cdot q + p \cdot r = p \cdot (q + r)$

(b) By the definition of a scalar product:

$\qquad\qquad p \cdot q = pq \cos \theta$, where θ is the angle between p and q.

$\therefore \qquad k\,(p \cdot q) \;=\; kpq\,\cos\theta$

Similarly, $(kp) \cdot q \;=\; (kp)\,q\,\cos\theta = kpq\,\cos\theta$

and $\qquad p \cdot (kq) \;=\; p\,(kq)\,\cos\theta = kpq\,\cos\theta$

Hence $\qquad k\,(p \cdot q) \;=\; (kp) \cdot q \;\;=\; p \cdot (kq)$

Problem 2. Find vector r joining points A and B when point A has co-ordinates $(3, -4, 5)$ and point B has co-ordinates $(2, 1, 0)$. Also find $|r|$, the magnitude of r.

Let O be the origin, i.e. its co-ordinates are $(0, 0, 0)$.
The position vectors of A and B are given by:
$$\mathbf{OA} = 3i - 4j + 5k$$
and $\mathbf{OB} = 2i + j$

By the addition laws of vectors $\mathbf{OA} + \mathbf{AB} = \mathbf{OB}$, hence
$$r = \mathbf{AB} = \mathbf{OB} - \mathbf{OA}$$
i.e. $r = \mathbf{AB} = (2i + j) - (3i - 4j + 5k)$
$$= -i + 5j - 5k$$

From equation (2), the magnitude of r, $|r| = \sqrt{(a^2 + b^2 + c^2)}$
$$= \sqrt{((-1)^2 + 5^2 + (-5)^2)}$$
$$= \sqrt{51}$$

Problem 3. If $p = i + 3j - k$ and $q = 2i + j + 4k$, determine:

(a) $p \cdot q$
(b) $p + q$
(c) $|p + q|$
(d) $|p| + |q|$
and (e) the angle between p and q

(a) $p \cdot q$

From equation (3), if $p = a_1 i + a_2 j + a_3 k$
and $q = b_1 i + b_2 j + b_3 k$,
then $p \cdot q = a_1 b_1 + a_2 b_2 + a_3 b_3$.
When $p = i + 3j - k$, $a_1 = 1, a_2 = 3, a_3 = -1$
and when $q = 2i + j + 4k$, $b_1 = 2, b_2 = 1, b_3 = 4$.
Hence $\qquad p \cdot q = (1)(2) + (3)(1) + (-1)(4)$
i.e. $\qquad p \cdot q = 1$

(b) $p + q$

If the unit vector i is considered, p contributes one unit of length in this direction and q two units of length in this direction. Hence, the resulting distance in the i direction is $1 + 2$, i.e. $3i$. Similarly, the resulting distance in the j direction is $3 + 1$, i.e. $4j$, and that in the k direction is $(-1) + 4$, i.e. $3k$.
Hence, when $p = i + 3j - k$ and $q = 2i + j + 4k$
$$p + q = (1 + 2)i + (3 + 1)j + (-1 + 4)k$$
$$= 3i + 4j + 3k.$$

(c) $|p + q|$

The magnitude or modulus of $p + q$, i.e. $|p + q| = |3i + 4j + 3k|$ from part (b). From equation (2)

$|p + q| = \sqrt{(3^2 + 4^2 + 3^2)} = \sqrt{34}$.

(d) $|p| + |q|$

From equation (2), if $a = a_1i + a_2j + a_3k$, $|a| = \sqrt{(a_1^2 + a_2^2 + a_3^2)}$.
Thus, when $p = i + 3j - k$, $|p| = \sqrt{(1^2 + 3^2 + (-1)^2)}$
$= \sqrt{11}$.
Similarly, $|q| = |2i + j + 4k| = \sqrt{(2^2 + 1^2 + 4^2)} = \sqrt{21}$.
Hence, $|p| + |q| = \sqrt{11} + \sqrt{21} = 7.899$, correct to three decimal places.

(e) The angle between p and q.

If θ is the angle between p and q, then from equation (4):

$$\cos \theta = \frac{p \cdot q}{|p| \, |q|}$$

From part (a), the value of the numerator is 1 and from part (d) the value of the denominator is $\sqrt{11} \sqrt{21}$.

$$\therefore \cos \theta = \frac{1}{\sqrt{11} \sqrt{21}}$$

$= 0.06580$
Hence $\theta = 86.23° = 86° \ 14'$.

Problem 4. Find the work done by a force of F newtons acting at point A on a body, when A is displaced to point B, the co-ordinates of A and B being $(3, 1, -2)$ and $(4, -1, 0)$ metres respectively, and when $F = -i -2j -k$ newtons.

If a vector displacement from A to B is d, then the work done is $F \cdot d$ newton metres or joules. The position vector **OA** is $3i + j - 2k$, (see Worked Problem 2) and **OB** is $4i - j$.

AB $= d =$ **OB** $-$ **OA** (see Worked Problem 2). Thus
$\quad d = (4i - j) - (3i + j - 2k)$
$\quad = i - 2j + 2k$.
Work done $= F \cdot d = (-1)(1) + (-2)(-2) + (-1)(2)$, from equation (3)
$\quad = 1$ N m or joule.

Further problems on scalar products and unit vectors may be found in Section 5 (Problems 1 to 10), page 78.

3 Vector products

Definition

The **vector** or **cross product** of a and b is c, where the magnitude of c is ab $\sin \theta$, θ being the acute angle between the position direction of a and b. The direction of c is perpendicular to both a and b, such that a, b and c form a right-handed system of vectors. Thus if a right-handed screw is screwed along c with its head at the origin of the vectors, the head rotates from a to b, as shown in Fig. 10. Typical applications of vector products as defined above to problems in mechanics are dealt with at the end of this section.

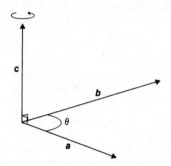

Fig. 10   Vector product: $c = a \cdot b$.

A vector product is shown as

$c = a \times b$, called a cross b

From the definition of a vector product,

$$|c| = |a \times b| = |a|\,|b| \sin \theta = ab \sin \theta \tag{5}$$

An alternative equation for $|a \times b|$ may be derived as follows:
Squaring equation (5) gives
$$\begin{aligned}
|c|^2 &= |a|^2\,|b|^2 \sin^2\theta \\
&= |a|^2\,|b|^2\,(1 - \cos^2\theta) \\
&= |a|^2\,|b|^2 - |a|^2\,|b|^2 \cos^2\theta
\end{aligned}$$

But from Section 2, equation (3)

$$a \cdot b = ab \cos \theta$$
hence $\quad a \cdot a = aa \cos \theta = a^2 \cos \theta = |a|^2 \cos \theta$

But for the vectors (a and a), $\theta = 0$, $\cos \theta = 1$, hence

$$a \cdot a = |a|^2$$
i.e. $\quad |a| = \sqrt{(a \cdot a)}$

Also, from Section 2, equation (4)

$$\cos \theta = \frac{a \cdot b}{ab}$$

Hence $(a^2\,b^2 \cos^2\theta) = \dfrac{a^2\,b^2\,(a \cdot b)^2}{a^2\,b^2} = (a \cdot b)^2$

Thus, substituting for $|a|^2$, $|b|^2$ and $a^2\,b^2 \cos^2\theta$ in:

$$|c|^2 = |a|^2\,|b|^2 - |a|^2\,|b|^2 \cos^2\theta$$
gives $\quad |c|^2 = (a \cdot a)\,(b \cdot b) - (a \cdot b)^2$
i.e. $\quad |a \times b| = \sqrt{[(a \cdot a)\,(b \cdot b) - (a \cdot b)^2]}$ $\qquad\qquad$ (6)

Basic relationships

In Section 2, it is shown that the scalar product of two vectors a and b is such that $a \cdot b = b \cdot a$, that is, scalar products are said to be commutative. However, the vector product of a and b, $a \times b$, is **not** commutative, i.e. $a \times b \neq b \times a$. The magnitudes of both $a \times b$ and $b \times a$ are the same, that is, $ab \sin \theta$, but from the definition of the direction of c, where $a \times b = c$, it follows that $a \times b = -b \times a$, and the vectors are as shown in Fig. 11.

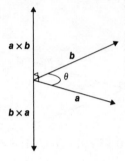

Fig. 11 $a \cdot b = -b \cdot a$.

The other basic relationships of vector products are similar to those of scalar products, i.e.

$$a \times (b + c) = a \times b + a \times c$$
$$\text{and } k\,(a \times b) = (ka) \times b = a \times (kb)$$

Unit vector relationships

The unit vector relationships for vector products are as follows:

(a) $i \times i$ (and $j \times j, k \times k$)

$i \times i$ corresponds to $a \times b$ where $a = 1i + 0j + 0k$
and $b = 1i + 0j + 0k$

Hence, $i \times i = ab \sin \theta = (1)(1)(\sin 0) = 0$, since the angle θ is 0.
Thus $i \times i = j \times j = k \times k = 0$.

(b) $i \times j$ (and $j \times k, k \times i, j \times i, \ldots$)

$i \times j$ corresponds to $a \times b$ where $a = 1i + 0j + 0k$
and $b = 0i + 1j + 0k$

Hence $i \times j = ab \sin \theta = (1)(1)(\sin 90) = 1$, since angle θ is $90°$. The direction of $i \times j$ is perpendicular to both i and j, that is, along k and of magnitude 1. It follows that $i \times k = k$. Similarly it may be shown that $j \times k = i$, and that $k \times i = j$, i.e.

$i \times j = k, j \times k = i, k \times i = j$.

This cyclic order must be adhered to, since vector products are not commutative, i.e. if $i \times j = k$, then $j \times i = -k$, and so on.

The vector product in terms of the unit vectors

For vectors $a = a_1 i + a_2 j + a_3 k$ and

$\qquad b = b_1 i + b_2 j + b_3 k$, the vector product

$\qquad a \times b = (a_1 i + a_2 j + a_3 k) \times (b_1 i + b_2 j + b_3 k)$

i.e., $a \times b = a_1 b_1\, i \times i + a_1 b_2\, i \times j + a_1 b_3\, i \times k + a_2 b_1\, j \times i + a_2 b_2\, j \times j$
$\qquad\qquad\qquad + a_2 b_3\, j \times k + a_3 b_1\, k \times i + a_3 b_2\, k \times j + a_3 b_3\, k \times k.$

Since $i \times i = 0$ and $i \times j = k$ and so on,

then $a \times b = a_1 b_2\, k - a_1 b_3\, j - a_2 b_1\, k + a_2 b_3\, i + a_3 b_1\, j - a_3 b_2\, i$
i.e. $a \times b = (a_2 b_3 - a_3 b_2) i + (a_3 b_1 - a_1 b_3) j + (a_1 b_2 - a_2 b_1) k.$

This relationship is best remembered in determinant form, introduced in Chapter 9. In determinant form:

$$a \times b = \begin{vmatrix} i & j & k \\ a_1 & a_2 & a_3 \\ b_1 & b_2 & b_3 \end{vmatrix} \qquad\qquad (7)$$

Thus for $a = 2i - j + 3k$ and $b = i + 4j - 2k$

$$a \times b = \begin{vmatrix} i & j & k \\ 2 & -1 & 3 \\ 1 & 4 & -2 \end{vmatrix}$$

Using the 'cover-up' rule and, say, a first row expansion, gives

$$a \times b = i \begin{vmatrix} -1 & 3 \\ 4 & -2 \end{vmatrix} - j \begin{vmatrix} 2 & 3 \\ 1 & -2 \end{vmatrix} + k \begin{vmatrix} 2 & -1 \\ 1 & 4 \end{vmatrix}$$

i.e. $a \times b = [(-1)(-2) - (3)(4)] i - [(2)(-2) - (3)(1)] j$
$\qquad\qquad\qquad\qquad\qquad\qquad\qquad + [(2)(4) - (-1)(1)] k$

That is, $a \times b = -10i + 7j + 9k.$

Typical applications of vector products

Typical applications of vector products are to moments and to angular velocity.

(a) **Moments.** If M is the moment vector about a point 0 of a force vector F which has position vector r from 0, as shown in Fig. 12, then

$\qquad M = r \times F$

By the definition of the modulus of a vector given in equation (5)
$|M| = |r|\,|F| \sin \theta$, where θ is the angle between the positive directions of r and F.
But with reference to Fig. 12, $|r| = OP$ and $\sin \theta = \sin(180 - \theta)$
$\qquad\qquad\qquad\qquad\qquad\qquad\qquad\qquad = \sin \phi$

Hence $|M| = |F|\,(OP \sin \theta)$

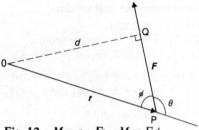

Fig. 12 $M = r \cdot F$; $M = Fd$

From triangle OPQ, $\sin \phi = \dfrac{OQ}{OP} = \dfrac{d}{OP}$, i.e. OP $\sin \phi = d$

Hence $|M| = |F| \, d = Fd$, which is the basic concept of the moment of a force in mechanics.

(b) **Angular velocity.** If v is the velocity vector of a point P on a body rotating about a fixed axis, $\boldsymbol{\omega}$ is its angular velocity vector and r is the position vector of P, then

$$v = \boldsymbol{\omega} \times r$$

The angular velocity vector, $\boldsymbol{\omega}$, has a magnitude of ω and its direction is along the axis. With reference to Fig. 13, if $QP = d$ and angle QOP, between $\boldsymbol{\omega}$ and r, is θ, then

$$|v| = |\boldsymbol{\omega}| \, |r| \sin \theta.$$

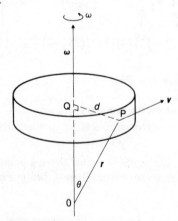

Fig. 13 $v = \omega \cdot r$

But $|r| = $ OP and $\sin \theta = \dfrac{QP}{OP} = \dfrac{d}{OP}$, hence $|r| \sin \theta = d$.

Hence $|v| = \omega \, d$

The relationship between linear and angular velocity, i.e. $v = \boldsymbol{\omega} \, d$, is a basic concept in dynamics.

Worked problems on vector products

Problem 1. If $p = 3i + 2j - k$, $q = i - j + \dfrac{k}{2}$ and $r = i - k$, find

(a) $p \times q$, (b) $|p \times r|$, (c) $(2p - q) \times r$ and (d) $p \times (2r \times q)$.

(a) $p \times q$.

From equation (7), when $a = a_1 i + a_2 j + a_3 k$ and
$b = b_1 i + b_2 j + b_3 k$, then

$$a \times b = \begin{vmatrix} i & j & k \\ a_1 & a_2 & a_3 \\ b_1 & b_2 & b_3 \end{vmatrix}$$

Thus, when $p = 3i + 2j - k$ and $q = i - j + \tfrac{1}{2}k$

$$p \times q = \begin{vmatrix} i & j & k \\ 3 & 2 & -1 \\ 1 & -1 & \tfrac{1}{2} \end{vmatrix}$$

i.e. $p \times q = i \begin{vmatrix} 2 & -1 \\ -1 & \tfrac{1}{2} \end{vmatrix} - j \begin{vmatrix} 3 & -1 \\ 1 & \tfrac{1}{2} \end{vmatrix} + k \begin{vmatrix} 3 & 2 \\ 1 & -1 \end{vmatrix}$

$$= i(1 - 1) - j(\tfrac{3}{2} - (-1)) + k(-3 - 2)$$
$$= -\tfrac{5}{2}j - 5k.$$

(b) $|p \times r|$

From equation (6), $|a \times b| = \sqrt{[(a \cdot a)(b \cdot b) - (a \cdot b)^2]}$
Hence $|p \times r| = \sqrt{[(p \cdot p)(r \cdot r) - (p \cdot r)^2]}$,

where $p = 3i + 2j - k$ and $r = i - k$.
Then $p \cdot p = (3)(3) + (2)(2) + (-1)(-1)$ from equation (3)
$= 9 + 4 + 1 = 14$
$r \cdot r = (1)(1) + (0)(0) + (-1)(-1) = 2$
and $p \cdot r = (3)(1) + (2)(0) + (-1)(-1) = 4$
Hence $|p \times r| = \sqrt{[(14)(2) - (4)^2]}$
$= \sqrt{12}$

(c) $(2p - q) \times r = [2(3i + 2j - k) - (i - j + \tfrac{1}{2}k)] \times (i - k)$
$= (5i + 5j - 2\tfrac{1}{2}k) \times (i - k)$

i.e. $(2p - q) \times r = \begin{vmatrix} i & j & k \\ 5 & 5 & -2\tfrac{1}{2} \\ 1 & 0 & -1 \end{vmatrix}$

$$= i(-5 - 0) - j(-5 + 2\tfrac{1}{2}) + k(0 - 5)$$
$$= -5i + 2\tfrac{1}{2}j - 5k$$

(d) $p \times (2r \times q)$

$$2r \times q = (2i - 2k) \times (i - j + \tfrac{1}{2}k)$$

$$= \begin{vmatrix} i & j & k \\ 2 & 0 & -2 \\ 1 & -1 & \tfrac{1}{2} \end{vmatrix}$$

$$= i(0 - 2) - j(1 + 2) + k(-2 - 0)$$

$$= -2i - 3j - 2k$$

Hence $p \times (2r \times q) = (3i + 2j - k) \times (-2i - 3j - 2k)$

$$= \begin{vmatrix} i & j & k \\ 3 & 2 & -1 \\ -2 & -3 & -2 \end{vmatrix}$$

$$= i(-4 - 3) - j(-6 - 2) + k(-9 + 4)$$

$$= -7i + 8j - 5k.$$

Problem 2. A force of $F = 2i + j - k$ newtons acts on a line passing through a point A. Find moment M and its magnitude M, of the force F about a point B when A has co-ordinates $(1, 2, -3)$ metres and B has co-ordinates $(0, 1, -1)$ metres.

Let r be the vector **BA**. From Worked Problem 2 in Section 2,

$$\textbf{OA} = i + 2j - 3k \text{ and } \textbf{OB} = j - k.$$
Then $\textbf{BA} = \textbf{OA} - \textbf{OB}.$
i.e. $r = (i + 2j - 3k) - (j - k) = i + j - 2k.$

Moment, $M = r \times F$

$$= (i + j - 2k) \times (2i + j - k)$$

$$= \begin{vmatrix} i & j & k \\ 1 & 1 & -2 \\ 2 & 1 & -1 \end{vmatrix}$$

$$= i(-1 + 2) - j(-1 + 4) + k(1 - 2)$$

$$= i - 3j - k \text{ N m.}$$

From equation (6) $M = |M| = |r \times F|$

$$= \sqrt{[(r \cdot r)(F \cdot F) - (r \cdot F)^2]}$$

$$r \cdot r = (1)(1) + (1)(1) + (-2)(-2) = 6$$

$$F \cdot F = (2)(2) + (1)(1) + (-1)(-1) = 6$$

$$r \cdot F = (1)(2) + (1)(1) + (-2)(-1) = 5$$

Hence, $M = \sqrt{[(6)(6) - 5^2]} = \sqrt{11}. \text{ N m.}$

Problem 3. A sphere of radius 10 centimetres with its centre at the origin, $(0, 0, 0)$, rotates about the z-axis with an angular velocity vector $\omega = i - 5j + 4k$ radians per second. Determine the velocity vector v and its magnitude for point A in the sphere when the position vector of A is $3i - j + 2k$ centimetres.

The velocity vector $v = \omega \times r$ where ω is the angular velocity vector and r is the position vector. Hence

$$v = \omega \times r = (i - 5j + 4k) \times (3i - j + 2k)$$

$$\text{i.e. } v = \begin{vmatrix} i & j & k \\ 1 & -5 & 4 \\ 3 & -1 & 2 \end{vmatrix}$$

$$= i(-10 + 4) - j(2 - 12) + k(-1 + 15)$$

$$= -6i + 10j + 14k.$$

From equation (6):

$$v = |\omega \times r| = \sqrt{[(\omega \cdot \omega)(r \cdot r) - (\omega \cdot r)^2]}$$

$$\omega \cdot \omega = 1^2 + (-5)^2 + 4^2 = 42$$

$$r \cdot r = 3^2 + (-1)^2 + 2^2 = 14$$

$$\omega \cdot r = (1)(3) + (-5)(-1) + (4)(2) = 16$$

$$\text{Hence, } v = \sqrt{[(42)(14) - 16^2]}$$

$$= 18.22 \text{ cm s}^{-1}$$

Further problems on vector products may be found in Section 5 (Problems 11 to 20), page 79.

4 Summary of scalar and vector product formulae

The formulae given below refers to two vectors:

$a = a_1 i + a_2 j + a_3 k$ and $b = b_1 i + b_2 j + b_3 k$, having an angle of θ between the positive directions of the vectors.

Scalar or dot products

By definition, the scalar or dot product is given by:

$$a \cdot b = |a| \, |b| \cos \theta = ab \cos \theta \qquad (1)$$
The modulus of a, $|a| = a = \sqrt{(a_1^2 + a_2^2 + a_3^2)}$ $\qquad (2)$

The value of the scalar product is:

$$a \cdot b = a_1 b_1 + a_2 b_2 + a_3 b_3 \qquad (3)$$

The angle between two vectors is given by.

$$\cos \theta = \frac{a \cdot b}{|a|\,|b|} = \frac{a \cdot b}{ab} = \frac{a_1 b_1 + a_2 b_2 + a_3 b_3}{\sqrt{(a_1^2 + a_2^2 + a_3^2)}\,\sqrt{(b_1^2 + b_2^2 + b_3^2)}} \tag{4}$$

Vector or cross products

The modulus of a vector or cross product is given by:

$$|a \times b| = |a|\,|b|\sin\theta = ab\sin\theta \tag{5}$$

Alternatively, in terms of scalar products.

$$|a \times b| = \sqrt{[(a \cdot a)(b \cdot b) - (a \cdot b)^2]} \tag{6}$$

The vector resulting from a vector product is given by:

$$a \times b = \begin{vmatrix} i & j & k \\ a_1 & a_2 & a_3 \\ b_1 & b_2 & b_3 \end{vmatrix} \tag{7}$$

5 Further problems

Scalar products and unit vectors

In Problems 1 to 6: $p = i + 3k$
$q = 2i + j - k$
$r = -i + 4j + 5k$

Determine the quantities stated.

1. (a) $p \cdot q$ $[-1]$
    (b) $p \cdot r$ $[14]$
    (c) $q \cdot r$ $[-3]$

2. (a) $|p|$ $[\sqrt{10}]$
    (b) $|q|$ $[\sqrt{6}]$
    (c) $|r|$ $[\sqrt{42}]$

3. (a) $q \cdot (p + r)$ $[-4]$
    (b) $r \cdot (p + q)$ $[11]$

4. $3p \cdot (q + 2r)$ $[81]$

5. (a) $|p + q|$ $[\sqrt{14}]$
    (b) $|p| + |q|$ $[\sqrt{10} + \sqrt{6} \simeq 5.612]$

6. The angle between

    (a) p and q $[97° \ 25']$
    (b) p and r $[46° \ 55']$
    (c) p and $-q$ $[82° \ 35']$
    (d) q and $(p - r)$ $[80° \ 24']$

In Problems 7 to 10, find the work done for the data given by a constant force of F newtons acting on a body at point A, when A is displaced to point B, in metres.

7. $F = 2i - 3k$, A : $(0, 1, 2)$, B : $(3, 0, -4)$ [24 N m]

8. $F = -i + 2j - k$, A : $(2, 0, 0)$, B : $(-1, -2, -3)$ [2 N m]

9. $F = 4i - 5j$, A : $(2, 1, -3)$, B : $(4, 4, 1)$ [−7 N m]

10. $F = 3i - j + k$, A : $(0, 1, 0)$, B : $(4, 2, 1)$ [12 N m]

Vector products

In Problems 11 to 16: $p = 2i + 3k$
                 $q = i - 2j + k$
                 $r = -2i + 4j - 3k$

Determine the quantities stated.

11. (a) $p \times q$ $[6i + j - 4k]$
   (b) $q \times p$ $[-6i - j + 4k]$

12. (a) $|p \times r|$ [14.42]
   (b) $|r \times q|$ [2.236]

13. (a) $2p \times 4r$ $[32(-3i + 2k)]$
   (b) $(p + q) \times r$ $[-10i + j + 8k]$

14. (a) $p \times (q \times r)$ $[-3i + 6j + 2k]$
   (b) $(p + 2q) \times r$ $[2(-4i + j + 4k]$

15. (a) $(r - 2p) \times \dfrac{q}{2}$ $[-7i - 1\frac{1}{2}j + 4k]$

   (b) $(2p \times 3r) \times q$ $[8(12i + 15j + 18k)]$

16. Prove that $p \times (q \times r) + q \times (r \times p) + r \times (p \times q) = 0$.

In Problems 17 and 18, a force F acts on a line through point P. Find the moment vector M and its modulus M of F about point Q, where:

17. $F = i + j$, P has co-ordinates $(2, 1, 1)$ and Q has co-ordinates $(3, -2, 4)$ $[M = 3i - 3j - 4k; M = 5.831]$

18. $F = 2i - j + k$, P has co-ordinates $(0, 3, 1)$ and Q has co-ordinates $(4, 0, -1)$ $[M = 5i + 8j - 2k; M = 9.644]$

In Problems 19 and 20, a circular cylinder with its centre at the origin rotates about the z-axis with vector angular velocity ω. Determine the vector velocity v and its magnitude v of point P on the cylinder for the data given and when $\omega = -5i + 2j - 7k$.

19. The position vector of P is $i + 2j$ $[v = 14i - 7j - 12k; v = 19.72]$

20. The position vector of P is $i - j + 2k$. $[v = 3(-i + j + k); v = 5.196]$

Chapter 6

Triple products of vectors (DE 17)

1 Triple scalar products of vectors

Definition

Products of three vectors of the form $a \cdot (b \times c)$ or $(a \times b) \cdot c$ are called **triple scalar products** of vectors a, b and c. The terms $b \times c$ and $a \times b$ are shown in brackets to indicate that vector products, $(\times)$, take precedence over scalar products, $(\cdot)$. Thus, for $a \cdot b \times c$, the $b \times c$ vector is determined before the scalar product is calculated. For $a \times b \cdot c$, the $a \times b$ vector is found first and then the scalar product is determined.

It is shown in Chapter 5, Section 3, that when $b = b_1 i + b_2 j + b_3 k$ and $c = c_1 i + c_2 j + c_3 k$, then the vector product $b \times c$ is given by:

$$b \times c = \begin{vmatrix} i & j & k \\ b_1 & b_2 & b_3 \\ c_1 & c_2 & c_3 \end{vmatrix}$$

$$= i(b_2 c_3 - b_3 c_2) - j(b_1 c_3 - b_3 c_1) + k(b_1 c_2 - b_2 c_1).$$

When $a = a_1 i + a_2 j + a_3 k$, then from Chapter 5, Section 2, the scalar product of a and $(b \times c)$ is given by:

$(a_1 i + a_2 j + a_3 k) \cdot [(b_2 c_3 - b_3 c_2)i - (b_1 c_3 - b_3 c_1)j + (b_1 c_2 - b_2 c_1)k]$,
i.e. $a \cdot (b \times c) = (a_1)(b_2 c_3 - b_3 c_2) - (a_2)(b_1 c_3 - b_3 c_1) + (a_3)(b_1 c_2 - b_2 c_1)$,
from equation (3), Chapter 5.

This relationship is best remembered in determinant form, i.e.

$$a \cdot (b \times c) = \begin{vmatrix} a_1 & a_2 & a_3 \\ b_1 & b_2 & b_3 \\ c_1 & c_2 & c_3 \end{vmatrix} \tag{1}$$

Properties

(i) The value of a triple scalar product remains unaltered if the dot and the cross are interchanged, i.e. $a \cdot (b \times c) = (a \times b) \cdot c$. However, when calculating these quantities '$\times$' still takes precedence over '$\cdot$'. This property may be verified as follows:

When $a = a_1 i + a_2 j + a_3 k$ and $b = b_1 i + b_2 j + b_3 k$, then from Chapter 5, Section 3

$$a \times b = \begin{vmatrix} i & j & k \\ a_1 & a_2 & a_3 \\ b_1 & b_2 & b_3 \end{vmatrix}$$

$$= i(a_2 b_3 - a_3 b_2) - j(a_1 b_3 - a_3 b_1) + k(a_1 b_2 - a_2 b_1).$$

When $c = c_1 i + c_2 j + c_3 k$, then the scalar product of $(a \times b)$ and c is:

$$[(a_2 b_3 - a_3 b_2)i - (a_1 b_3 - a_3 b_1)j + (a_1 b_2 - a_2 b_1)k] \cdot (c_1 i + c_2 j + c_3 k)$$

i.e. $(a \times b) \cdot c = (a_2 b_3 - a_3 b_2)(c_1) - (a_1 b_3 - a_3 b_1)(c_2) + (a_1 b_2 - a_2 b_1)(c_3)$

$$= \begin{vmatrix} c_1 & c_2 & c_3 \\ a_1 & a_2 & a_3 \\ b_1 & b_2 & b_3 \end{vmatrix}$$

When a row or a column of a determinant are interchanged, the sign changes, hence, interchanging the first and second rows and then interchanging the second and third rows gives:

$$\begin{vmatrix} c_1 & c_2 & c_3 \\ a_1 & a_2 & a_3 \\ b_1 & b_2 & b_3 \end{vmatrix} = -\begin{vmatrix} a_1 & a_2 & a_3 \\ c_1 & c_2 & c_3 \\ b_1 & b_2 & b_3 \end{vmatrix} = +\begin{vmatrix} a_1 & a_2 & a_3 \\ b_1 & b_2 & b_3 \\ c_1 & c_2 & c_3 \end{vmatrix}$$

i.e. $(a \times b) \cdot c = \begin{vmatrix} a_1 & a_2 & a_3 \\ b_1 & b_2 & b_3 \\ c_1 & c_2 & c_3 \end{vmatrix}$

But this determinant is the same as the determinant for $a \cdot (b \times c)$ given in equation (1), thus it follows that

$$a \cdot (b \times c) = (a \times b) \cdot c.$$

(ii) The value of a triple scalar product remains unaltered by the cyclic interchange of the three vectors, i.e.

$$a \cdot (b \times c) = c \cdot (a \times b) = b \cdot (c \times a)$$

This property may be verified as follows. From equation (1):

$$a \cdot (b \times c) = \begin{vmatrix} a_1 & a_2 & a_3 \\ b_1 & b_2 & b_3 \\ c_1 & c_2 & c_3 \end{vmatrix}$$

Thus, $c \cdot (a \times b) = \begin{vmatrix} c_1 & c_2 & c_3 \\ a_1 & a_2 & a_3 \\ b_1 & b_2 & b_3 \end{vmatrix}$ and $b \cdot (c \times a) = \begin{vmatrix} b_1 & b_2 & b_3 \\ c_1 & c_2 & c_3 \\ a_1 & a_2 & a_3 \end{vmatrix}$

Interchanging the first and second rows and then the second and third rows of the determinant for $c \cdot (a \times b)$ gives:

$$c \cdot (a \times b) = \begin{vmatrix} c_1 & c_2 & c_3 \\ a_1 & a_2 & a_3 \\ b_1 & b_2 & b_3 \end{vmatrix} = - \begin{vmatrix} a_1 & a_2 & a_3 \\ c_1 & c_2 & c_3 \\ b_1 & b_2 & b_3 \end{vmatrix} = \begin{vmatrix} a_1 & a_2 & a_3 \\ b_1 & b_2 & b_3 \\ c_1 & c_2 & c_3 \end{vmatrix}$$

Similarly, interchanging the second and third rows and then the first and second rows of the determinant for $b \cdot (c \times a)$, gives:

$$b \cdot (c \times a) = \begin{vmatrix} b_1 & b_2 & b_3 \\ c_1 & c_2 & c_3 \\ a_1 & a_2 & a_3 \end{vmatrix} = - \begin{vmatrix} b_1 & b_2 & b_3 \\ a_1 & a_2 & a_3 \\ c_1 & c_2 & c_3 \end{vmatrix} = \begin{vmatrix} a_1 & a_2 & a_3 \\ b_1 & b_2 & b_3 \\ c_1 & c_2 & c_3 \end{vmatrix}$$

i.e. $a \cdot (b \times c) = c \cdot (a \times b) = b \cdot (c \times a) = \begin{vmatrix} a_1 & a_2 & a_3 \\ b_1 & b_2 & b_3 \\ c_1 & c_2 & c_3 \end{vmatrix}$

Applications

(i) One of the geometrical applications of triple scalar products is in determining the volumes of certain regular shapes. For example, the volume of the parallelepiped (see Fig. 1), with a. b and c as adjacent edges, is the value of the triple scalar product $a \cdot (b \times c)$. This may be shown as follows:

From Chapter 5, Section 2, $a \cdot b = |a| \, |b| \cos \theta$, where θ is the angle between a and b. Hence, with reference to Fig. 1:

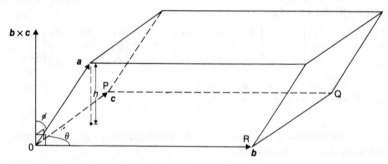

Fig. 1 Volume $= a \cdot (b \times c)$

$a \cdot (b \times c) = |a| \, |b \times c| \cos \phi$

But $|a| \cos \phi = a \cos \phi = h$, the perpendicular height of the parallelepiped. Also, from Chapter 5, Section 3, $|b \times c| = |b| \, |c| \sin \theta$
$$= b c \sin \theta = \text{the area of } OPQR.$$
Hence, $a \cdot (b \times c) =$ (perpendicular height of parallelepiped) $\times$ (area of base)
$$= \text{volume of the parallelepiped.} \tag{2}$$

By considering the parallelepiped shown in Fig. 1, it is possible to deduce that if three vectors are coplanar, that is, say, they all act in the b-c plane, then the triple scalar product is zero. When the vectors are coplanar, either the height or the width or the depth of the parallelepiped is zero, and if this is so, it has no volume, i.e. $a \cdot (b \times c) = 0$.

(ii) In mechanics, it is possible to determine the component of a vector product, such as the moment M of a force, or the linear velocity v of a point on a rotating body, about any axis by using triple scalar products. For example, in Fig. 2, a force of F is shown acting on a body at point P. OP is the position vector, r, of P from the origin. It is required to find the component of the moment vector of F about axis ON, lying in the i-k plane and at an angle of $\theta°$ to i.

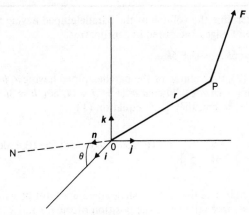

Fig. 2 $M_1 = (r \times F) \cdot n.$

The moment of F about an axis through 0 normal to the plane of F and r is given by $M = r \times F$ (see Ch. 5, Section 3). The component of M along the ON-axis is $M \cdot n$ (see Ch. 5, Section 2), where n is the unit vector along the ON-axis. Thus if M_1 is the modulus of the moment of the component of M acting along ON

$$M_1 = (r \times F) \cdot n, \tag{3}$$

i.e. a triple scalar product.

Worked problems on triple scalar products of vectors

Problem 1. (a) Find the triple scalar product, $p \cdot (q \times r)$ when $p = i$, $q = 2j$ and $r = 3k$.

(b) Find $(p \times q) \cdot r$ when $p = i + j$, $q = 4i - 6j + 7k$ and $r = -5j + 2k$.

(a) $p = 1i + 0j + 0k$, $q = 0i + 2j + 0k$ and $r = 0i + 0j + 3k$.
From equation (1):

$$p \cdot (q \times r) = \begin{vmatrix} 1 & 0 & 0 \\ 0 & 2 & 0 \\ 0 & 0 & 3 \end{vmatrix} = 1(6 - 0) - 0(0 - 0) + 0(0 - 0)$$

i.e. $p \cdot (q \times r) = 6$

(b) It is shown in Section 1 that $(p \times q) \cdot r = p \cdot (q \times r)$, hence equation (1) can be used to determine the value of $(p \times q) \cdot r$ when $p = 1i + 1j + 0k$, $q = 4i - 6j + 7k$ and $r = 0i - 5j + 2k$, then

$$(p \times q) \cdot r = p \cdot (q \times r) = \begin{vmatrix} 1 & 1 & 0 \\ 4 & -6 & 7 \\ 0 & -5 & 2 \end{vmatrix} = \begin{aligned} & 1(-12 + 35) - 1(8 - 0) \\ & \qquad + 0(-20 - 0) \end{aligned}$$

i.e. $(p \times q) \cdot r = 15$

Problem 2. Determine the volume of the parallelepiped having the following vectors as adjacent edges, measured in centimetres:

$i + j, \quad 3i - 5j + 6k, \quad -4j + 5k.$

From equation (2), the volume of the parallelepiped having a, b and c as adjacent edges is $a \cdot (b \times c)$. When $a = 1i + 1j + 0k$ cm, $b = 3i - 5j + 6k$ cm and $c = 0i - 4j + 5k$ cm, then from equation (1)

$$a \cdot (b \times c) = \begin{vmatrix} 1 & 1 & 0 \\ 3 & -5 & 6 \\ 0 & -4 & 5 \end{vmatrix} = 1(-25 + 24) - 1(15 - 0) + 0(-12 + 0)$$

$$= -16.$$

The minus sign indicates that the parallelepiped does not lie wholly in the first quadrant of space relative to the position of the i, j and k unit vectors. Hence, volume of parallelepiped is 16 cm^3.

Problem 3. A force of $F = i + 2j + 5k$ newtons acts at P, as shown in Fig. 2(a). Position vector $r = 2i + 4j - 3k$ metres. Determine the magnitude of the component of the moment vector of F about axis ON. ON lies in the i-k plane at an angle of 30° to i. Determine also the vector equation of this component in terms of i and k.

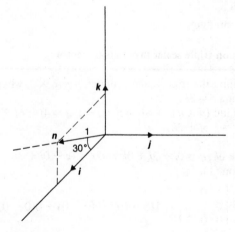

Fig. 3

The unit vector, n, can be expressed in terms of unit vectors i, j and k. From Fig. 3, the component of n along i is $(1)(\cos 30°)$, i.e. $\dfrac{\sqrt{3}}{2}$, the component along j is 0 and the component along k is $(1)(\sin 30°)$, i.e. $\dfrac{1}{2}$. Thus

$$n = \frac{\sqrt{3}}{2} i + 0j + \frac{1}{2} k.$$

From equation (3), the magnitude of the component of M about axis ON, M_1, is given by

$$M_1 = (r \times F) \cdot n = r \cdot (F \times n)$$

From equation (1)

$$r \cdot (F \times n) = \begin{vmatrix} 2 & 4 & -3 \\ 1 & 2 & 5 \\ \frac{\sqrt{3}}{2} & 0 & \frac{1}{2} \end{vmatrix} = 2(1) - 4\left(\frac{1}{2} - \frac{5\sqrt{3}}{2}\right) - 3(-\sqrt{3})$$

$$= 2 - 2 + 10\sqrt{3} + 3\sqrt{3}$$

$$\text{Hence,} \; M_1 = 13\sqrt{3} \; \text{N m}.$$

The component of M_1 acting along i is $13\sqrt{3} \cos 30°$, i.e. $\dfrac{39}{2}i$.

The component of M_1 acting along j is zero.

The component of M_1 acting along k is $13\sqrt{3} \sin 30°$, i.e. $\dfrac{13\sqrt{3}}{2} k$.

Thus in vector form $M_1 = \dfrac{39}{2}i + \dfrac{13\sqrt{3}}{2} k$ N m.

Further problems on triple scalar products of vectors may be found in Section 4 (Problems 1 to 8), page 89.

2 Triple vector products

If a, b and c are three vectors, the vectors formed by determining $a \times (b \times c)$ and $(a \times b) \times c$ are called the **triple vector products** of a, b and c. The relationship between triple vector products and scalar products is:

$$a \times (b \times c) = (a \cdot c)b - (a \cdot b)c \tag{4}$$

This may be shown as follows. In order to simplify the number and size of the algebraic term in the verification of equation (4), the cartesian co-ordinate system is chosen so that one axis, say, the x-axis coincides with a, i.e. $a = a_1 i$. The x-y plane is selected so that b lies in this plane, then $b = b_1 i + b_2 j$. Let c be $c_1 i + c_2 j + c_3 k$. By choosing the cartesian co-ordinate system in this way, it is 'lined-up' with the vector system, but places no restriction on the magnitude or direction of any of the vectors. The vector product of b and c

is given by

$$b \times c = \begin{vmatrix} i & j & k \\ b_1 & b_2 & 0 \\ c_1 & c_2 & c_3 \end{vmatrix} = i\,(b_2 c_3) - j(b_1 c_3) + k(b_1 c_2 - b_2 c_1)$$

$$a \times (b \times c) = (a_1 i + 0j + 0k) \times [(b_2 c_3)i - (b_1 c_3)j + (b_1 c_2 - b_2 c_1)k]$$

i.e. $a \times (b \times c) = \begin{vmatrix} i & j & k \\ a_1 & 0 & 0 \\ b_2 c_3 & -b_1 c_3 & b_1 c_2 - b_2 c_1 \end{vmatrix}$

$$= -j[a_1(b_1 c_2 - b_2 c_1)] + k(-a_1 b_1 c_3)$$

Also, when $a = a_1 i$, $b = b_1 i + b_2 j$ and $c = c_1 i + c_2 j + c_3 k$, then

$$(a \cdot c)b - (a \cdot b)c = [(a_1 i) \cdot (c_1 i + c_2 j + c_3 k)]\,(b_1 i + b_2 j)$$
$$- [(a_1 i) \cdot (b_1 i + b_2 j)]\,(c_1 i + c_2 j + c_3 k)$$
$$= [(a_1)(c_1) + (0)(c_2) + (0)(c_3)]\,(b_1 i + b_2 j)$$
$$- [(a_1)(b_1) + (0)(b_2) + (0)(0)]\,(c_1 i + c_2 j + c_3 k)$$
$$= a_1 c_1 (b_1 i + b_2 j) - a_1 b_1 (c_1 i + c_2 j + c_3 k)$$
$$= a_1 c_1 b_1 i + a_1 c_1 b_2 j - a_1 b_1 c_1 i - a_1 b_1 c_2 j - a_1 b_1 c_3 k$$
$$= -j[a_1 (b_1 c_2 - b_2 c_1)] + k(-a_1 b_1 c_3),$$

the same result as obtained for $a \times (b \times c)$. Thus:

$a \times (b \times c) = (a \cdot c)b - (a \cdot b)c$ as stated in equation (4).

By following the same procedure as that used above for finding $a \times (b \times c)$ in terms of scalar products, it can be shown that:

$$(a \times b) \times c = -c \times (a \times b) = -[(c \cdot b)a - (c \cdot a)b]$$

i.e. $(a \times b) \times c = (c \cdot a)b - (c \cdot b)a$ \hfill (5)

It can be seen that equation (5) is not the same as equation (4). Thus, in general $a \times (b \times c)$ is **not** equal to $(a \times b) \times c$.

Worked problems on triple vector products

Problem 1. If $a = 2i + 5k$, $b = i + 3j - 4k$, $c = -3j + 7k$ and $d = -3i + 4j - 2k$, find (a) $a \times (b \times c)$, (b) $(a \times b) \times c$, (c) $4a \times (2c \times 3d)$, (d) $(a \times b) \times (c \times d)$ and (e) $[(a - b) \times 3c] \times 2d$.

(a) From equation (4), $a \times (b \times c) = (a \cdot c)b - (a \cdot b)c$
When $a = 2i + 5k$, $b = i + 3j - 4k$ and $c = -3j + 7k$, then

$$a \times (b \times c) = [(2)(0) + (0)(-3) + (5)(7)]\,(i + 3j - 4k)$$
$$- [(2)(1) + (0)(3) + (5)(-4)]\,(-3j + 7k).$$
$$= 35(i + 3j - 4k) + 18(-3j + 7k)$$
$$= 35\,i + 51\,j - 14\,k.$$

(b) From equation (5),

$$(a \times b) \times c = (c \cdot a)b - (c \cdot b)a$$
$$= [(0)(2) + (-3)(0) + (7)(5)] \ (i + 3j - 4k)$$
$$- \ [(0)(1) + (-3)(3) + (7)(-4)] \ (2i + 5k)$$
$$= 35 \ (i + 3j - 4k) + 37(2i + 5k)$$
$$= 109i + 105j + 45k.$$

From the results of (a) and (b) it can be seen that $a \times (b \times c) \neq (a \times b) \times c$ in this case.

(c) From equation (4), $4a \times (2c \times 3d) = (4a \cdot 3d) \ 2c - (4a \cdot 2c) \ 3d$

$4a = 8i + 20k, \qquad 2c = -6j + 14k$ and $3d = -9i + 12j - 6k$.
Thus, $4a \times (2c \times 3d) = [(8)(-9) + (0)(12) + (20)(-6)] \ (-6j + 14k)$
$$- \ [(8)(0) + (0)(-6) + (20)(14)] \ (-9i + 12j - 6k)$$
$$= -192 \ (-6j + 14k) - 280 \ (-9i + 12j - 6k)$$
$$= 2\ 520i - 2\ 208j - 1\ 008k.$$

(d) The vector product $(a \times b)$ is given by:

$$(a \times b) = \begin{vmatrix} i & j & k \\ 2 & 0 & 5 \\ 1 & 3 & -4 \end{vmatrix}$$
$$= i(-15) - j(-8 - 5) + k(6)$$
$$= -15i + 13j + 6k.$$

From equation (4):

$$(a \times b) \times (c \times d) = [(a \times b) \cdot d]c - [(a \times b) \cdot c]d$$
$$= [(-15i + 13j + 6k) \cdot (-3i + 4j - 2k)] \ (-3j + 7k)$$
$$- \ [(-15i + 13j + 6k) \cdot (-3j + 7k)] \ (-3i + 4j - 2k)$$
$$= [(-15)(-3) + (13)(4) + (6)(-2)] \ (-3j + 7k)$$
$$- \ [(-15)(0) + (13)(-3) + (6)(7)] \ (-3i + 4j - 2k)$$
$$= 85 \ (-3j + 7k) - 3(-3i + 4j - 2k)$$
$$= 9i - 267j + 601k.$$

(e) $(a - b) = (2i + 5k) - (i + 3j - 4k)$
$$= i - 3j + 9k$$
$3c = 3(-3j + 7k) = -9j + 21k$
$2d = 2(-3i + 4j - 2k) = -6i + 8j - 4k.$

From equation (5)

$$[(a - b) \times 3c] \times 2d = [2d \cdot (a - b)] \ 3c - (2d \cdot 3c)(a - b)$$

$$= [(-6)(1) + (8)(-3) + (-4)(9)] \ (-9j + 21k)$$
$$\quad - [(-6)(0) + (8)(-9) + (-4)(21)] \ (i - 3j + 9k)$$
$$= -66 \ (-9j + 21k) + 156 \ (i - 3j + 9k)$$
$$= 156i + 126j + 18k.$$

Problem 2. The centrifugal force, in newtons, experienced by a body rotating with constant angualr velocity about a fixed axis is equal to $-m \, \omega \times (\omega \times r)$, where ω is the angular velocity vector in radians per second and r is the position vector, in metres, of the body relative to the origin. Determine the centrifugal force vector and its magnitude when $r = -2i + j - 4k$ metres and $\omega = 3i - 2j + 5k$ radians per second. The mass of the body is 12 kilograms.

$$-m \, \omega = -12 \, (3i - 2j + 5k) = -36i + 24j - 60k$$

Using equation (4), where, $a = -m \, \omega, b = \omega$ and $c = r$, gives

$$-m \, \omega \times (\omega \times r) = (-m \omega \cdot r)\omega - (-m \omega \cdot \omega)r$$
$$= [(-36)(-2) + (24)(1) + (-60)(-4)] \ (3i - 2j + 5k)$$
$$\quad - [(-36)(3) + (24)(-2) + (-60)(5)] \ (-2i + j - 4k)$$
$$= 336 \ (3i - 2j + 5k) + 456 \ (-2i + j - 4k)$$
$$= 96i - 216j - 144k.$$

i.e. the centrifugal force vector is $96i - 216j - 144k$ newtons.
From Chapter 5, Section 2, the magnitude of this vector is given by

$$\sqrt{[96^2 + (-216)^2 + (-144)^2]}$$

i.e. the magnitude of the centrifugal force is 276.8 newtons, correct to four significant figures.

Further problems on triple vector products may be found in Section 4 (Problems 9 to 20), page 90.

3 Summary of triple scalar and vector product formulae

Triple scalar products of vectors
Triple scalar products are of the form $a \cdot (b \times c)$ or $(a \times b) \cdot c$

$$a \cdot (b \times c) = \begin{vmatrix} a_1 & a_2 & a_3 \\ b_1 & b_2 & b_3 \\ c_1 & c_2 & c_3 \end{vmatrix} \qquad (1)$$

$$a \cdot (b \times c) = (a \times b) \cdot c$$
$$a \cdot (b \times c) = c \cdot (a \times b) \ = b \cdot (c \times a)$$

When a triple scalar product is determined, the result is a **scalar** quantity.

The volume of a parallelepiped $= a \cdot (b \times c)$, (2)

where a, b and c are adjacent edges of the parallelepiped.

The modulus of the component of moment vector M (see Fig. 2), acting along any axis ON, is given by M_1 where

$$M_1 = (r \times F) \cdot n \qquad (3)$$

where r is the position vector and n the unit vector along ON.

Triple vector products

Triple vector products are of the form $a \times (b \times c)$ or $(a \times b) \times c$.

$$a \times (b \times c) = (a \cdot c)b - (a \cdot b)c \qquad (4)$$

$$(a \times b) \times c = (c \cdot a)b - (c \cdot b)a \qquad (5)$$

Thus, in general, $a \times (b \times c) \neq (a \times b) \times c$.

When a triple vector product is determined, the result is a **vector** quantity.

4 Further problems

Triple scalar products of vectors

In Problems 1 to 4, $p = 3i - 7j + 4k$
$\qquad\qquad\qquad\quad q = -4i + 2k$
$\qquad\qquad\qquad\quad r = 3j - 5k$

Determine the values of the triple scalar products given.

1. $p \cdot q \times r$ [74]

2. $(q \times p) \cdot r$ [−74]

3. $r \cdot (p \times q)$ [74]

4. $(p \times r) \cdot q$ [−74]

In Problems 5 and 6, determine the volumes of the parallelepipeds having the vectors stated as adjacent edges. The vectors are measured in centimetres and the answers should be expressed correct to four significant figures.

5. $-3i + 4j - 10k$, $5i - 1.2j + 4.7k$ and $7i - 2.3j - 4.9k$. [210.5 cm^3]

6. $1.7i - 2.4j + 3k, 4.3j - 7.4k$ and $14.8i + 10.3k$. [147.2 cm^3]

In Problems 7 and 8, a force of $F = 6i - 4j + 5k$ newtons acts at P, as shown in Fig. 2. Determine the magnitude of the component of the moment vector about axis ON and its vector equation for the data given.

7. Position vector $r = -3i - 2j + k$ cm and axis ON lies in the i-j plane at an angle of $45°$ to i.

$$\left[M_1 = \frac{15}{\sqrt{2}} \text{ N cm}, \quad M_1 = \frac{15}{2}i + \frac{15}{2}j \text{ N cm} \right]$$

8. Position vector, $r = 1.7i + 2.3j - 4.4k$ centimetres and axis ON lies in the j-k plane at an angle of $50°$ to k.

$$\left[\begin{array}{l} M_1 = -39.98 \text{ N cm, correct to four significant figures.} \\ M_1 = -30.63j - 25.70k \text{ N cm, correct to four significant figures.} \end{array} \right]$$

Triple vector products

In Problems 9 to 15, $a = 4i + 3k$, $b = 3j - k$, $c = i - 2j + 5k$
and $d = -2i + j - 4k$. Determine the quantities stated.

9. (a) $a \times (b \times c)$ $[3i + 51j - 4k]$
   (b) $b \times (c \times d)$ $[-15i - 3j - 9k]$

10. (a) $(a \times c) \times d$ $[76i + 40j - 28k]$
    (b) $(b \times d) \times c$ $[22i + 61j + 20k]$

11. (a) $2a \times (3b \times c)$ $[18i + 306j - 24k]$
    (b) $(2b \times 3d) \times 4c$ $[528i + 1\,464j + 480k]$

12. (a) $(3a - 2b) \times (b \times 2c)$ $[58i + 358j + 132k]$
    (b) $(a + 4c) \times (2b \times \frac{1}{2}d)$ $[-94i - 301j - 72k]$

13. (a) $(a \times b) \times (c \times d)$ $[60i + 9j + 42k]$
    (b) $(a \times a) \times (b \times c)$ $[0]$

14. (a) $(a \times (b \times c)) \times (c \times a)$ $[476i + 357k]$
    (b) $(a \cdot (c \times b)) \times d$ $[86i - 43j + 172k]$

15. (a) $((a \times b) \cdot c) \times (b \times c)$ $[559i - 43j - 129k]$
    (b) $b \times (a \cdot (d \times c))$ $[-9j + 3k]$

16. Show that $(q \times r) \times (r \times p) = [p \cdot (q \times r)]\, r$

17. Prove that $(p \times q) \cdot [(q \times r) \times (r \times p)] = [p \cdot (q \times r)]^2$

18. Prove that $p \times [q \times (r \times s)] = p \cdot (r \times s)\, q - (p \cdot q)(r \times s)$

In Problems 19 and 20, determine the centrifugal force vector and its magnitude for the data given. The centrifugal force is given by $-m\, \boldsymbol{\omega} \times (\boldsymbol{\omega} \times r)$ newtons where m is the mass of the body in kilograms rotating at constant angular velocity $\boldsymbol{\omega}$ radians per second about a fixed axis. r is the position vector of the body from the origin.

19. $m = 3$ kg, $r = -2i + 5j + 7k$ metres and $\boldsymbol{\omega} = j + 2k$ radians per second.

$[-30i + 18j - 9k$ N; 36.12 N, correct to four significant figures]

20. $m = 3.14$ kg, $r = -2.0j + 3.7k$ metres and $\boldsymbol{\omega} = 2.4i - 1.8j + 6.3k$ radians per second.

$[-202.8i - 153.7j + 33.35k$ N; 256.6 N, both correct to
four significant figures]

Chapter 7

The notation of second order matrices and determinants (BC 7)

1 Introduction

Matrices are used in engineering and science for solving linear simultaneous equations.

Terms used in connection with matrices
Consider the linear simultaneous equations:

$$2x + 3y = 4 \tag{1}$$
$$\text{and } 5x - 6y = 7 \tag{2}$$

In **matrix** notation, the coefficients of x and y are written as $\begin{pmatrix} 2 & 3 \\ 5 & -6 \end{pmatrix}$,

that is, occupying the same relative positions as in equations (1) and (2) above. The grouping of the coefficients of x and y in this way is called an **array** and the coefficients forming the array are called the **elements** of the matrix.

If there are m rows across an array and n columns down an array, then the matrix is said to be of order $m \times n$, called 'm by n'. Thus for the equations

$$2x + 3y - 4z = 5 \tag{3}$$
$$6x - 7y + 8z = 9 \tag{4}$$

the matrix of the coefficients of x, y and z is $\begin{pmatrix} 2 & 3 & -4 \\ 6 & -7 & 8 \end{pmatrix}$ and is a

'2 by 3' matrix. A matrix having a single row is called a **row** matrix and one having a single column is called a **column** matrix. For example, in equation (3) above, the coefficients of x, y and z form a row matrix of $(2 \quad 3 \quad -4)$ and the coefficients of x in equations (3) and (4) form a column matrix of $\begin{pmatrix} 2 \\ 6 \end{pmatrix}$.

A matrix having the same number of rows as columns is called a **square** matrix. Thus the matrix for the coefficients of x and y in equations (1) and (2) above, i.e. $\begin{pmatrix} 2 & 3 \\ 5 & -6 \end{pmatrix}$, is a square matrix, and is called a second order matrix. Matrices are generally denoted by capital letters and if the matrix representing the coefficients of x and y in equations (1) and (2) above is A,

then $A = \begin{pmatrix} 2 & 3 \\ 5 & -6 \end{pmatrix}$.

2 Addition, subtraction and multiplication of second order matrices

In arithmetic, once the basic procedures associated with addition, subtraction, multiplication and division have been mastered, simple problems may be solved. With matrices, the various rules governing them have to be understood before they can be used to solve practical problems. In this section the basic laws of addition, subtraction and multiplication are introduced.

A matrix does not have a single numerical value and cannot be simplified to a particular answer. The main advantage of using matrices is that by applying the laws of matrices, given in this section, they can be simplified, and by comparing one matrix with another similar matrix, values of unknown elements can be determined. It will be seen in Chapter 8 that matrices can be used in this way for solving simultaneous equations. Matrices can be added, subtracted and multiplied and suitable definitions for these operations are formulated, so that they obey most of the laws which govern the algebra of numbers.

Addition

Only matrices of the same order may be added. Thus a 2 by 2 matrix can be added to a 2 by 2 matrix by adding corresponding elements, but a 3 by 2 matrix cannot be added to a 2 by 2 matrix, since some elements in one matrix do not have corresponding elements in the other. The sum of two matrices is the matrix obtained by adding the elements occupying corresponding positions in the matrix, and results in two matrices being simplified to a single matrix.

For example, the matrices $\begin{pmatrix} 1 & 3 \\ 2 & -4 \end{pmatrix}$ and $\begin{pmatrix} 2 & 5 \\ 6 & -7 \end{pmatrix}$

are added as follows:

$$\begin{pmatrix} 1 & 3 \\ 2 & -4 \end{pmatrix} + \begin{pmatrix} 2 & 5 \\ 6 & -7 \end{pmatrix} = \begin{pmatrix} 1+2 & 3+5 \\ 2+6 & (-4)+(-7) \end{pmatrix}$$

$$= \begin{pmatrix} 3 & 8 \\ 8 & -11 \end{pmatrix}$$

Subtraction

Only matrices of the same order can be subtracted and the difference between two matrices, say $A - B$, is the matrix obtained by subtracting the elements of matrix B from those occupying the corresponding positions in matrix A.

For example:

$$\begin{pmatrix} 1 & 3 \\ 2 & -4 \end{pmatrix} - \begin{pmatrix} 2 & 5 \\ 6 & -7 \end{pmatrix} = \begin{pmatrix} 1-2 & 3-5 \\ 2-6 & (-4)-(-7) \end{pmatrix}$$

$$= \begin{pmatrix} -1 & -2 \\ -4 & 3 \end{pmatrix}$$

By adding the single matrix obtained by adding A and B to, say, matrix C, the single matrix representing $A + B + C$ is obtained. By taking, say, matrix D from this single matrix, $A + B + C - D$ is obtained. Thus the laws of addition and subtraction can be applied to more than two matrices, providing that they are all of the same order.

Multiplication

(a) Scalar multiplication

When a matrix is multiplied by a number, the resultant matrix is one of the same order having each element multiplied by the number.

Thus, if matrix $A = \begin{pmatrix} 1 & 3 \\ 2 & -4 \end{pmatrix}$, then $2A = 2 \begin{pmatrix} 1 & 3 \\ 2 & -4 \end{pmatrix}$

$$= \begin{pmatrix} 2 \times 1 & 2 \times 3 \\ 2 \times 2 & 2 \times (-4) \end{pmatrix}$$

$$= \begin{pmatrix} 2 & 6 \\ 4 & -8 \end{pmatrix}$$

(b) Multiplication of matrices

Two matrices can only be multiplied together when the number of columns in the first one is equal to the number of rows in the second one. This is because the process of matrix multiplication depends on finding the sum of the products of the rows in one matrix with the columns in the other. Thus it is possible to multiply a 2 by 2 matrix by a column matrix having two elements or by another 2 by 2 matrix, but it is not possible to multiply it by a row matrix. Thus if:

$$A = \begin{pmatrix} 2 & 3 \\ 5 & 6 \end{pmatrix}, B = \begin{pmatrix} 1 \\ 8 \end{pmatrix} \text{ and } C = (4 \quad 9)$$

it is possible to find $A \times B$, since the number of columns in A is equal to the number of rows in B, but it is not possible to find $A \times C$ since there are two columns in A but only one row in C. If a 2 by 2 matrix A is multiplied by a column matrix, B, having two elements, the resulting matrix is a two element column matrix. The top element is the sum of the products obtained by taking the elements of the top row of A with B. The bottom element is the sum of the products obtained by taking the bottom elements of A with B.

If $A = \begin{pmatrix} a & b \\ c & d \end{pmatrix}$ and $B = \begin{pmatrix} p \\ q \end{pmatrix}$

then $A \times B = \begin{pmatrix} a & b \\ c & d \end{pmatrix} \times \begin{pmatrix} p \\ q \end{pmatrix} = \begin{pmatrix} ap + bq \\ cp + dq \end{pmatrix}$

For example, to multiply the matrices, say, $\begin{pmatrix} 2 & -5 \\ 4 & 3 \end{pmatrix}$ and $\begin{pmatrix} 1 \\ 6 \end{pmatrix}$ gives

$$\begin{pmatrix} 2 & -5 \\ 4 & 3 \end{pmatrix} \times \begin{pmatrix} 1 \\ 6 \end{pmatrix} = \begin{pmatrix} 2 \times 1 + (-5) \times 6 \\ 4 \times 1 + 3 \times 6 \end{pmatrix} = \begin{pmatrix} 2 - 30 \\ 4 + 18 \end{pmatrix}$$

$$= \begin{pmatrix} -28 \\ 22 \end{pmatrix}$$

If a 2 by 2 matrix, say, A, is multiplied by a 2 by 2 matrix, say B, the resulting matrix is a 2 by 2 matrix, say C. The top elements of C are the sum of the products obtained by taking the top row of A with the columns of B. The bottom elements of C are the sum of the products obtained by taking the bottom row of A with the columns of B.

For example:

$$\begin{pmatrix} 1 & 3 \\ -2 & 4 \end{pmatrix} \times \begin{pmatrix} 5 & 0 \\ 7 & -6 \end{pmatrix} = \begin{pmatrix} 1 \times 5 + 3 \times 7 & 1 \times 0 + 3 \times (-6) \\ -2 \times 5 + 4 \times 7 & -2 \times 0 + 4 \times (-6) \end{pmatrix}$$

$$= \begin{pmatrix} 26 & -18 \\ 18 & -24 \end{pmatrix}$$

In general, when a matrix of dimension (m by n) is multiplied by a matrix of dimension (n by q), the resulting matrix is one of dimension (m by q).

Although the laws of matrices are so formulated that they follow most of the laws which govern the algebra of numbers, frequently in the multiplication of matrices

$A \times B \neq B \times A$,

It is shown above that

$$A \times B = \begin{pmatrix} 1 & 3 \\ -2 & 4 \end{pmatrix} \times \begin{pmatrix} 5 & 0 \\ 7 & -6 \end{pmatrix} = \begin{pmatrix} 26 & -18 \\ 18 & -24 \end{pmatrix}$$

However, $B \times A = \begin{pmatrix} 5 & 0 \\ 7 & -6 \end{pmatrix} \times \begin{pmatrix} 1 & 3 \\ -2 & 4 \end{pmatrix} =$

$$= \begin{pmatrix} 5 \times 1 + 0 \times (-2) & 5 \times 3 + 0 \times 4 \\ 7 \times 1 + (-6) \times (-2) & 7 \times 3 + (-6) \times 4 \end{pmatrix}$$

i.e. $B \times A = \begin{pmatrix} 5 & 15 \\ 19 & -3 \end{pmatrix}$

That is, $A \times B \neq B \times A$ in this case. The results are said to be *non-commutative* (i.e., they are not in agreement).

Worked problems on the addition, subtraction and multiplication of second order matrices

---

Problem 1. If $A = \begin{pmatrix} 1 & 4 \\ -3 & 2 \end{pmatrix}$, $B = \begin{pmatrix} 5 & -1 \\ 0 & 1 \end{pmatrix}$ and $C = \begin{pmatrix} 3 & -4 \\ 7 & 2 \end{pmatrix}$

determine the single matrix for (a) $A + C$, (b) $A - C$, and (c) $A + B - C$.

(a) $A + C = \begin{pmatrix} 1 & 4 \\ -3 & 2 \end{pmatrix} + \begin{pmatrix} 3 & -4 \\ 7 & 2 \end{pmatrix} = \begin{pmatrix} 1+3 & 4+(-4) \\ (-3)+7 & 2+2 \end{pmatrix}$

$$= \begin{pmatrix} 4 & 0 \\ 4 & 4 \end{pmatrix}$$

(b) $A - C = \begin{pmatrix} 1 & 4 \\ -3 & 2 \end{pmatrix} - \begin{pmatrix} 3 & -4 \\ 7 & 2 \end{pmatrix} = \begin{pmatrix} 1-3 & 4-(-4) \\ (-3)-7 & 2-2 \end{pmatrix}$

$$= \begin{pmatrix} -2 & 8 \\ -10 & 0 \end{pmatrix}$$

(c) From part (b), $A - C = \begin{pmatrix} -2 & 8 \\ -10 & 0 \end{pmatrix}$

Hence $A + B - C = \begin{pmatrix} -2 & 8 \\ -10 & 0 \end{pmatrix} + \begin{pmatrix} 5 & -1 \\ 0 & 1 \end{pmatrix}$

$$= \begin{pmatrix} -2+5 & 8+(-1) \\ -10+0 & 0+1 \end{pmatrix}$$

$$= \begin{pmatrix} 3 & 7 \\ -10 & 1 \end{pmatrix}$$

Problem 2. Determine the single matrix for (a) $A \cdot C$ and (b) $A \cdot B$, where

$A = \begin{pmatrix} 2 & 4 \\ 1 & -3 \end{pmatrix}$, $B = \begin{pmatrix} 3 & -7 \\ 4 & -5 \end{pmatrix}$ and $C = \begin{pmatrix} 2 \\ -5 \end{pmatrix}$

(a) $A \cdot C = \begin{pmatrix} 2 & 4 \\ 1 & -3 \end{pmatrix} \times \begin{pmatrix} 2 \\ -5 \end{pmatrix} = \begin{pmatrix} 2 \times 2 + 4 \times (-5) \\ 1 \times 2 + (-3) \times (-5) \end{pmatrix}$

$$= \begin{pmatrix} 4-20 \\ 2+15 \end{pmatrix} = \begin{pmatrix} -16 \\ 17 \end{pmatrix}$$

(b) $A \cdot B = \begin{pmatrix} 2 & 4 \\ 1 & -3 \end{pmatrix} \times \begin{pmatrix} 3 & -7 \\ 4 & -5 \end{pmatrix}$

$= \begin{pmatrix} [2 \times 3 + 4 \times 4] & [2 \times (-7) + 4 \times (-5)] \\ [1 \times 3 + (-3) \times 4] & [1 \times (-7) + (-3) \times (-5)] \end{pmatrix}$

$= \begin{pmatrix} 6 + 16 & -14 + (-20) \\ 3 + (-12) & -7 + 15 \end{pmatrix}$

$= \begin{pmatrix} 22 & -34 \\ -9 & 8 \end{pmatrix}$

Further problems on the addition, subtraction and multiplication of matrices may be found in Section 5 (Problems 1 to 10), page 98.

3 The unit matrix

A unit matrix is one in which the values of the elements in the leading diagonal, ($\diagdown$), are 1, the remaining elements being 0. Thus, a 2 by 2 unit

matrix is $\begin{pmatrix} 1 & 0 \\ 0 & 1 \end{pmatrix}$, and is usually denoted by the symbol I. If A is a

square matrix and I the unit matrix, then $A \times I = I \times A$, that is, this is one case in matrices where the law $A \times B = B \times A$ of the algebra of numbers is true. The unit matrix is analogous to the number 1 in ordinary algebra.

4 Second order determinants

The solution of the linear simultaneous equations:

$$a_1 x + b_1 y + c_1 = 0 \tag{1}$$
$$a_2 x + b_2 y + c_2 = 0 \tag{2}$$

may be found by the elimination method of solving simultaneous equations. To eliminate y:

Equation (1) $\times b_2$: $a_1 b_2 x + b_1 b_2 y + c_1 b_2 = 0$
Equation (2) $\times b_1$: $a_2 b_1 x + b_1 b_2 y + c_2 b_1 = 0$

Subtracting: $(a_1 b_2 - a_2 b_1) x + (c_1 b_2 - c_2 b_1) = 0$

Thus, $x = \dfrac{-(c_1 b_2 - c_2 b_1)}{a_1 b_2 - a_2 b_1}$

i.e. $x = \dfrac{(b_1 c_2 - b_2 c_1)}{a_1 b_2 - a_2 b_1}$ \hfill (3)

Similarly, to eliminate x:

Equation (1) $\times a_2$: $a_1a_2x + a_2b_1y + a_2c_1 = 0$
Equation (2) $\times a_1$: $a_1a_2x + a_1b_2y + a_1c_2 = 0$

Subtracting: $(a_2b_1 - a_1b_2)y + (a_2c_1 - a_1c_2) = 0$

Thus, $y = \dfrac{-(a_2c_1 - a_1c_2)}{(a_2b_1 - a_1b_2)} = \dfrac{(a_1c_2 - a_2c_1)}{(a_2b_1 - a_1b_2)} = \dfrac{(a_1c_2 - a_2c_1)}{-(a_1b_2 - a_2b_1)}$

i.e. $-y = \dfrac{(a_1c_2 - a_2c_1)}{(a_1b_2 - a_2b_1)}$ \hfill (4)

Equations (3) and (4) can be written in the form:

$$\frac{x}{b_1c_2 - b_2c_1} = \frac{-y}{a_1c_2 - a_2c_1} = \frac{1}{a_1b_2 - a_2b_1} \hfill (5)$$

The denominators of equation (5) are all of the general form:

$pq - rs$.

Although as stated in Section 2 a matrix does not have a single numerical value and cannot be simplified to a particular answer, coefficients written in this form may be expressed as a special matrix, denoted by an array within vertical lines, rather than brackets. In this case:

$\begin{vmatrix} a & b \\ c & d \end{vmatrix} = ad - bc$, and is called a second order **determinant**.

It is shown in Chapter 8 following that determinants can be used to solve linear simultaneous equations such as those given in equations (1) and (2) above.

Worked problem on second order determinants

Problem 1. Evaluate the determinants: (a) $\begin{vmatrix} 3 & -1 \\ 4 & 2 \end{vmatrix}$ and (b) $\begin{vmatrix} a & -2b \\ 2a & -3b \end{vmatrix}$

By the definition of a determinant, $\begin{vmatrix} a & b \\ c & d \end{vmatrix} = ad - bc$, hence,

(a) $\begin{vmatrix} 3 & -1 \\ 4 & 2 \end{vmatrix} = (3 \times 2) - ((-1) \times 4) = 6 + 4 = \mathbf{10}.$

(b) $\begin{vmatrix} a & -2b \\ 2a & -3b \end{vmatrix} = (a \times (-3b)) - ((-2b) \times 2a) = -3ab + 4ab$

$\qquad\qquad\qquad\qquad\qquad\qquad\qquad\qquad\qquad\qquad = \mathbf{ab}.$

Further problems on second order determinants may be found in Section 5 following (Problems 11 to 15), page 99.

98

5 Further problems

Addition, subtraction and multiplication of second order matrices

In Problems 1 to 5, matrices A, B, C and D are given by:

$$A = \begin{pmatrix} 1 & 4 \\ -3 & 2 \end{pmatrix}, B = \begin{pmatrix} 2 & 7 \\ -1 & 0 \end{pmatrix},$$

$$C = \begin{pmatrix} 5 & -1 \\ 0 & 1 \end{pmatrix}, \text{ and } D = \begin{pmatrix} 3 & -4 \\ 7 & 2 \end{pmatrix}.$$

Determine the single matrix for the expressions given.

1. (a) $A + B$ $\left[\begin{pmatrix} 3 & 11 \\ -4 & 2 \end{pmatrix}\right.$

    (b) $A + C$ $\left.\begin{pmatrix} 6 & 3 \\ -3 & 3 \end{pmatrix}\right]$

2. (a) $C + D$ $\left[\begin{pmatrix} 8 & -5 \\ 7 & 3 \end{pmatrix}\right.$

    (b) $B + D$ $\left.\begin{pmatrix} 5 & 3 \\ 6 & 2 \end{pmatrix}\right]$

3. (a) $B - A$ $\left[\begin{pmatrix} 1 & 3 \\ 2 & -2 \end{pmatrix}\right.$

    (b) $D - B$ $\left.\begin{pmatrix} 1 & -11 \\ 8 & 2 \end{pmatrix}\right]$

4. (a) $C - A$ $\left[\begin{pmatrix} 4 & -5 \\ 3 & -1 \end{pmatrix}\right.$

    (b) $B - C$ $\left.\begin{pmatrix} -3 & 8 \\ -1 & -1 \end{pmatrix}\right]$

5. (a) $A + B + C$ $\left[\begin{pmatrix} 8 & 10 \\ -4 & 3 \end{pmatrix}\right.$

    (b) $A - B + C - D$ $\left.\begin{pmatrix} 1 & 0 \\ -9 & 1 \end{pmatrix}\right]$

In Problems 6 to 10, matrices A, B, C, D and E are given by:

$$A = \begin{pmatrix} 3 & 1 \\ -2 & 4 \end{pmatrix}, B = \begin{pmatrix} 2 & -5 \\ 0 & 1 \end{pmatrix}, C = \begin{pmatrix} -1 & 6 \\ 3 & 0 \end{pmatrix}$$

$$D = \begin{pmatrix} 2 \\ 3 \end{pmatrix} \text{ and } E = \begin{pmatrix} -1 \\ 4 \end{pmatrix}.$$

Determine the single matrix for the expressions given.

6. (a) $A \cdot D$ $\begin{bmatrix} \begin{pmatrix} 9 \\ 8 \end{pmatrix} \end{bmatrix}$

    (b) $B \cdot E$ $\begin{bmatrix} \begin{pmatrix} -22 \\ 4 \end{pmatrix} \end{bmatrix}$

7. (a) $C \cdot D$ $\begin{bmatrix} \begin{pmatrix} 16 \\ 6 \end{pmatrix} \end{bmatrix}$

    (b) $A \cdot E$ $\begin{bmatrix} \begin{pmatrix} 1 \\ 18 \end{pmatrix} \end{bmatrix}$

8. (a) $A \cdot C$ $\begin{bmatrix} \begin{pmatrix} 0 & 18 \\ 14 & -12 \end{pmatrix} \end{bmatrix}$

    (b) $C \cdot B$ $\begin{bmatrix} \begin{pmatrix} -2 & 12 \\ 6 & -15 \end{pmatrix} \end{bmatrix}$

9. (a) $A \cdot B$ $\begin{bmatrix} \begin{pmatrix} 6 & -14 \\ -4 & 14 \end{pmatrix} \end{bmatrix}$

    (b) $B \cdot A$ $\begin{bmatrix} \begin{pmatrix} 16 & -18 \\ -2 & 4 \end{pmatrix} \end{bmatrix}$

(Note that $A \cdot B \neq B \cdot A$)

10. (a) $B \cdot C$ $\begin{bmatrix} \begin{pmatrix} -17 & 12 \\ 3 & 0 \end{pmatrix} \end{bmatrix}$

    (b) $C \cdot A$ $\begin{bmatrix} \begin{pmatrix} -15 & 23 \\ 9 & 3 \end{pmatrix} \end{bmatrix}$

Second order determinants

In Problems 11 to 15, evaluate the determinants given.

11. (a) $\begin{vmatrix} 2 & 3 \\ 4 & 5 \end{vmatrix}$ (b) $\begin{vmatrix} -1 & -1 \\ 7 & 2 \end{vmatrix}$ (a) $[-2]$ (b) $[5]$

12. (a) $\begin{vmatrix} 3 & -1 \\ 4 & 7 \end{vmatrix}$ (b) $\begin{vmatrix} 5 & -2 \\ 3 & 1 \end{vmatrix}$ (a) $[25]$ (b) $[11]$

13. (a) $\begin{vmatrix} -2 & 4 \\ 3 & 1 \end{vmatrix}$ (b) $\begin{vmatrix} 1 & -4 \\ 5 & 1 \end{vmatrix}$ (a) $[-14]$ (b) $[21]$

14. (a) $\begin{vmatrix} x & 2x \\ -3x & 5 \end{vmatrix}$ (b) $\begin{vmatrix} y^2 & 3y \\ 4y^2 & -2y \end{vmatrix}$

(a) $[5x + 6x^2]$ (b) $[-14y^3]$

15. (a) $\begin{vmatrix} c & -2b \\ 4c & -3b \end{vmatrix}$ (b) $\begin{vmatrix} a & c \\ 2a & 4c \end{vmatrix}$

(a) $[5bc]$ (b) $[2ac]$

Chapter 8

The solution of simultaneous equations having two unknowns using matrices and determinants (BC 8)

1 The solution of simultaneous equations having two unknowns using determinants

When introducing determinants in Chapter 7, Section 4, the simultaneous equations

$$a_1 x + b_1 y + c_1 = 0 \tag{1}$$
$$a_2 x + b_2 y + c_2 = 0 \tag{2}$$

are solved using the elimination method of solving simultaneous equations, and it is shown that

$$\frac{x}{b_1 c_2 - b_2 c_1} = \frac{-y}{a_1 c_2 - a_2 c_1} = \frac{1}{a_1 b_2 - a_2 b_1} \tag{3}$$

It is also stated that the denominators of equation (3) are all of the general form:

$$pq - rs.$$

This algebraic expression is denoted by a special matrix having its array within vertical lines and is called a **determinant**. Thus

$$\begin{vmatrix} a & b \\ c & d \end{vmatrix} = ad - bc,$$

and is called a second order determinant.

The denominators of equation (3) can be written in determinant form, giving:

$$\frac{x}{\begin{vmatrix} b_1 & c_1 \\ b_2 & c_2 \end{vmatrix}} = \frac{-y}{\begin{vmatrix} a_1 & c_1 \\ a_2 & c_2 \end{vmatrix}} = \frac{1}{\begin{vmatrix} a_1 & b_1 \\ a_2 & b_2 \end{vmatrix}} \tag{4}$$

This expression is used to solve simultaneous equations by determinants and can be remembered by the 'cover-up' rule. In this rule:

  (i) the equations are written in the form $a_1x + b_1y + c_1 = 0$.

 (ii) If equation (4) is written in the form:

$$\frac{x}{|D_1|} = \frac{-y}{|D_2|} = \frac{1}{|D|}, \text{ then}$$

(iii) $|D_1|$ is obtained by covering-up the x-column and writing down the remaining coefficients in determinant form in positions corresponding to the positions they occupy in the equations.

(iv) $|D_2|$ is obtained by covering-up the y-column and treating the coefficients as in (iii) above.

 (v) $|D|$ is obtained by covering-up the constants-column and treating the coefficients as in (iii) above.

For example, to solve the equations:

$$2x + 3y = 11 \tag{5}$$
$$4x + 2y = 10 \tag{6}$$

by using determinants:

  (i) the equations are written as:

$$2x + 3y - 11 = 0$$
$$4x + 2y - 10 = 0$$

 (ii) $\dfrac{x}{|D_1|} = \dfrac{-y}{|D_2|} = \dfrac{1}{|D|}$

(iii) $|D_1| = \begin{vmatrix} 3 & -11 \\ 2 & -10 \end{vmatrix}$, obtained by covering-up the x-column in (i) above.

(iv) $|D_2| = \begin{vmatrix} 2 & -11 \\ 4 & -10 \end{vmatrix}$, obtained by covering-up the y-column in (i) above.

 (v) $|D| = \begin{vmatrix} 2 & 3 \\ 4 & 2 \end{vmatrix}$, obtained by covering-up the constants-column in (i) above.

Thus,

$$\frac{x}{\begin{vmatrix} 3 & -11 \\ 2 & -10 \end{vmatrix}} = \frac{-y}{\begin{vmatrix} 2 & -11 \\ 4 & -10 \end{vmatrix}} = \frac{1}{\begin{vmatrix} 2 & 3 \\ 4 & 2 \end{vmatrix}}$$

i.e.
$$\frac{x}{3 \times (-10) - 2 \times (-11)} = \frac{-y}{2 \times (-10) - 4 \times (-11)} = \frac{1}{2 \times 2 - 4 \times 3}$$

$$\frac{x}{-8} = \frac{-y}{24} = \frac{1}{-8}$$

giving: $x = \dfrac{-8}{-8} = 1$ and $-y = \dfrac{24}{-8}$, i.e. $y = 3$.

Checking in the original equations:

L.H.S. of equation (5) $= 2 \times 1 + 3 \times 3 = 11 =$ R.H.S.
L.H.S. of equation (6) $= 4 \times 1 + 2 \times 3 = 10 =$ R.H.S.

Hence, $x = 1, y = 3$ is the correct solution.

If, in a determinant of the form $D = \begin{vmatrix} a_1 & b_1 \\ a_2 & b_2 \end{vmatrix}$, $a_1 b_2 = a_2 b_1$,

then $a_1 b_2 - a_2 b_1 = 0$, i.e. $D = 0$. This means that in the simultaneous equations on which the determinant is based (say of the form $ax + by = c$), a_1 and b_1 are each multiplied by the same constant to give a_2 and b_2. So essentially there is only one equation with two unknown quantities, which is

not capable of solution. An example is the determinant $\begin{vmatrix} 2 & 1 \\ 10 & 5 \end{vmatrix}$ which

arises when trying to solve the equations, say,

$$2x + y = 3$$
and $10x + 5y = 15,$

and in this case, $D = 2 \times 5 - 10 \times 1 = 0$, i.e. the equations cannot be solved.

The matrix of the coefficients of x and y, i.e. $\begin{pmatrix} 2 & 1 \\ 10 & 5 \end{pmatrix}$, is called a

singular matrix when the determinant of the matrix is equal to 0.

Worked problems on the solution of simultaneous equations having two unknowns using determinants

Problem 1. Solve the simultaneous equations:

$$\tfrac{3}{2}p - 2q = \tfrac{1}{2} \tag{1}$$
$$p + \tfrac{3}{2}q = 6 \tag{2}$$

by using determinants.

(i) Writing the equations in the form $ax + by + c = 0$ gives:

$$\tfrac{3}{2}p - 2q - \tfrac{1}{2} = 0$$

$$p + \tfrac{3}{2}q - 6 = 0$$

(ii) $\dfrac{p}{|D_1|} = \dfrac{-q}{|D_2|} = \dfrac{1}{|D|}$ (note the signs are $+, -, +$)

(iii) Covering-up the p-column gives: $|D_1| = \begin{vmatrix} -2 & -\frac{1}{2} \\ \frac{3}{2} & -6 \end{vmatrix}$

(iv) Covering-up the q-column gives: $|D_2| = \begin{vmatrix} \frac{3}{2} & -\frac{1}{2} \\ 1 & -6 \end{vmatrix}$

(v) Covering-up the constants-column gives: $|D| = \begin{vmatrix} \frac{3}{2} & -2 \\ 1 & \frac{3}{2} \end{vmatrix}$

Thus,

$$\dfrac{p}{\begin{vmatrix} -2 & -\frac{1}{2} \\ \frac{3}{2} & -6 \end{vmatrix}} = \dfrac{-q}{\begin{vmatrix} \frac{3}{2} & -\frac{1}{2} \\ 1 & -6 \end{vmatrix}} = \dfrac{1}{\begin{vmatrix} \frac{3}{2} & -2 \\ 1 & \frac{3}{2} \end{vmatrix}}$$

i.e. $\dfrac{p}{(-2) \times (-6) - (\frac{3}{2}) \times (-\frac{1}{2})} = \dfrac{-q}{(\frac{3}{2}) \times (-6) - (1) \times (-\frac{1}{2})}$

$$= \dfrac{1}{(\frac{3}{2}) \times (\frac{3}{2}) - (1) \times (-2)}$$

$$\dfrac{p}{12\frac{3}{4}} = \dfrac{-q}{-8\frac{1}{2}} = \dfrac{1}{-4\frac{1}{4}}$$

Hence, $p = \dfrac{12\frac{3}{4}}{4\frac{1}{4}} = 3$ and $q = \dfrac{8\frac{1}{2}}{4\frac{1}{4}} = 2$.

Checking in the original equations:

L.H.S. of equation (1) $= \frac{3}{2} \times 3 - 2 \times 2 = \frac{1}{2} =$ R.H.S.

L.H.S. of equation (2) $= 3 + \frac{3}{2} \times 2 = 6 =$ R.H.S.

Hence $p = 3, q = 2$ is the correct solution.

Problem 2. Use determinants to solve the simultaneous equations:

$$-0.5f + 0.4g = 0.7 \tag{1}$$
$$1.2f - 0.3g = 3.6 \tag{2}$$

Writing the equations in the form $ax + by + c = 0$ gives:

$$-0.5f + 0.4g - 0.7 = 0$$
$$1.2f - 0.3g - 3.6 = 0$$

$$\dfrac{f}{|D_1|} = \dfrac{-g}{|D_2|} = \dfrac{1}{|D|}$$

Covering-up the f-column gives $|D_1| = \begin{vmatrix} 0.4 & -0.7 \\ -0.3 & -3.6 \end{vmatrix}$

$$= (0.4) \times (-3.6) - (-0.3) \times (-0.7)$$
$$= -1.44 - 0.21 = -1.65$$

Covering-up the g-column, gives $|D_2| = \begin{vmatrix} -0.5 & -0.7 \\ 1.2 & -3.6 \end{vmatrix}$

$$= (-0.5) \times (-3.6) - (1.2) \times (-0.7)$$

$$= 1.8 - (-0.84) = 2.64.$$

Covering-up the constants column, gives $|D|$ $\quad = \begin{vmatrix} -0.5 & 0.4 \\ 1.2 & -0.3 \end{vmatrix}$

$$= (-0.5) \times (-0.3) - (1.2) \times (0.4)$$
$$= 0.15 - 0.48 = -0.33$$

Hence, $\dfrac{f}{-1.65} = \dfrac{-g}{2.64} = \dfrac{1}{-0.33}$

i.e. $f = \dfrac{-1.65}{-0.33} = 5$ and $g = \dfrac{-2.64}{-0.33} = 8.$

Checking:

L.H.S. of equation (1) $= -0.5 \times 5 + 0.4 \times 8 = 0.7 = $ R.H.S.
L.H.S. of equation (2) $= 1.2 \times 5 - 0.3 \times 8 = 3.6 = $ R.H.S.

Thus, $f = 5, g = 8$ is the correct solution.

Problem 3. Use determinants to solve the simultaneous equations:

$$\frac{10}{a} - \frac{4}{b} = 3 \tag{1}$$

$$\frac{6}{a} + \frac{8}{b} = 7 \tag{2}$$

Let $x = \dfrac{1}{a}$ and $y = \dfrac{1}{b}$, then

$$10x - 4y = 3$$
$$6x + 8y = 7.$$

Writing in the $ax + by + c = 0$ form gives:

$$10x - 4y - 3 = 0$$
$$6x + 8y - 7 = 0$$

Applying the cover-up rule gives:

$$\frac{x}{\begin{vmatrix} -4 & -3 \\ 8 & 7 \end{vmatrix}} = \frac{-y}{\begin{vmatrix} 10 & -3 \\ 6 & -7 \end{vmatrix}} = \frac{1}{\begin{vmatrix} 10 & -4 \\ 6 & 8 \end{vmatrix}}$$

i.e. $\dfrac{x}{28+24} = \dfrac{-y}{-70+18} = \dfrac{1}{80+24}$

$x = \frac{52}{104} = \frac{1}{2}, y = \frac{52}{104} = \frac{1}{2}$

Since $x = \dfrac{1}{a}$, $a = 2$ and since $y = \dfrac{1}{b}$, $b = 2$.

Checking:

L.H.S. of equation (1) $= \frac{10}{2} - \frac{4}{2} = 3 =$ R.H.S.

L.H.S. of equation (2) $= \frac{6}{2} + \frac{8}{2} = 7 =$ R.H.S.

Hence, $a = 2$, $b = 2$ is the correct solution.

Problem 4. The forces acting on a bolt are resolved horizontally and vertically, giving the simultaneous equations shown below. Use determinants to find the values of F_1 and F_2, correct to three significant figures.

$$3.4F_1 - 0.83F_2 = 3.9 \tag{1}$$
$$0.7F_1 + 1.47F_2 = -2.05 \tag{2}$$

Writing the equations in the $ax + by + c = 0$ form, gives:

$3.4F_1 - 0.83F_2 - 3.9 = 0$
$0.7F_1 + 1.47F_2 + 2.05 = 0$

Applying the cover-up rule gives:

$$\frac{F_1}{\begin{vmatrix} -0.83 & -3.9 \\ 1.47 & 2.05 \end{vmatrix}} = \frac{-F_2}{\begin{vmatrix} 3.4 & -3.9 \\ 0.7 & 2.05 \end{vmatrix}} = \frac{1}{\begin{vmatrix} 3.4 & -0.83 \\ 0.7 & 1.47 \end{vmatrix}}$$

Hence,

$$\frac{F_1}{(-0.83) \times 2.05 - 1.47 \times (-3.9)} = \frac{-F_2}{3.4 \times 2.05 - 0.7 \times (-3.9)}$$

$$= \frac{1}{3.4 \times 1.47 - 0.7 \times (-0.83)}$$

that is: $\dfrac{F_1}{-1.701\,5 + 5.733} = \dfrac{-F_2}{6.970 + 2.730} = \dfrac{1}{4.998 + 0.581}$

i.e. $\dfrac{F_1}{4.032} = \dfrac{-F_2}{0.700} = \dfrac{1}{5.579}$

Thus, $F_1 = \dfrac{4.032}{5.579} = 0.723$, correct to three significant figures, and $\quad F_2 = \dfrac{-9.700}{5.579} = -1.74$, correct to three significant figures.

Checking:

L.H.S. of equation (1) $= 3.4 \times 0.723 - 0.83 \times (-1.74)$
$\qquad\qquad\qquad\quad = 3.90$ correct to three significant figures.
L.H.S. of equation (2) $= 0.7 \times 0.723 + 1.47 \times (-1.74)$
$\qquad\qquad\qquad\quad = -2.05$, correct to three significant figures.

Hence, $F_1 = 0.723, F_2 = -1.74$ is the correct solution.

Further problems on solving simultaneous equations having two unknowns using determinants may be found in Section 4 (Problems 1 to 15) page 113.

2 The inverse of a matrix

The inverse or reciprocal of matrix A is the matrix A^{-1}, such that

$A \cdot A^{-1} = I = A^{-1} \cdot A$, where I is the unit matrix, introduced in Chapter 7, Section 3.

The process of inverting a matrix makes division possible. If three matrices, A, B and X, are such that

$$A \cdot X = B$$

then $\quad X = \dfrac{B}{A} = A^{-1} \cdot B.$

Let the inverse of matrix A be $A^{-1} = \begin{pmatrix} a & b \\ c & d \end{pmatrix}$, and let matrix A be, say,

$\begin{pmatrix} 2 & 3 \\ -1 & 1 \end{pmatrix}$. By the definition of the inverse of a matrix,

$\begin{pmatrix} 2 & 3 \\ -1 & 1 \end{pmatrix} \times \begin{pmatrix} a & b \\ c & d \end{pmatrix} = \begin{pmatrix} 1 & 0 \\ 0 & 1 \end{pmatrix}$, the unit matrix.

Multiplying the matrices on the left-hand side gives:

$$\begin{pmatrix} 2a + 3c & 2b + 3d \\ -a + c & -b + d \end{pmatrix} = \begin{pmatrix} 1 & 0 \\ 0 & 1 \end{pmatrix} \tag{1}$$

Since these two matrices are equal to one another, the corresponding elements are equal to one another, hence

$-a + c = 0$, that is, $a = c$
$2b + 3d = 0$, that is, $b = -\tfrac{3}{2}d$

Substituting for a and b in equation (1) above gives:

$\begin{pmatrix} 5c & 0 \\ 0 & \dfrac{5d}{2} \end{pmatrix} = \begin{pmatrix} 1 & 0 \\ 0 & 1 \end{pmatrix}$

Thus, $5c = 1$, that is, $c = \frac{1}{5}$

$$\frac{5d}{2} = 1, \text{ that is, } d = \frac{2}{5}$$

Since $a = c$, $a = \frac{1}{5}$ and since $b = -\frac{3}{2}d$, $b = -\frac{3}{5}$.

Thus the inverse of matrix $\begin{pmatrix} 2 & 3 \\ -1 & 1 \end{pmatrix}$ is $\begin{pmatrix} \frac{1}{5} & -\frac{3}{5} \\ \frac{1}{5} & \frac{2}{5} \end{pmatrix}$

There is an alternative method of finding the inverse of a matrix. If the inverses of many matrices are determined and the inverses of the matrices are compared with the matrices, a relationship is seen to exist between matrices and their inverses. This relationship for a matrix of the form

$\begin{pmatrix} a & b \\ c & d \end{pmatrix}$ is that in the inverse:

  (i) the position of the a and d elements are interchanged,
  (ii) the sign of both the b and c elements is changed, and

  (iii) the matrix is multiplied by $\dfrac{1}{ad - bc}$, i.e. the reciprocal of the deter-

     minant of the matrix.

Thus, the inverse of matrix $\begin{pmatrix} a & b \\ c & d \end{pmatrix}$ is $\dfrac{1}{ad - bc} \begin{pmatrix} d & -b \\ -c & a \end{pmatrix}$

For the matrix $\begin{pmatrix} 2 & 3 \\ -1 & 1 \end{pmatrix}$ considered previously the inverse is

$$\frac{1}{2 \times 1 - 3 \times (-1)} \begin{pmatrix} 1 & -3 \\ 1 & 2 \end{pmatrix} = \frac{1}{5} \begin{pmatrix} 1 & -3 \\ 1 & 2 \end{pmatrix} = \begin{pmatrix} \frac{1}{5} & -\frac{3}{5} \\ \frac{1}{5} & \frac{2}{5} \end{pmatrix} \text{ as shown}$$

previously.

Worked problem on the inverse of a matrix

Problem 1. Determine the inverse of the matrix, $A = \begin{pmatrix} 5 & -3 \\ -2 & 1 \end{pmatrix}$

Let the inverse matrix be $A^{-1} = \begin{pmatrix} a & b \\ c & d \end{pmatrix}$

Since $A \cdot A^{-1} = I$, the unit matrix, then

$$\begin{pmatrix} 5 & -3 \\ -2 & 1 \end{pmatrix} \times \begin{pmatrix} a & b \\ c & d \end{pmatrix} = \begin{pmatrix} 1 & 0 \\ 0 & 1 \end{pmatrix}$$

Applying the multiplication law to the left-hand side gives:

$$\begin{pmatrix} 5a - 3c & 5b - 3d \\ -2a + c & -2b + d \end{pmatrix} = \begin{pmatrix} 1 & 0 \\ 0 & 1 \end{pmatrix} \tag{1}$$

Equating corresponding elements gives.

$-2a + c = 0$, i.e. $a = \dfrac{c}{2}$ and $5b - 3d = 0$, i.e. $b = \tfrac{3}{5}d$.

Substituting in equation (1):

$$\begin{pmatrix} -\dfrac{c}{2} & 0 \\ 0 & -\dfrac{d}{5} \end{pmatrix} = \begin{pmatrix} 1 & 0 \\ 0 & 1 \end{pmatrix}$$

i.e. $-\dfrac{c}{2} = 1$, $c = -2$ and $-\dfrac{d}{5} = 1$, $d = -5$.

Since $a = \tfrac{c}{2}$, $a = -1$ and since $b = \tfrac{3}{5}d$, $b = -3$.

Thus the inverse matrix of $\begin{pmatrix} 5 & -3 \\ -2 & 1 \end{pmatrix}$ is $\begin{pmatrix} -1 & -3 \\ -2 & -5 \end{pmatrix}$

The solution may be checked, using $A \cdot A^{-1} = I$. Thus,

$$\begin{pmatrix} 5 & -3 \\ -2 & 1 \end{pmatrix} \times \begin{pmatrix} -1 & -3 \\ -2 & -5 \end{pmatrix} = \begin{pmatrix} -5 + 6 & -15 + 15 \\ 2 - 2 & 6 - 5 \end{pmatrix}$$

$$= \begin{pmatrix} 1 & 0 \\ 0 & 1 \end{pmatrix}, \text{ the inverse matrix.}$$

Hence the solution $\begin{pmatrix} -1 & -3 \\ -2 & -5 \end{pmatrix}$ is correct.

Alternatively, the relationship that the inverse of matrix

$\begin{pmatrix} a & b \\ c & d \end{pmatrix}$ is $\dfrac{1}{ad - bc} \begin{pmatrix} d & -b \\ -c & a \end{pmatrix}$ could have been applied.

The inverse of $\begin{pmatrix} 5 & -3 \\ -2 & 1 \end{pmatrix} = \dfrac{1}{5 \times 1 - (-2) \times (-3)} \begin{pmatrix} 1 & 3 \\ 2 & 5 \end{pmatrix}$

$$= \dfrac{1}{-1} \begin{pmatrix} 1 & 3 \\ 2 & 5 \end{pmatrix}$$

Applying the law for scalar multiplication gives $\begin{pmatrix} -1 & -3 \\ -2 & -5 \end{pmatrix}$, as obtained

previously. The alternative method of applying a formula is the easiest method of determining the inverse of a matrix.

Further problems on the inverse of a matrix may be found in Section 4 (Problems 16 to 20), page 115. .

3 The solution of simultaneous equations having two unknowns using matrices

Matrices may be used to solve linear simultaneous equations. For equations having two unknown quantities there is no advantage in using a matrix method. However, for equations having three or more unknown quantities, solution by a matrix method can usually be performed more quickly and accurately.

Two linear simultaneous equations, such as:

$$2x + 3y = 4 \tag{1}$$
$$x - 5y = 6 \tag{2}$$

may be written in matrix form, as:

$$\begin{pmatrix} 2 & 3 \\ 1 & -5 \end{pmatrix} \begin{pmatrix} x \\ y \end{pmatrix} = \begin{pmatrix} 4 \\ 6 \end{pmatrix} \tag{3}$$

The inverse of the matrix $\begin{pmatrix} 2 & 3 \\ 1 & -5 \end{pmatrix}$ is obtained as shown in Section 2

and is $\begin{pmatrix} \frac{5}{13} & \frac{3}{13} \\ \frac{1}{13} & \frac{-2}{13} \end{pmatrix}$. Multiplying both sides of equation (3) by this inversed

matrix gives:

$$\begin{pmatrix} 1 & 0 \\ 0 & 1 \end{pmatrix} \begin{pmatrix} x \\ y \end{pmatrix} = \begin{pmatrix} \frac{5}{13} & \frac{3}{13} \\ \frac{1}{13} & \frac{-2}{13} \end{pmatrix} \begin{pmatrix} 4 \\ 6 \end{pmatrix} = \begin{pmatrix} \frac{20}{13} + \frac{18}{13} \\ \frac{4}{13} - \frac{12}{13} \end{pmatrix}$$

$$= \begin{pmatrix} \frac{38}{13} \\ \frac{-8}{13} \end{pmatrix}$$

i.e. $\begin{pmatrix} x \\ y \end{pmatrix} = \begin{pmatrix} \frac{38}{13} \\ \frac{-8}{13} \end{pmatrix}$

Equating corresponding elements gives:

$x = \frac{38}{13}$ and $y = -\frac{8}{13}$

Check: L.H.S. of equation (1) is $2 \times \frac{38}{13} + 3 \times (-\frac{8}{13}) = 4 =$ R.H.S.

L.H.S. of equation (2) is $\frac{38}{13} - 5(-\frac{8}{13}) = 6 =$ R.H.S.

Hence, $x = \frac{38}{13}$, $y = -\frac{8}{13}$ is the correct solution.

Summary

To solve linear simultaneous equations with two unknown quantities by using matrices:

(i) write the equations in the standard form

$$ax + by = c$$
$$dx + ey = f$$

(ii) write this in matrix form, i.e.

$$\begin{pmatrix} a & b \\ d & e \end{pmatrix} \begin{pmatrix} x \\ y \end{pmatrix} = \begin{pmatrix} c \\ f \end{pmatrix}$$

(iii) determine the inverse of matrix $\begin{pmatrix} a & b \\ d & e \end{pmatrix}$,

(iv) multiply each side of (ii) by the inversed matrix, and express in the

form $\begin{pmatrix} x \\ y \end{pmatrix} = \begin{pmatrix} g \\ h \end{pmatrix}$

(v) solve for x and y by equating corresponding elements,
and (vi) check the solution in the original equations.

Worked problems on solving simultaneous equations having two unknowns using matrices

Problem 1. Use matrices to solve the simultaneous equations:

$$4a - 3b = 18 \tag{1}$$
$$a + 2b = -1 \tag{2}$$

With reference to the above summary:

(i) The equations are in the standard form.

(ii) The matrices are $\begin{pmatrix} 4 & -3 \\ 1 & 2 \end{pmatrix} \begin{pmatrix} a \\ b \end{pmatrix} = \begin{pmatrix} 18 \\ -1 \end{pmatrix}$

(iii) The inverse matrix of $\begin{pmatrix} 4 & -3 \\ 1 & 2 \end{pmatrix}$ is $\dfrac{1}{4 \times 2 - (-3) \times 1} \begin{pmatrix} 2 & 3 \\ -1 & 4 \end{pmatrix}$

i.e. $\frac{1}{11} \begin{pmatrix} 2 & 3 \\ -1 & 4 \end{pmatrix}$, that is, $\begin{pmatrix} \frac{2}{11} & \frac{3}{11} \\ -\frac{1}{11} & \frac{4}{11} \end{pmatrix}$

(iv) Multiplying each side of (ii) by this inversed matrix, gives:

$$\begin{pmatrix} a \\ b \end{pmatrix} = \begin{pmatrix} \frac{2}{11} & \frac{3}{11} \\ -\frac{1}{11} & \frac{4}{11} \end{pmatrix} \begin{pmatrix} 18 \\ -1 \end{pmatrix}$$

$$\begin{pmatrix} a \\ b \end{pmatrix} = \begin{pmatrix} \frac{36}{11} + (-\frac{3}{11}) \\ -\frac{18}{11} + (-\frac{4}{11}) \end{pmatrix} = \begin{pmatrix} 3 \\ -2 \end{pmatrix}$$

(v) Thus, $a = 3$, $b = -2$.
(vi) Checking: L.H.S. of equation (1) is $4 \times 3 - 3(-2) = 18 =$ R.H.S.
    L.H.S. of equation (2) is $3 + 2(-2)$ $= -1 =$ R.H.S.

Hence $a = 3$, $b = -2$ is the correct solution.

Problem 2. Solve the simultaneous equations:

$$\frac{3}{x} - \frac{2}{y} = 0 \tag{1}$$

$$\frac{1}{x} + \frac{4}{y} = 14 \tag{2}$$

by using matrices.

With reference to the summary:

(i) The equations may be expressed in standard form by letting

$\dfrac{1}{x}$ be p and $\dfrac{1}{y}$ be q. Thus, equations (1) and (2)

become $3p - 2q = 0$
$\qquad\quad p + 4q = 14$

(ii) The matrices are $\begin{pmatrix} 3 & -2 \\ 1 & 4 \end{pmatrix} \begin{pmatrix} p \\ q \end{pmatrix} = \begin{pmatrix} 0 \\ 14 \end{pmatrix}$

(iii) The inverse of $\begin{pmatrix} 3 & -2 \\ 1 & 4 \end{pmatrix}$ is $\frac{1}{14}\begin{pmatrix} 4 & 2 \\ -1 & 3 \end{pmatrix}$

(iv) Multiplying each side of (ii) by (iii) gives:

$$\begin{pmatrix} p \\ q \end{pmatrix} = \tfrac{1}{14}\begin{pmatrix} 4 & 2 \\ -1 & 3 \end{pmatrix}\begin{pmatrix} 0 \\ 14 \end{pmatrix}$$

i.e. $\begin{pmatrix} p \\ q \end{pmatrix} = \tfrac{1}{14}\begin{pmatrix} 4 \times 0 + 2 \times 14 \\ -1 \times 0 + 3 \times 14 \end{pmatrix} = \tfrac{1}{14}\begin{pmatrix} 28 \\ 42 \end{pmatrix} = \begin{pmatrix} \frac{1}{14} \times 28 \\ \frac{1}{14} \times 42 \end{pmatrix}$

$$= \begin{pmatrix} 2 \\ 3 \end{pmatrix}$$

(v) Thus $p = 2$ and $q = 3$, i.e. $x = \frac{1}{2}$, $y = \frac{1}{3}$.

(vi) Checking: L.H.S. of equation (1) is $\frac{3}{1/2} - \frac{2}{1/3} = 6-6 = 0 = $ R.H.S.

L.H.S. of equation (2) is $\frac{1}{1/2} + \frac{4}{1/3} = 2+12 = 14 = $ R.H.S.

Hence $x = \frac{1}{2}$, $y = \frac{1}{3}$ is the correct solution.

Problem 3. A force system is analysed, and by resolving the forces horizontally and vertically the following equations are obtained:

$$6F_1 - F_2 = 5 \tag{1}$$
$$5F_1 + 2F_2 = 7 \tag{2}$$

Use matrices to solve for F_1 and F_2.

The matrices are $\begin{pmatrix} 6 & -1 \\ 5 & 2 \end{pmatrix} \begin{pmatrix} F_1 \\ F_2 \end{pmatrix} = \begin{pmatrix} 5 \\ 7 \end{pmatrix}$

The inverse of $\begin{pmatrix} 6 & -1 \\ 5 & 2 \end{pmatrix}$ is $\frac{1}{17} \begin{pmatrix} 2 & 1 \\ -5 & 6 \end{pmatrix}$

Hence $\begin{pmatrix} F_1 \\ F_2 \end{pmatrix} = \frac{1}{17} \begin{pmatrix} 2 & 1 \\ -5 & 6 \end{pmatrix} \begin{pmatrix} 5 \\ 7 \end{pmatrix}$

$= \frac{1}{17} \begin{pmatrix} 10 + 7 \\ -25 + 42 \end{pmatrix} = \begin{pmatrix} 1 \\ 1 \end{pmatrix}$

Thus, $F_1 = 1, F_2 = 1$.

Checking: L.H.S. of equation (1) is $6 - 1 = 5 =$ R.H.S.
L.H.S. of equation (2) is $5 + 2 = 7 =$ R.H.S.

Thus, $F_1 = 1$ and $F_2 = 1$ is the correct solution.

Any technical problems, such as the equations formed by the resolution of vector quantities or the equations relating load and effort in machines, which were previously solved by using simultaneous equations, may be solved either by using determinants or by using matrices.

Further problems on the solution of simultaneous equations having two unknowns using matrices may be found in Section 4 following (Problems 21 to 30), page 115.

4 Further problems

The solution of simultaneous equations having two unknowns using determinants

In Problems 1 to 11, use determinants to solve the simultaneous equations given.

1. $4v_1 - 3v_2 = 18$
    $v_1 + 2v_2 = -1$ $[v_1 = 3, v_2 = -2]$

2. $3m - 2n = -4.5$
    $4m + 3n = 2.5$ $[m = -\frac{1}{2}, n = 1\frac{1}{2}]$

3. $\dfrac{a}{3} + \dfrac{b}{4} = 8$

    $\dfrac{a}{6} - \dfrac{b}{8} = 1$ $[a = 15, b = 12]$

4. $s + t = 17$

    $\dfrac{s}{5} - \dfrac{t}{7} = 1$ $[s = 10, t = 7]$

5. $\dfrac{c}{5} + \dfrac{d}{3} = \dfrac{43}{30}$

   $\dfrac{c}{9} - \dfrac{d}{6} = -\dfrac{1}{12}$ $[c = 3, d = 2\frac{1}{2}]$

6. $0.5i_1 - 1.2i_2 = -13$
   $0.8i_1 + 0.3i_2 = 12.5$ $[i_1 = 10, i_2 = 15]$

7. $1.25L_1 - 0.75L_2 = 1$
   $0.25L_1 + 1.25L_2 = 17$ $[L_1 = 8.0, L_2 = 12.0]$

8. $\dfrac{1}{2a} + \dfrac{3}{5b} = 7$

   $\dfrac{4}{a} + \dfrac{1}{2b} = 13$ $[a = \frac{1}{2}, b = \frac{1}{10}]$

9. $\dfrac{3}{v_1} - \dfrac{2}{v_2} = \frac{1}{2}$

   $\dfrac{5}{v_1} + \dfrac{3}{v_2} = \dfrac{29}{12}$ $[v_1 = 3, v_2 = 4]$

10. $\dfrac{4}{p_1 - p_2} = \dfrac{16}{21}$

    $\dfrac{3}{p_1 + p_2} = \dfrac{4}{9}$ $[p_1 = 6, p_2 = \frac{3}{4}]$

11. $\dfrac{2x + 1}{5} - \dfrac{1 - 4y}{2} = \dfrac{5}{2}$

    $\dfrac{1 - 3x}{7} + \dfrac{2y - 3}{5} + \dfrac{32}{35} = 0$ $[x = 2, y = 1]$

12. A vector system to determine the shortest distance between two moving bodies is analysed, producing the following equations:

    $11S_1 - 10S_2 = 30$
    $21S_2 - 20S_1 = -40$

    Use determinants to find the values of S_1 and S_2.
    $$[S_1 = 7.42, S_2 = 5.16]$$

13. The power in a mechanical device is given by $p = aN + \dfrac{b}{N}$ where a and

    b are constants. Use determinants to find the value of a and b if $p = 13$ when $N = 3$ and $p = 12$ when $N = 2$.
    $$[a = 3, b = 12]$$

14. The law connecting friction F and load L for an experiment to find the friction force between two surfaces is of the type $F = aL + b$, where a and b are constants.

When $F = 6.0$, $L = 7.5$ and when $F = 2.7$, $L = 2.0$.
Find the values of a and b by using determinants.

$$[a = 0.60, b = 1.5]$$

15. The length L metres of an alloy at temperature $t\,°C$ is given by:
    $L = L_0\,(1 + \alpha\,t)$, where L_0 and α are constants.
    Use determinants to find the values of L_0 and α if $L = 20$ m when t is
    $52°C$ and $L = 21$ m when $t = 100°C$.

$$[L_0 = 18.92 \text{ m}, \alpha = 0.001\ 1]$$

The inverse of a matrix
In Problems 16 to 20, find the inverse of the matrices given.

16. $\begin{pmatrix} 2 & -1 \\ -5 & -1 \end{pmatrix}$ $\quad \left[\begin{pmatrix} \frac{1}{7} & -\frac{1}{7} \\ -\frac{5}{7} & -\frac{2}{7} \end{pmatrix} \right]$

17. $\begin{pmatrix} 1 & -3 \\ 1 & 7 \end{pmatrix}$ $\quad \left[\begin{pmatrix} \frac{7}{10} & \frac{3}{10} \\ -\frac{1}{10} & \frac{1}{10} \end{pmatrix} \right]$

18. $\begin{pmatrix} 3 & 5 \\ -2 & 1 \end{pmatrix}$ $\quad \left[\begin{pmatrix} \frac{1}{13} & -\frac{5}{13} \\ \frac{2}{13} & \frac{3}{13} \end{pmatrix} \right]$

19. $\begin{pmatrix} -2 & -1 \\ 4 & 3 \end{pmatrix}$ $\quad \left[\begin{pmatrix} -\frac{3}{2} & -\frac{1}{2} \\ 2 & 1 \end{pmatrix} \right]$

20. $\begin{pmatrix} -4 & -3 \\ 5 & 3 \end{pmatrix}$ $\quad \left[\begin{pmatrix} 1 & 1 \\ -\frac{5}{3} & -\frac{4}{3} \end{pmatrix} \right]$

The solution of simultaneous equations having two unknowns using matrices
In Problems 21 to 26, solve the simultaneous equations given by using
matrices.

21. $p + 3q = 11$
    $p + 2q = \ 8$ $\qquad [p = 2, q = 3]$

22. $3a + 4b - 5 = 0$
    $12 = 5b - 2a$ $\qquad [a = -1, b = 2]$

23. $\dfrac{m}{3} + \dfrac{n}{4} = 6$

    $\dfrac{m}{6} - \dfrac{n}{8} = 0$ $\qquad [m = 9, n = 12]$

24. $4a - 6b + 2.5 = 0$
    $7a - 5b + 0.25 = 0$ $\qquad [a = \frac{1}{2}, b = \frac{3}{4}]$

25. $\dfrac{x}{8} + \dfrac{5}{2} = y$

    $13 - \dfrac{y}{3} - 3x = 0$ $\qquad [x = 4, y = 3]$

26. $\dfrac{a-1}{3} + \dfrac{b+2}{2} = 3$

$\dfrac{1-a}{6} + \dfrac{4-b}{2} = \frac{1}{2}$ $[a = 4, b = 2]$

27. When determining the relative velocity of a system, the following equations are produced:

$3.0 = 0.10\,v_1 + (v_1 - v_2)$
$-2.0 = 0.05\,v_2 - (v_1 - v_2)$

Use matrices to find the values of v_1 and v_2.

$[v_1 = 7.42, v_2 = 5.16]$

28. Applying Newton's laws of motion to a mechanical system gives the following equations:

$14 = 0.2u + 2u + 8\,(u - v)$
$0 = -8\,(u - v) + 2v + 10v$

Use matrices to find the values of u and v.

$[u = 2.0, v = 0.8]$

29. Equations connecting the lens system in a position transducer are:

$\dfrac{4}{u_1} + \dfrac{6}{v_1} + \dfrac{9}{v_2} = 6$

$\dfrac{15}{u_1} + \dfrac{11}{v_1} + \dfrac{2}{v_2} = 8\frac{1}{12}$

If $v_1 = v_2$, use matrices to find the values of u_1, v_1 and v_2.

$[u_1 = 4, v_1 = v_2 = 3]$

30. When an effort E is applied to the gearbox on a diesel motor it is found that a resistance R can be overcome and that E and R are connected by a formula: $E = a + b R$ where a and b are constants. An effort of 3.5 newtons overcomes a resistance of 5 newtons and an effort of 5.3 newtons overcomes a resistance of 8 newtons. Use matrices to find the values of a and b.

$[a = 0.50, b = 0.60]$

Chapter 9

Matrix arithmetic and the determinant of a matrix (BC 11)

1 Introduction

Second-order matrices are introduced in Chapters 7 and 8. In this chapter, the arithmetic of matrices is extended to third order matrices, i.e. 3 by 3 matrices, and the arithmetic operations of addition, subtraction, multiplication and inversion are introduced.

The terms used in connection with third-order matrices are the same as those used for second-order matrices. For the linear simultaneous equations, say,

$$2x + 3y + 4z = 7 \qquad (1)$$
$$6x - 4y + 5z = 8 \qquad (2)$$
$$7x + 5y - 6z = 9 \qquad (3)$$

the coefficients of x, y and z are written in matrix notation as

$$\begin{pmatrix} 2 & 3 & 4 \\ 6 & -4 & 5 \\ 7 & 5 & -6 \end{pmatrix},$$

that is, occupying the same relative positions as in equations (1), (2) and (3). The grouping of the coefficients of x, y and z in this way is called an **array** and the coefficients forming the array are called the **elements** of the matrix.

If there are m rows across an array and n columns down an array, the matrix is said to be of order 'm by n'. A matrix having a single row is called a **row matrix** and one having a single column is called a **column matrix**. For example, in equation (1) above the coefficients of x, y and z form a row matrix of (2 3 4) and the coefficients of x in equations (1), (2) and (3) form a column matrix of $\begin{pmatrix} 2 \\ 6 \\ 7 \end{pmatrix}$.

A matrix having the same number of rows as columns is called a **square matrix** and a square matrix having three rows and three columns is called a third-order matrix. Matrices are generally denoted by capital letters and if the matrix representing the coefficients of x, y and z in equations (1), (2) and (3) above is A, then

$$A = \begin{pmatrix} 2 & 3 & 4 \\ 6 & -4 & 5 \\ 7 & 5 & -6 \end{pmatrix}$$

To refer to a particular element in a matrix, the notation a_{ij} is used, where i refers to the row number and j the column number. Thus in matrix A above, the element a_{22} is -4 and element a_{31} is 7, and so on.

2. Addition, subtraction and multiplication of third order matrices

A matrix does not have a single numerical value and cannot be simplified to a particular numerical answer. However, by applying the laws of matrices given in this section, they can be simplified and often the number of matrices can be reduced, for example, by adding two matrices, a single matrix is produced. Also, by comparing one matrix with another similar matrix, values of unknown elements can be determined. Matrices are of very little value in themselves but an understanding of the laws of matrices can lead to a better understanding of determinants and also enables determinants to be simplified before evaluation, saving the number of arithmetic operations and reducing their size.

Addition

As with second-order matrices, only matrices having the same number of rows and columns may be added. The sum of two matrices is the matrix obtained by adding the elements occupying corresponding positions in the two matrices. When two matrices are added, the result is a single matrix of the same order as those being added. Thus,

$$\begin{pmatrix} p & q & r \\ s & t & u \\ v & w & x \end{pmatrix} + \begin{pmatrix} a & b & c \\ d & e & f \\ g & h & i \end{pmatrix} = \begin{pmatrix} p+a & q+b & r+c \\ s+d & t+e & u+f \\ v+g & w+h & x+i \end{pmatrix}$$

This principle may be applied to more than two m by n matrices being added, the resulting matrix being a single m by n matrix.

Subtraction

As with second-order matrices, only matrices having the same number of rows and columns may be subtracted. The difference between two matrices, say $A - B$, is the matrix obtained by subtracting the elements of matrix B from those occupying the corresponding position in matrix A. Thus,

$$\begin{pmatrix} p & q & r \\ s & t & u \\ v & w & x \end{pmatrix} - \begin{pmatrix} a & b & c \\ d & e & f \\ g & h & i \end{pmatrix} = \begin{pmatrix} p-a & q-b & r-c \\ s-d & t-e & u-f \\ v-g & w-h & x-i \end{pmatrix}$$

This principle can be applied to the difference of two m by n matrices, the resulting matrix being a single m by n matrix.

Multiplication

(a) Scalar multiplication

When a matrix is multiplied by a number, the resulting matrix is one of the same order, having each element multiplied by the number. Thus if,

$$A = \begin{pmatrix} p & q & r \\ s & t & u \\ v & w & x \end{pmatrix} \text{, then } kA = \begin{pmatrix} kp & kq & kr \\ ks & kt & ku \\ kv & kw & kx \end{pmatrix}$$

where k is a constant.

(b) Multiplication of matrices

When a 3 by 3 matrix, say A, is multiplied by a 3 by 3 matrix, say B, the resulting matrix is a 3 by 3 matrix, say C. The a_{11} element of C is the sum of the products of the first row of A and the first column of B, taken element by element. Thus, if $A \times B = C$

$$\begin{pmatrix} p & q & r \\ - & - & - \\ - & - & - \end{pmatrix} \times \begin{pmatrix} a & - & - \\ d & - & - \\ g & - & - \end{pmatrix} = \begin{pmatrix} pa + qd + rg & - & - \\ - & & - & - \\ - & & - & - \end{pmatrix}$$

The a_{12} element of C is the sum of the products obtained by taking the first row of A with the second column of B, element by element. Thus

$$\begin{pmatrix} p & q & r \\ - & - & - \\ - & - & - \end{pmatrix} \times \begin{pmatrix} - & b & - \\ - & e & - \\ - & h & - \end{pmatrix} = \begin{pmatrix} - & pb + qe + rh & - \\ - & - & - \\ - & - & - \end{pmatrix}$$

The remaining elements of C are calculated in a similar way, giving

$$\begin{pmatrix} p & q & r \\ s & t & u \\ v & w & x \end{pmatrix} \times \begin{pmatrix} a & b & c \\ d & e & f \\ g & h & i \end{pmatrix} = \begin{pmatrix} (pa + qd + rg) & (pb + qe + rh) & (pc + qf + ri) \\ (sa + td + ug) & (sb + te + uh) & (sc + tf + ui) \\ (va + wd + xg) & (vb + we + xh) & (vc + wf + xi) \end{pmatrix}$$

Although the laws of matrices are so formulated that they follow most of the laws which govern the algebra of numbers, frequently in the multiplication of two matrices $A \times B$ is **not** equal to $B \times A$. This is shown in worked problem 4 following.

Worked problems on addition, subtraction and multiplication of third-order matrices

In Problems 1 to 4 following, matrices A, B and C refer to:

$$A = \begin{pmatrix} 2 & 3 & -1 \\ -2 & 0 & 4 \\ 3 & 1 & 5 \end{pmatrix} B = \begin{pmatrix} 4 & 3 & 0 \\ 5 & -1 & 2 \\ 5 & 3 & 4 \end{pmatrix} \text{ and } C = \begin{pmatrix} 1 & -3 & 1 \\ -5 & 7 & 4 \\ 7 & 0 & 6 \end{pmatrix}$$

Problem 1. Determine the matrix D, where $D = A + B$.

$$A + B = \begin{pmatrix} 2 & 3 & -1 \\ -2 & 0 & 4 \\ 3 & 1 & 5 \end{pmatrix} + \begin{pmatrix} 4 & 3 & 0 \\ 5 & -1 & 2 \\ 5 & 3 & 4 \end{pmatrix} = \begin{pmatrix} 2 + 4 & 3 + 3 & -1 + 0 \\ -2 + 5 & 0 + (-1) & 4 + 2 \\ 3 + 5 & 1 + 3 & 5 + 4 \end{pmatrix}$$

$$D = \begin{pmatrix} 6 & 6 & -1 \\ 3 & -1 & 6 \\ 8 & 4 & 9 \end{pmatrix}$$

Problem 2. Determine the matrix E, where $E = B - A$.

$$B - A = \begin{pmatrix} 4 & 3 & 0 \\ 5 & -1 & 2 \\ 5 & 3 & 4 \end{pmatrix} - \begin{pmatrix} 2 & 3 & -1 \\ -2 & 0 & 4 \\ 3 & 1 & 5 \end{pmatrix} = \begin{pmatrix} [4 - 2] & [3 - 3] & [0 - (-1)] \\ [5 - (-2)] & [-1 - 0] & [2 - 4] \\ [5 - 3] & [3 - 1] & [4 - 5] \end{pmatrix}$$

$$E = \begin{pmatrix} 2 & 0 & 1 \\ 7 & -1 & -2 \\ 2 & 2 & -1 \end{pmatrix}$$

Problem 3. Determine matrix F, where $F = A - B + C$.

This problem may be done in two parts, in which case the single matrix representing $A - B$ can be determined as shown in Problem 2 above, and then this single matrix can be added to matrix C, as shown in Problem 1 above. Alternatively, it may be done in one step, the elements of F being obtained from the corresponding elements of A, B and C as shown below.

$$A - B + C = \begin{pmatrix} 2 & 3 & -1 \\ -2 & 0 & 4 \\ 3 & 1 & 5 \end{pmatrix} - \begin{pmatrix} 4 & 3 & 0 \\ 5 & -1 & 2 \\ 5 & 3 & 4 \end{pmatrix} + \begin{pmatrix} 1 & -3 & 1 \\ -5 & 7 & 4 \\ 7 & 0 & 6 \end{pmatrix}$$

$$= \begin{pmatrix} [2 - 4 + 1] & [3 - 3 + (-3)] & [-1 - 0 + 1] \\ [-2 - 5 + (-5)] & [0 - (-1) + 7] & [4 - 2 + 4] \\ [3 - 5 + 7] & [1 - 3 + 0] & [5 - 4 + 6] \end{pmatrix}$$

$$F = \begin{pmatrix} -1 & -3 & 0 \\ -12 & 8 & 6 \\ 5 & -2 & 7 \end{pmatrix}$$

Problem 4. (i) Determine matrix G, where $G = A \times B$.
  (ii) Show that $A \times B$ is not equal to $B \times A$, in this case.

(i) The element g_{11} is obtained from the sum of the products of the first row of A and the first column of B, taken element by element. Thus $g_{11} = a_{11} \times b_{11} + a_{12} \times b_{21} + a_{13} \times b_{31}$.

Since $A = \begin{pmatrix} 2 & 3 & -1 \\ -2 & 0 & 4 \\ 3 & 1 & 5 \end{pmatrix}$ and $B = \begin{pmatrix} 4 & 3 & 0 \\ 5 & -1 & 2 \\ 5 & 3 & 4 \end{pmatrix}$

then $\quad g_{11} = 2 \times 4 + 3 \times 5 + (-1) \times 5 = 8 + 15 - 5 = 18$

Similarly, $g_{12} = a_{11} \times b_{12} + a_{12} \times b_{22} + a_{13} \times b_{32}$
$\qquad\qquad = 2 \times 3 + 3 \times (-1) + (-1) \times 3 = 6 - 3 - 3 = 0$

The remaining seven elements are found in a similar way, giving:

$$A \times B = \begin{pmatrix} [18] & [0] \\ [-2 \times 4 + 0 \times 5 + 4 \times 5] & [-2 \times 3 + 0 \times (-1) + 4 \times 3] \\ [3 \times 4 + 1 \times 5 + 5 \times 5] & [3 \times 3 + 1 \times (-1) + 5 \times 3] \end{pmatrix}$$

$$\begin{pmatrix} [2 \times 0 + 3 \times 2 + (-1) \times 4] \\ [-2 \times 0 + 0 \times 2 + 4 \times 4] \\ [3 \times 0 + 1 \times 2 + 5 \times 4] \end{pmatrix}$$

$$G = A \times B = \begin{pmatrix} 18 & 0 & 2 \\ 12 & 6 & 16 \\ 42 & 23 & 22 \end{pmatrix}$$

(ii)
$$B \times A = \begin{pmatrix} 4 & 3 & 0 \\ 5 & -1 & 2 \\ 5 & 3 & 4 \end{pmatrix} \times \begin{pmatrix} 2 & 3 & -1 \\ -2 & 0 & 4 \\ 3 & 1 & 5 \end{pmatrix}$$

$$= \begin{pmatrix} [4 \times 2 + 3 \times (-2) + 0 \times 3] & [4 \times 3 + 3 \times 0 + 0 \times 1] \\ [5 \times 2 + (-1) \times (-2) + 2 \times 3] & [5 \times 3 + (-1) \times 0 + 2 \times 1] \\ [5 \times 2 + 3 \times (-2) + 4 \times 3] & [5 \times 3 + 3 \times 0 + 4 \times 1] \end{pmatrix}$$

$$\begin{pmatrix} [4 \times (-1) + 3 \times 4 + 0 \times 5] \\ [5 \times (-1) + (-1) \times 4 + 2 \times 5] \\ [5 \times (-1) + 3 \times 4 + 4 \times 5] \end{pmatrix}$$

i.e. $B \times A = \begin{pmatrix} 2 & 12 & 8 \\ 18 & 17 & 1 \\ 16 & 19 & 27 \end{pmatrix}$

Since the matrix obtained for $A \times B$ is different to that obtained for $B \times A$, this shows that in this case $A \times B \neq B \times A$.

Problem 5.

If $P = \begin{pmatrix} 2 & 3 & -1 & 4 \\ 4 & 2 & 7 & 8 \\ 3 & -2 & 10 & 0 \end{pmatrix}$ and $Q = \begin{pmatrix} 7 \\ 4 \\ 3 \\ 1 \end{pmatrix}$,

find matrix R, where $R = P \times Q$.

The previous problem dealt with multiplication of a square, 3×3 matrix. However, the principles introduced can be extended to any m by n matrix, provided the number of elements in the rows of matrix P are equal to the number of elements in the columns of matrix Q. Using the same principle as that used in Problem 4, gives:

$$R = P \times Q = \begin{pmatrix} 2 & 3 & -1 & 4 \\ 4 & 2 & 7 & 8 \\ 3 & -2 & 10 & 0 \end{pmatrix} \times \begin{pmatrix} 7 \\ 4 \\ 3 \\ 1 \end{pmatrix}$$

$$= \begin{pmatrix} 2 \times 7 + 3 \times 4 + (-1) \times 3 + 4 \times 1 \\ 4 \times 7 + 2 \times 4 + 7 \times 3 + 8 \times 1 \\ 3 \times 7 + (-2) \times 4 + 10 \times 3 + 0 \times 1 \end{pmatrix}$$

i.e. $R = \begin{pmatrix} 14 + 12 - 3 + 4 \\ 28 + 8 + 21 + 8 \\ 21 - 8 + 30 + 0 \end{pmatrix} = \begin{pmatrix} 27 \\ 65 \\ 43 \end{pmatrix}$

In general, when multiplying a matrix of dimension (m by n) by a matrix of dimension (n by q), the resulting matrix has a dimension of (m by q), i.e. (**rows** of first $\times$ **columns** of second). In this problem, matrix P has 3 **rows** and matrix Q has 1 **column**, hence the matrix resulting from the multiplication of these two must be a matrix having 3 rows and 1 column.

Further problems on addition, subtraction and multiplication of third-order matrices may be found in Section 5 (Problems 1 to 8), page 129.

3 The determinant of a 3 by 3 matrix

For a 2 by 2 matrix, say $\begin{pmatrix} a & b \\ c & d \end{pmatrix}$, the quantity $(ad - bc)$ is called the **determinant** of a matrix.

The determinant of a matrix differs from the matrix itself in that it can be evaluated to a single numerical result. To distinguish it from the matrix, it is written as $\begin{vmatrix} a & b \\ c & d \end{vmatrix}$, vertical lines being used instead of brackets.

For a 3 by 3 matrix, say $\begin{pmatrix} a_1 & b_1 & c_1 \\ a_2 & b_2 & c_2 \\ a_3 & b_3 & c_3 \end{pmatrix}$, the **minor** of element a_1 is the determinant obtained if the row and column containing element a_1 are considered to be covered up, i.e. row a_1, b_1, c_1 and column $\begin{pmatrix} a_1 \\ a_2 \\ a_3 \end{pmatrix}$ are covered up, i.e. the minor of element a_1 is found from

matrix $\begin{pmatrix} a_1 & b_1 & c_1 \\ a_2 & b_2 & c_2 \\ a_3 & b_3 & c_3 \end{pmatrix}$ and is $\begin{vmatrix} b_2 & c_2 \\ b_3 & c_3 \end{vmatrix}$

Similarly, to find the minor of element, say c_2, row a_2, b_2, c_2 and column c_2 are considered to be covered, leaving exposed the determinant $\begin{vmatrix} a_1 & b_1 \\ a_3 & b_3 \end{vmatrix}$. It follows that each of the nine elements of a 3 by 3 matrix has its own minor.

The sign of the minor of an element depends on the position of the element within the matrix. Each minor has a + or a − sign, the signs being such that they are alternatively + then − in both rows and columns, element a_{11} always being +. Thus for a 3 by 3 matrix, the sign pattern is:

$$\begin{pmatrix} + & - & + \\ - & + & - \\ + & - & + \end{pmatrix}$$

It follows that the sign of the minor of element, say c_1 is +, but the sign of the element, say b_3, is −.

A signed − minor is called the **cofactor** of an element. Thus

the cofactor of element b_2, B_2 $= +\begin{vmatrix} a_1 & c_1 \\ a_3 & c_3 \end{vmatrix} = a_1 c_3 - a_3 c_1$, and

the cofactor of element b_3, B_3 $= -\begin{vmatrix} a_1 & c_1 \\ a_2 & c_2 \end{vmatrix} = -(a_1 c_2 - a_2 c_1)$

$$= a_2 c_1 - a_1 c_2.$$

The value of a 3 by 3 determinant for matrix A is written as $|A|$ and is given by the sum of the products of elements and their cofactors of **any** row or **any** column. Thus, for the matrix

$$A = \begin{pmatrix} a_1 & b_1 & c_1 \\ a_2 & b_2 & c_2 \\ a_3 & b_3 & c_3 \end{pmatrix}$$

the value of the determinant $|A|$ is given by any of the six expressions:

$$|A| = a_1 A_1 + b_1 B_1 + c_1 C_1 \text{ or } a_2 A_2 + b_2 B_2 + c_2 C_2$$
$$\text{or } a_3 A_3 + b_3 B_3 + c_3 C_3 \text{ or } a_1 A_1 + a_2 A_2 + a_3 A_3$$
$$\text{or } b_1 B_1 + b_2 B_2 + b_3 B_3 \text{ or } c_1 C_1 + c_2 C_2 + c_3 C_3 .$$

Using, say, the first row elements,

$$|A| = a_1 \begin{vmatrix} b_2 & c_2 \\ b_3 & c_3 \end{vmatrix} - b_1 \begin{vmatrix} a_2 & c_2 \\ a_3 & c_3 \end{vmatrix} + c_1 \begin{vmatrix} a_2 & b_2 \\ a_3 & b_3 \end{vmatrix}$$

$$= a_1 (b_2 c_3 - b_3 c_2) - b_1 (a_2 c_3 - a_3 c_2) + c_1 (a_2 b_3 - a_3 b_2)$$

$$= a_1 b_2 c_3 - a_1 b_3 c_2 - a_2 b_1 c_3 + a_3 b_1 c_2 + a_2 b_3 c_1 - a_3 b_2 c_1$$

Using, say, the first column elements,

$$|A| = a_1 \begin{vmatrix} b_2 & c_2 \\ b_3 & c_3 \end{vmatrix} - a_2 \begin{vmatrix} b_1 & c_1 \\ b_3 & c_3 \end{vmatrix} + a_3 \begin{vmatrix} b_1 & c_1 \\ b_2 & c_2 \end{vmatrix}$$

$$= a_1 (b_2 c_3 - b_3 c_2) - a_2 (b_1 c_3 - b_3 c_1) + a_3 (b_1 c_2 - b_2 c_1)$$
$$= a_1 b_2 c_3 - a_1 b_3 c_2 - a_2 b_1 c_3 + a_2 b_3 c_1 + a_3 b_1 c_2 - a_3 b_2 c_1.$$

This result is the same as that obtained by using the first row elements, and any of the six expressions for the value of a determinant given above will give the same result. When finding the value of a determinant containing one or more 0's as elements, it is usually easier to select the row or column containing the most 0's as the basis of the expansion, since the cofactor of a zero element is zero and need not be calculated.

Worked problems on the determinant of a 3 by 3 matrix

Problem 1. Determine the value of (a) the minors of elements $-1, 0$ and 5, and (b) the cofactors of elements $7, 3$ and -2 for the matrix.

$$\begin{pmatrix} 2 & -1 & 7 \\ 3 & 4 & -2 \\ 0 & 8 & 5 \end{pmatrix}$$

(a) The minor of the element -1 is obtained by writing the 2 by 2 determinant left when the row and column containing the element -1 are covered up (i.e. the first row and the second column).

Thus the value of the minor of the element $-1 = \begin{vmatrix} 3 & -2 \\ 0 & 5 \end{vmatrix} = 3 \times 5 - 0 \times (-2)$
$$= 15$$

Similarly, the value of the minor of element $0 = \begin{vmatrix} -1 & 7 \\ 4 & -2 \end{vmatrix} = (-1) \times (-2) - 4 \times 7$
$$= -26$$

The value of the minor of element $5 = \begin{vmatrix} 2 & -1 \\ 3 & 4 \end{vmatrix} = 2 \times 4 - 3 \times (-1)$
$$= 11$$

(b) Cofactors are signed — minors, the signs following the pattern

$$\begin{pmatrix} + & - & + \\ - & + & - \\ + & - & + \end{pmatrix}.$$ For the element 7, the cofactor $= + \begin{vmatrix} 3 & 4 \\ 0 & 8 \end{vmatrix} = 3 \times 8 - 0 \times 4$
$$= 24$$

For the element 3, the cofactor $= - \begin{vmatrix} -1 & 7 \\ 8 & 5 \end{vmatrix} = -((-1) \times 5 - 8 \times 7)$
$$= 61$$

For the element -2, the cofactor $= - \begin{vmatrix} 2 & -1 \\ 0 & 8 \end{vmatrix} = -(2 \times 8 - 0 \times (-1))$
$$= -16$$

Problem 2. Find the value of the determinant $|A|$ of the matrix

$$A = \begin{pmatrix} 3 & 2 & -5 \\ 0 & 1 & 7 \\ 4 & 5 & 2 \end{pmatrix}$$

The value of the determinant is the sum of the products of the elements and their cofactors of any row or any column. Since element a_{21} is 0, the arithmetic is kept to a minimum by selecting either the second row or the first column as the basis for expansion. Selecting, say, the first column, i.e. $|A| = a_1 A_1 + a_2 A_2 + a_3 A_3$, gives

$$|A| = \begin{vmatrix} 3 & 2 & -5 \\ 0 & 1 & 7 \\ 4 & 5 & 2 \end{vmatrix} = 3 \begin{vmatrix} 1 & 7 \\ 5 & 2 \end{vmatrix} - 0 \begin{vmatrix} 2 & -5 \\ 5 & 2 \end{vmatrix} + 4 \begin{vmatrix} 2 & -5 \\ 1 & 7 \end{vmatrix}$$

$$= 3 (1 \times 2 - 5 \times 7) - (0) + 4 (2 \times 7 - 1 \times (-5))$$

$$= 3 (-33) + 4 (19) = -23$$

Further problems on the determinant of a 3 by 3 matrix may be found in Section 5 (Problems 9 to 16) page 130.

4 The inverse or reciprocal of a matrix

To determine the inverse or reciprocal of a matrix, two new operations associated with matrices are introduced below. These refer to finding the transpose of a matrix and to finding the adjoint of a matrix.

The **transpose** of a matrix A, usually denoted by A^T, is obtained by writing the rows of matrix A as the columns of matrix A^T. Thus if

$$A = \begin{pmatrix} a_1 & b_1 & c_1 \\ a_2 & b_2 & c_2 \\ a_3 & b_3 & c_3 \end{pmatrix} \text{, then } A^T = \begin{pmatrix} a_1 & a_2 & a_3 \\ b_1 & b_2 & b_3 \\ c_1 & c_2 & c_3 \end{pmatrix}$$

It can be seen that the elements on the leading diagonal, (\ ), of A and A^T are the same.

The algebraic expression for the determinant of A, using the first row as the basis for the expansion, is:

$$|A| = a_1 \begin{vmatrix} b_2 & c_2 \\ b_3 & c_3 \end{vmatrix} - b_1 \begin{vmatrix} a_2 & c_2 \\ a_3 & c_3 \end{vmatrix} + c_1 \begin{vmatrix} a_2 & b_2 \\ a_2 & b_3 \end{vmatrix}$$

$$= a_1 b_2 c_3 - a_1 b_3 c_2 - a_2 b_1 c_3 + a_3 b_1 c_2 + a_2 b_3 c_1 - a_3 b_2 c_1$$

and for A^T, using the first column as the basis for the expansion, is:

$$|A^T| = a_1 \begin{vmatrix} b_2 & b_3 \\ c_2 & c_3 \end{vmatrix} - b_1 \begin{vmatrix} a_2 & a_3 \\ c_2 & c_3 \end{vmatrix} + c_1 \begin{vmatrix} a_2 & a_3 \\ b_2 & b_3 \end{vmatrix}$$

$$= a_1\,b_2\,c_3 - a_1\,b_3\,c_2 - a_2\,b_1\,c_3 + a_3\,b_1\,c_2 + a_2\,b_3\,c_1 - a_3\,b_2\,c_1$$

Since the algebraic equations for $|A|$ and $|A^T|$ are the same, it follows that:

$$|A| = |A^T|$$

The **adjoint** matrix of matrix A, usually abbreviated to 'adj A', is obtained by:

  (i) determining the nine cofactors of matrix A,
 (ii) forming a matrix of these cofactors, and
(iii) transposing the matrix formed in (ii) above.

Thus, to determine the adjoint matrix of matrix A, where

$$A = \begin{pmatrix} 1 & 2 & 1 \\ 2 & 0 & 2 \\ 1 & 1 & 0 \end{pmatrix}\text{, and following the steps outlined above, gives:}$$

 (i) the cofactors of the first row are:

$$+\begin{vmatrix} 0 & 2 \\ 1 & 0 \end{vmatrix} = -2, -\begin{vmatrix} 2 & 2 \\ 1 & 0 \end{vmatrix} = 2 \text{ and } +\begin{vmatrix} 2 & 0 \\ 1 & 1 \end{vmatrix} = 2,$$

the cofactors of the second row are:

$$-\begin{vmatrix} 2 & 1 \\ 1 & 0 \end{vmatrix} = 1, +\begin{vmatrix} 1 & 1 \\ 1 & 0 \end{vmatrix} = -1 \text{ and } -\begin{vmatrix} 1 & 2 \\ 1 & 1 \end{vmatrix} = 1$$

and the cofactors of the third row are:

$$+\begin{vmatrix} 2 & 1 \\ 0 & 2 \end{vmatrix} = 4, -\begin{vmatrix} 1 & 1 \\ 2 & 2 \end{vmatrix} = 0 \text{ and } +\begin{vmatrix} 1 & 2 \\ 2 & 0 \end{vmatrix} = -4.$$

 (ii) the matrix of these cofactors is $\begin{pmatrix} -2 & 2 & 2 \\ 1 & -1 & 1 \\ 4 & 0 & -4 \end{pmatrix}$

(iii) transposing the matrix of cofactors, gives

$$\text{adj } A = \begin{pmatrix} -2 & 1 & 4 \\ 2 & -1 & 0 \\ 2 & 1 & -4 \end{pmatrix}$$

The **unit matrix** I for a 2 by 2 matrix has been previously defined in Chapter 7 and is the matrix having elements of value 1 on its leading diagonal, all other elements being 0. It is analogous to the number 1 in ordinary algebra. For a 3 by 3 matrix, the unit matrix is

$$I = \begin{pmatrix} 1 & 0 & 0 \\ 0 & 1 & 0 \\ 0 & 0 & 1 \end{pmatrix}$$

It has also been shown previously that the inverse or reciprocal of matrix A

is the matrix A^{-1}, such that

$$A \cdot A^{-1} = I$$

i.e. $A^{-1} = \dfrac{I}{A}$ \hfill (4)

The inverse of matrix A can also be expressed in terms of the adjoint of matrix A and the determinant of matrix A, the relationship being:

$$A^{-1} = \frac{\text{adj } A}{|A|}$$ \hfill (5)

Equation (5) may be verified for matrix A, where A is, say,

$$\begin{pmatrix} 1 & 2 & 1 \\ 2 & 0 & 2 \\ 1 & 1 & 0 \end{pmatrix}, \text{ as follows.}$$

It is shown above that adj $A = \begin{pmatrix} -2 & 1 & 4 \\ 2 & -1 & 0 \\ 2 & 1 & -4 \end{pmatrix}$

The product of A and adj A is

$$A \times \text{adj } A = \begin{pmatrix} 1 & 2 & 1 \\ 2 & 0 & 2 \\ 1 & 1 & 0 \end{pmatrix} \times \begin{pmatrix} -2 & 1 & 4 \\ 2 & -1 & 0 \\ 2 & 1 & -4 \end{pmatrix}$$

$$= \begin{pmatrix} 4 & 0 & 0 \\ 0 & 4 & 0 \\ 0 & 0 & 4 \end{pmatrix} = 4 \begin{pmatrix} 1 & 0 & 0 \\ 0 & 1 & 0 \\ 0 & 0 & 1 \end{pmatrix}$$

(Extracting a common factor is dealt with in Chapter 10 following.)

i.e. $A \times \text{adj } A = 4I$.

Also, the determinant of A, $|A| = 1 \begin{vmatrix} 0 & 2 \\ 1 & 0 \end{vmatrix} - 2 \begin{vmatrix} 2 & 2 \\ 1 & 0 \end{vmatrix} + 1 \begin{vmatrix} 2 & 0 \\ 1 & 1 \end{vmatrix}$

$$= 4$$

i.e. $A \times \text{adj } A \qquad = |A| \times I$.

But from equation (4), $A \times A^{-1} = I$

Hence, $A \times \text{adj } A \qquad = |A| \times A \times A^{-1}$

i.e. $A^{-1} \qquad\qquad\qquad = \dfrac{\text{adj } A}{A}$

Although this relationship has been verified for a specific matrix, the proof can be done, following the same steps, for the general matrix

$$\begin{pmatrix} a_1 & b_1 & c_1 \\ a_2 & b_2 & c_2 \\ a_3 & b_3 & c_3 \end{pmatrix}$$

In this case, the algebraic expressions produced become numerous and contain a large number of factors and hence become difficult to follow.

It is now possible, using equation (5), to find the reciprocal of a matrix and since dividing by, say a, is the same as multiplying by $\dfrac{1}{a}$, the inverse matrix may be used for the operation of matrix division,

i.e. $\dfrac{A}{B} = A \times \dfrac{1}{B} = AB^{-1}$

Worked problems on the inverse or reciprocal of a matrix

Problem 1. Find the adjoint matrix of matrix A, where

$$A = \begin{pmatrix} 4 & 2 & 0 \\ 0 & 1 & 1 \\ 3 & 2 & 0 \end{pmatrix}$$

Using the procedure given in the text for determining the adjoint of a matrix:

(i) determine the nine cofactors of A.

First row: $\quad + \begin{vmatrix} 1 & 1 \\ 2 & 0 \end{vmatrix} = -2, - \begin{vmatrix} 0 & 1 \\ 3 & 0 \end{vmatrix} = 3, + \begin{vmatrix} 0 & 1 \\ 3 & 2 \end{vmatrix} = -3.$

Second row: $- \begin{vmatrix} 2 & 0 \\ 2 & 0 \end{vmatrix} = 0, + \begin{vmatrix} 4 & 0 \\ 3 & 0 \end{vmatrix} = 0, - \begin{vmatrix} 4 & 2 \\ 3 & 2 \end{vmatrix} = -2.$

Third row: $\quad + \begin{vmatrix} 2 & 0 \\ 1 & 1 \end{vmatrix} = 2, - \begin{vmatrix} 4 & 0 \\ 0 & 1 \end{vmatrix} = -4, + \begin{vmatrix} 4 & 2 \\ 0 & 1 \end{vmatrix} = 4.$

(ii) form the matrix of these cofactors, i.e. $\begin{pmatrix} -2 & 3 & -3 \\ 0 & 0 & -2 \\ 2 & -4 & 4 \end{pmatrix}$

(iii) transpose the matrix of cofactors, giving

$$\text{adj } A = \begin{pmatrix} -2 & 0 & 2 \\ 3 & 0 & -4 \\ -3 & -2 & 4 \end{pmatrix}$$

Problem 2. Determine the inverse or reciprocal matrix, A^{-1}, for the matrix A given in Problem 1 above.

From equation (5), $A^{-1} = \dfrac{\text{adj } A}{|A|}$

It is shown in Problem 1 above that the adjoint of matrix A is:

$$\text{adj } A = \begin{pmatrix} -2 & 0 & 2 \\ 3 & 0 & -4 \\ -3 & -2 & 4 \end{pmatrix}$$

The determinant of matrix A, $|A|$, using the third column as the basis of the expansion, since it contains two 0's, is: $0 - 1 \begin{vmatrix} 4 & 2 \\ 3 & 2 \end{vmatrix} + 0$

i.e. $|A| = -2$

Hence, $A^{-1} = -\dfrac{1}{2} \begin{pmatrix} -2 & 0 & 2 \\ 3 & 0 & -4 \\ -3 & -2 & 4 \end{pmatrix}$

Further problems on the inverse or reciprocal of a matrix may be found in the following Section (5) (Problems 17 to 25), page 130.

5 Further problems

Addition, subtraction and multiplication of third-order matrices

In problems 1 to 4, matrix $M = \begin{pmatrix} 1 & 2 & 1 \\ 3 & 2 & 1 \\ 2 & 3 & 2 \end{pmatrix}$ and

matrix $N = \begin{pmatrix} 3 & 2 & 1 \\ 1 & 1 & 2 \\ 1 & 2 & 1 \end{pmatrix}$

1. Determine matrix L, where $L = M + N$. $\left[\begin{pmatrix} 4 & 4 & 2 \\ 4 & 3 & 3 \\ 3 & 5 & 3 \end{pmatrix} \right]$

2. Prove that $M + N = N + M$.

3. Determine the matrix $M \times N$. $\left[\begin{pmatrix} 6 & 6 & 6 \\ 12 & 10 & 8 \\ 11 & 11 & 10 \end{pmatrix} \right]$

4. Prove that $M \times N \neq N \times M$. $\left[N \times M = \begin{pmatrix} 11 & 13 & 7 \\ 8 & 10 & 6 \\ 9 & 9 & 5 \end{pmatrix} \right]$

In Problems 5 to 8, $A = \begin{pmatrix} 2 & 5 & 4 & 3 \\ 3 & 1 & 4 & 5 \\ 1 & 8 & 3 & 5 \end{pmatrix}, B = \begin{pmatrix} 1 & 2 & 3 & 1 \\ 0 & 1 & 2 & -2 \\ 1 & -3 & 1 & 0 \end{pmatrix}$

$C = \begin{pmatrix} 2 & 2 & 1 \\ 4 & 5 & 0 \\ 3 & 5 & 7 \\ 5 & 3 & 5 \end{pmatrix}$ and $Q = \begin{pmatrix} 2 \\ 4 \\ 3 \\ 5 \end{pmatrix}$

5. Determine $A + B$. $\left[\begin{pmatrix} 3 & 7 & 7 & 4 \\ 3 & 2 & 6 & 3 \\ 2 & 5 & 4 & 5 \end{pmatrix} \right]$

6. Determine $A \times Q$.
$$\left[\begin{pmatrix} 51 \\ 47 \\ 68 \end{pmatrix} \right]$$

7. Determine $A \times C$.
$$\left[\begin{pmatrix} 51 & 58 & 45 \\ 47 & 46 & 56 \\ 68 & 72 & 47 \end{pmatrix} \right]$$

8. Determine $C \times A$.
$$\left[\begin{pmatrix} 11 & 20 & 19 & 21 \\ 23 & 25 & 36 & 37 \\ 28 & 76 & 53 & 69 \\ 24 & 68 & 47 & 55 \end{pmatrix} \right]$$

The determinant of a 3 by 3 matrix

Problems 9 to 12 refer to the determinant $\begin{vmatrix} a & b & c \\ m & n & p \\ x & y & z \end{vmatrix}$

9. Write down the minors for elements a, p and y.
$$[nz - yp, \; ay - xb, \; ap - mc]$$

10. Write down the cofactors for elements c, m and x.
$$[my - xn, \; yc - bz, \; bp - nc]$$

11. Find the algebraic expression for the determinant based on the first row elements. $\quad [a\,(nz - yp) - b\,(mz - xp) + c\,(my - xn)]$

12. Find the algebraic expression for the determinant based on the second column elements. $\quad [b\,(mz - xp) - n\,(az - xc) + y\,(ap - mc)]$

In problems 13 to 16, determine the values of the determinants for the matrices given.

13. $\begin{pmatrix} 2 & -1 & 3 \\ -2 & 3 & -2 \\ 2 & 1 & 5 \end{pmatrix}$ [4]

14. $\begin{pmatrix} 3 & 5 & 1 \\ -2 & 3 & 1 \\ 4 & -2 & 1 \end{pmatrix}$ [37]

15. $\begin{pmatrix} 1 & 2 & -3 \\ 2 & -3 & 4 \\ 3 & 4 & 5 \end{pmatrix}$ [−78]

16. $\begin{pmatrix} -13 & 3 & 2 \\ 32 & -6 & 3 \\ 12 & -4 & 1 \end{pmatrix}$ [−178]

The inverse or reciprocal of a matrix

17. Write down the transpose matrix of the matrix $\begin{pmatrix} 2 & -7 & 4 \\ 3 & 2 & 5 \\ -4 & 1 & 0 \end{pmatrix}$

$$\left[\begin{pmatrix} 2 & 3 & -4 \\ -7 & 2 & 1 \\ 4 & 5 & 0 \end{pmatrix} \right]$$

18. Write down the transpose matrix of the matrix $\begin{pmatrix} 4 & 8 & 3 \\ 1 & 4 & 7 \\ -3 & -2 & 6 \end{pmatrix}$

$$\left[\begin{pmatrix} 4 & 1 & -3 \\ 8 & 4 & -2 \\ 3 & 7 & 6 \end{pmatrix} \right]$$

In Problems 19 to 21, determine the adjoint matrices for the matrices given.

19. $\begin{pmatrix} 3 & 2 & 1 \\ 4 & 2 & 2 \\ 1 & 3 & 1 \end{pmatrix}$ $\left[\begin{pmatrix} -4 & 1 & 2 \\ -2 & 2 & -2 \\ 10 & -7 & -2 \end{pmatrix} \right]$

20. $\begin{pmatrix} 1 & -3 & 0 \\ 2 & 0 & 1 \\ 4 & 1 & 3 \end{pmatrix}$ $\left[\begin{pmatrix} -1 & 9 & -3 \\ -2 & 3 & -1 \\ 2 & -13 & 6 \end{pmatrix} \right]$

21. $\begin{pmatrix} 1 & 2 & 3 \\ 2 & -1 & 4 \\ 0 & -1 & 1 \end{pmatrix}$ $\left[\begin{pmatrix} 3 & -5 & 11 \\ -2 & 1 & 2 \\ -2 & 1 & -5 \end{pmatrix} \right]$

In Problems 22 to 25, determine the reciprocals of the matrices given. Show, in each case, that the product of the matrix and its reciprocal is the unit matrix.

22. $\begin{pmatrix} -1 & 2 & -3 \\ 2 & -1 & 4 \\ 3 & 4 & 1 \end{pmatrix}$ $\left[\frac{1}{4}\begin{pmatrix} -17 & -14 & 5 \\ 10 & 8 & -2 \\ 11 & 10 & -3 \end{pmatrix} \right]$

23. $\begin{pmatrix} 0 & 0 & 1 \\ 0 & 1 & 0 \\ 1 & 0 & 0 \end{pmatrix}$ $\left[-1\begin{pmatrix} 0 & 0 & -1 \\ 0 & -1 & 0 \\ -1 & 0 & 0 \end{pmatrix} \right]$

24. $\begin{pmatrix} 2 & 1 & 1 \\ 1 & 2 & 2 \\ 1 & 3 & 2 \end{pmatrix}$ $\left[-\frac{1}{3}\begin{pmatrix} -2 & 1 & 0 \\ 0 & 3 & -3 \\ 1 & -5 & 3 \end{pmatrix} \right]$

25. $\begin{pmatrix} 1 & 2 & 1 \\ 2 & 4 & 6 \\ 3 & 1 & 2 \end{pmatrix}$ $\left[\frac{1}{20}\begin{pmatrix} 2 & -3 & 8 \\ 14 & -1 & -4 \\ -10 & 5 & 0 \end{pmatrix} \right]$

Chapter 10

The general properties of 3 by 3 determinants and the solution of simultaneous equations (BC 12)

1 The properties of third-order determinants

The amount of arithmetic used to determine the value of a determinant can often be reduced by simplifying the determinant before evaluation. The principal properties of a determinant used in the simplifying process are given below.

(a) Changing rows or columns

If all the elements in a row or a column of a determinant are interchanged with the corresponding elements in another row or column, the value of the determinant obtained by doing this is (-1) multiplied by the value of the original determinant. It is shown in Chapter 9 that, using the first-row elements:

$$\begin{vmatrix} a_1 & b_1 & c_1 \\ a_2 & b_2 & c_2 \\ a_3 & b_3 & c_3 \end{vmatrix} = a_2\,(b_2c_3 - b_3c_2) - b_1\,(a_2c_3 - a_3c_2) + c_1\,(a_2b_3 - a_3b_2)$$

$$= a_1b_2c_3 - a_1b_3c_2 - a_2b_1c_3 + a_3b_1c_2 + a_2b_3c_1 - a_3b_2c_1$$

Suppose now that the first and third columns are interchanged. The determinant becomes:

$$\begin{vmatrix} c_1 & b_1 & a_1 \\ c_2 & b_2 & a_2 \\ c_3 & b_3 & a_3 \end{vmatrix}$$

Evaluating, using the first row elements, gives:

$$c_1(b_2a_3 - a_2b_3) - b_1(c_2a_3 - c_3a_2) + a_1(c_2b_3 - c_3b_2)$$
$$= a_3b_2c_1 - a_2b_3c_1 - a_3b_1c_2 + a_2b_1c_3 + a_1b_3c_2 - a_1b_2c_3$$

Comparing these terms with those obtained for the determinant $\begin{vmatrix} a_1 & b_1 & c_1 \\ a_2 & b_2 & c_2 \\ a_3 & b_3 & c_3 \end{vmatrix}$

shows that each term in the second expression is (-1) multiplied by its corresponding term in the first expression. Thus

$$\begin{vmatrix} a_1 & b_1 & c_1 \\ a_2 & b_2 & c_2 \\ a_3 & b_3 & c_3 \end{vmatrix} = (-1) \times \begin{vmatrix} c_1 & b_1 & a_1 \\ c_2 & b_2 & a_2 \\ c_3 & b_3 & a_3 \end{vmatrix}$$

Although this has been proved by interchanging the first and third columns, it may be shown in the same way that it applies when interchanging any row with any other row and when interchanging any column with any other column.

(b) The value of a determinant if two rows or two columns are equal

Let $|A|$ be a determinant in which, say, the first and second columns are equal,

i.e. $|A| = \begin{vmatrix} a_1 & a_1 & c_1 \\ a_2 & a_2 & c_2 \\ a_3 & a_3 & c_3 \end{vmatrix}$

Let $|B|$ be the determinant $|A|$ in which the first and second columns

have been interchanged, then $|B| = \begin{vmatrix} a_1 & a_1 & c_1 \\ a_2 & a_2 & c_2 \\ a_3 & a_3 & c_3 \end{vmatrix}$

Since the first and second columns of $|A|$ are interchanged in $|B|$, then from (a) above

$$|A| = -|B|$$

But it can be seen that $|A| = |B|$ from the elements obtained. These two statements can only be true if $|A| = 0$. It follows that the value of a determinant in which two rows or two columns are equal is 0.

This statement can also be verified by obtaining the algebraic expression for a determinant in which, say, the first and second columns are equal. For the determinant $|A|$ given above, using the first-row elements as the basis for the expansion:

$$|A| = a_1 \begin{vmatrix} a_2 & c_2 \\ a_3 & c_3 \end{vmatrix} - a_1 \begin{vmatrix} a_2 & c_2 \\ a_3 & c_3 \end{vmatrix} + c_1 \begin{vmatrix} a_2 & a_2 \\ a_3 & a_3 \end{vmatrix}$$

i.e. $|A| = 0$.

(c) Common factors

All elements in a row or a column having a common factor can be divided by this factor. This factor then becomes a factor of the determinant. Thus

$$\begin{vmatrix} a_1 & b_1 & c_1 \\ \lambda a_2 & \lambda b_2 & \lambda c_2 \\ a_3 & b_3 & c_3 \end{vmatrix} = \lambda \begin{vmatrix} a_1 & b_1 & c_1 \\ a_2 & b_2 & c_2 \\ a_2 & b_3 & c_3 \end{vmatrix}$$

This property can be verified by obtaining the algebraic expression for the determinant. Using the first-row elements to obtain this, gives:

$$\begin{vmatrix} a_1 & b_1 & c_1 \\ \lambda a_2 & \lambda b_2 & \lambda c_2 \\ a_3 & b_3 & c_3 \end{vmatrix} = a_1 \begin{vmatrix} \lambda b_2 & \lambda c_2 \\ b_3 & c_3 \end{vmatrix} - b_1 \begin{vmatrix} \lambda a_2 & \lambda c_2 \\ a_3 & c_3 \end{vmatrix} + c_1 \begin{vmatrix} \lambda a_2 & \lambda b_2 \\ a_3 & b_3 \end{vmatrix}$$

$$= a_1(\lambda b_2 c_3 - \lambda b_3 c_2) - b_1(\lambda a_2 c_3 - \lambda a_3 c_2)$$
$$+ c_1(\lambda a_2 b_3 - \lambda a_3 b_2)$$

$$= \lambda[a_1(b_2 c_3 - b_3 c_2) - b_1(a_2 c_3 - a_3 c_2)$$
$$+ c_1(a_2 b_3 - a_3 b_2)]$$

$$= \lambda \begin{vmatrix} a_1 & b_1 & c_1 \\ a_2 & b_2 & c_2 \\ a_3 & b_3 & c_3 \end{vmatrix}$$

(d) Addition or subtraction of rows or columns

If a multiple of the elements of any row or column are added to the corresponding elements of any other row or column, then the value of the determinant so obtained is equal to the value of the original determinant. Thus multiplying the third-row elements by λ and adding them to the corresponding elements in the first row, gives:

$$\begin{vmatrix} a_1 & b_1 & c_1 \\ a_2 & b_2 & c_2 \\ a_3 & b_3 & c_3 \end{vmatrix} = \begin{vmatrix} a_1 + \lambda c_1 & b_1 & c_1 \\ a_2 + \lambda c_2 & b_2 & c_2 \\ a_3 + \lambda c_3 & b_3 & c_3 \end{vmatrix}$$

This property may be verified as follows. Expanding the determinant $\begin{vmatrix} a_1 + \lambda c_1 & b_1 & c_1 \\ a_2 + \lambda c_2 & b_2 & c_2 \\ a_3 + \lambda c_3 & b_3 & c_3 \end{vmatrix}$, using the first-row elements, gives:

$$(a_1 + \lambda c_1)(b_2 c_3 - b_3 c_2) - b_1[(a_2 + \lambda c_2)c_3 - (a_3 + \lambda c_3)c_2]$$
$$+ c_1[(a_2 + \lambda c_2)b_3 - (a_3 + \lambda c_3)b_2]$$

i.e. $a_1(b_2 c_3 - b_3 c_2) - b_1(a_2 c_3 - a_3 c_2) + c_1(a_2 b_3 - a_3 b_2)$

$$+ \lambda(c_1 b_2 c_3 - c_1 b_3 c_2 - b_1 c_2 c_3 + b_1 c_3 c_2 + c_1 c_2 b_3 - c_1 c_3 b_2)$$

All the terms containing λ cancel out and the remaining terms are equal to $\begin{vmatrix} a_1 & b_1 & c_1 \\ a_2 & b_2 & c_2 \\ a_3 & b_3 & c_3 \end{vmatrix}$. Thus

$$\begin{vmatrix} a_1 + \lambda c_1 & b_1 & c_1 \\ a_2 + \lambda c_2 & b_2 & c_2 \\ a_3 + \lambda c_3 & b_3 & c_3 \end{vmatrix} = \begin{vmatrix} a_1 & b_1 & c_1 \\ a_2 & b_2 & c_2 \\ a_3 & b_3 & c_3 \end{vmatrix}$$

When λ is equal to 1, then the elements of a row or column are being added to the corresponding elements of another row or column. When λ = −1, then the elements of a row or column are being subtracted from the corresponding elements of another row or column.

The four properties introduced above are used mainly in two ways. These are (i) to reduce the magnitude of the elements in a row or column, making subsequent arithmetic easier, and (ii) to introduce as many 0's as practical into a determinant before evaluating it. The way in which this is done is shown in the worked problems following.

Worked problems on the properties of third order determinants

Problem 1. Use the properties of determinants to simplify the determinant

$\begin{vmatrix} 3 & 4 & 5 \\ 6 & 7 & 8 \\ 1 & 11 & 6 \end{vmatrix}$, and hence find its value.

There are many ways of simplifying this determinant and one method is shown below.

Taking the first row elements from the corresponding second-row elements (property (d) above) and then taking out a common factor of 3 from the second row elements (property (c) above) gives:

$$\begin{vmatrix} 3 & 4 & 5 \\ 6 & 7 & 8 \\ 1 & 11 & 6 \end{vmatrix} = \begin{vmatrix} 3 & 4 & 5 \\ 3 & 3 & 3 \\ 1 & 11 & 6 \end{vmatrix} = 3 \begin{vmatrix} 3 & 4 & 5 \\ 1 & 1 & 1 \\ 1 & 11 & 6 \end{vmatrix}$$

To introduce some elements whose values are 0, property (d) is applied again. The second-row elements are taken from the corresponding third-row elements, then the first-column elements are taken from the third-column elements and finally the first-column elements from the corresponding second-column elements. These three steps are shown below.

$$3 \begin{vmatrix} 3 & 4 & 5 \\ 1 & 1 & 1 \\ 1 & 11 & 6 \end{vmatrix} = 3 \begin{vmatrix} 3 & 4 & 5 \\ 1 & 1 & 1 \\ 0 & 10 & 5 \end{vmatrix} = 3 \begin{vmatrix} 3 & 4 & 2 \\ 1 & 1 & 0 \\ 0 & 10 & 5 \end{vmatrix} = 3 \begin{vmatrix} 3 & 1 & 2 \\ 1 & 0 & 0 \\ 0 & 10 & 5 \end{vmatrix}$$

It is now fairly easy to evaluate this determinant by using the second row elements, since this row contains two 0's.

Hence $3 \begin{vmatrix} 3 & 1 & 2 \\ 1 & 0 & 0 \\ 0 & 10 & 5 \end{vmatrix} = 3 \left\{ -1 \begin{vmatrix} 1 & 2 \\ 10 & 5 \end{vmatrix} + 0 + 0 \right\} = -3(1 \times 5 - 10 \times 2)$

i.e $\begin{vmatrix} 3 & 4 & 5 \\ 6 & 7 & 8 \\ 1 & 11 & 6 \end{vmatrix} = 45$

Problem 2. Evaluate the determinant $\begin{vmatrix} 3 & 5 & 7 \\ 11 & 9 & 13 \\ 15 & 17 & 19 \end{vmatrix}$

To reduce the magnitude of the elements in the second and third rows, the second-row elements are taken from the corresponding third-row elements, then the first-row elements are taken from the corresponding second-row elements, giving:

$$\begin{vmatrix} 3 & 5 & 7 \\ 11 & 9 & 13 \\ 15 & 17 & 19 \end{vmatrix} = \begin{vmatrix} 3 & 5 & 7 \\ 11 & 9 & 13 \\ 4 & 8 & 6 \end{vmatrix} = \begin{vmatrix} 3 & 5 & 7 \\ 8 & 4 & 6 \\ 4 & 8 & 6 \end{vmatrix}$$

To introduce 0's, the second-row elements are taken from the corresponding third-row elements and then the first-column elements are added to the corresponding second-column elements. This gives:

$$\begin{vmatrix} 3 & 5 & 7 \\ 8 & 4 & 6 \\ 4 & 8 & 6 \end{vmatrix} = \begin{vmatrix} 3 & 5 & 7 \\ 8 & 4 & 6 \\ -4 & 4 & 0 \end{vmatrix} = \begin{vmatrix} 3 & 8 & 7 \\ 8 & 12 & 6 \\ -4 & 0 & 0 \end{vmatrix}$$

Evaluating, using the third-row elements, gives

$$\begin{vmatrix} 3 & 8 & 7 \\ 8 & 12 & 6 \\ -4 & 0 & 0 \end{vmatrix} = -4 \begin{vmatrix} 8 & 7 \\ 12 & 6 \end{vmatrix} = -4\,(48 - 84)$$

i.e. $\begin{vmatrix} 3 & 5 & 7 \\ 11 & 9 & 13 \\ 15 & 17 & 19 \end{vmatrix} = 144$

Further problems on the properties of third order determinants may be found in Section 3 (Problems 1 to 7), page 141.

2 The solution of simultaneous equations having three unknowns

In engineering and science, the principal use of determinants is for the solution of simultaneous equations. Solving simultaneous equations having two unknowns by determinants is discussed in Chapter 8. This section deals with the solution of simultaneous equations having three unknowns by determinants.

Let the simultaneous equations to be solved be:

$$a_1x + b_1y + c_1z + d_1 = 0 \tag{1}$$
$$a_2x + b_2y + c_2z + d_2 = 0 \tag{2}$$
$$a_3x + b_3y + c_3z + d_3 = 0 \tag{3}$$

where a, b, c and d are constants.
The variables y and z may be eliminated from equations (1), (2) and (3) by the following procedure:

(i) multiplying equation (1) by $(b_2c_3 - b_3c_2)$,

(ii) multiplying equation (2) by $-(b_1 c_3 - b_3 c_1)$,

(iii) multiplying equation (3) by $(b_1 c_2 - b_2 c_1)$ and

(iv) adding the equations obtained in (i), (ii) and (iii) above.

(i) Equation (1) multiplied by $(b_2 c_3 - b_3 c_2)$ gives:

$$a_1(b_2 c_3 - b_3 c_2)x + b_1(b_2 c_3 - b_3 c_2)y + c_1(b_2 c_3 - b_3 c_2)z$$
$$+ d_1(b_2 c_3 - b_3 c_2) = 0$$

(ii) Equation (2) multiplied by $-(b_1 c_3 - b_3 c_1)$ gives:

$$-a_2(b_1 c_3 - b_3 c_1)x - b_2(b_1 c_3 - b_3 c_1)y - c_2(b_1 c_3 - b_3 c_1)z$$
$$- d_2(b_1 c_3 - b_3 c_1) = 0$$

(iii) Equation (3) multiplied by $(b_1 c_2 - b_2 c_1)$ gives:

$$a_3(b_1 c_2 - b_2 c_1)x + b_3(b_1 c_2 - b_2 c_1)y + c_3(b_1 c_2 - b_2 c_1)z$$
$$+ d_3(b_1 c_2 - b_2 c_1) = 0$$

(iv) Adding (i), (ii) and (iii) above gives:

$$x\{a_1(b_2 c_3 - b_3 c_2) - a_2(b_1 c_3 - b_3 c_1) + a_3(b_1 c_2 - b_2 c_1)\}$$
$$+ d_1(b_2 c_3 - b_3 c_2) - d_2(b_1 c_3 - b_3 c_1) + d_3(b_1 c_2 - b_2 c_1) = 0.$$

Writing these algebraic expressions in determinant form gives:

$$x \begin{vmatrix} a_1 & b_1 & c_1 \\ a_2 & b_2 & c_2 \\ a_3 & b_3 & c_3 \end{vmatrix} + \begin{vmatrix} b_1 & c_1 & d_1 \\ b_2 & c_2 & d_2 \\ b_3 & c_3 & d_3 \end{vmatrix} = 0$$

$$\text{Thus } x = \frac{- \begin{vmatrix} b_1 & c_1 & d_1 \\ b_2 & c_2 & d_2 \\ b_3 & c_3 & d_3 \end{vmatrix}}{\begin{vmatrix} a_1 & b_1 & c_1 \\ a_2 & b_2 & c_2 \\ a_3 & b_3 & c_3 \end{vmatrix}}$$

Equations (1), (2) and (3) may be written as:

$$b_1 y + a_1 x + c_1 z + d_1 = 0$$
$$b_2 y + a_2 x + c_2 z + d_2 = 0$$
$$b_3 y + a_3 x + c_3 z + d_3 = 0$$

By comparison with the determinant obtained for x:

$$y = \frac{- \begin{vmatrix} a_1 & c_1 & d_1 \\ a_2 & c_2 & d_2 \\ a_3 & c_3 & d_3 \end{vmatrix}}{\begin{vmatrix} b_1 & a_1 & c_1 \\ b_2 & a_2 & c_2 \\ b_3 & a_3 & c_3 \end{vmatrix}}$$

Using property (a) of determinants introduced in Section 1, the first and second columns of the denominator can be interchanged, giving

$$y = \dfrac{+\begin{vmatrix} a_1 & c_1 & d_1 \\ a_2 & c_2 & d_2 \\ a_3 & c_3 & d_3 \end{vmatrix}}{\begin{vmatrix} a_1 & b_1 & c_1 \\ a_2 & b_2 & c_2 \\ a_3 & b_3 & c_3 \end{vmatrix}}$$

Similarly, to find the determinant for z, equations (1), (2) and (3) may be written as:

$$c_1 z + b_1 y + a_1 x + d_1 = 0$$
$$c_2 z + b_2 y + a_2 x + d_2 = 0$$
$$c_3 z + b_3 y + a_3 x + d_3 = 0$$

By comparison with the determinant obtained for x:

$$z = \dfrac{-\begin{vmatrix} b_1 & a_1 & d_1 \\ b_2 & a_2 & d_2 \\ b_3 & a_3 & d_3 \end{vmatrix}}{\begin{vmatrix} c_1 & b_1 & a_1 \\ c_2 & b_2 & a_2 \\ c_3 & b_3 & a_3 \end{vmatrix}}$$

Using property (a) of determinants introduced in Section 1, the first and second columns of the numerator and the first and third columns of the denominator may be interchanged, giving

$$z = \dfrac{+\begin{vmatrix} a_1 & b_1 & d_1 \\ a_2 & b_2 & d_2 \\ a_3 & b_3 & d_3 \end{vmatrix}}{-\begin{vmatrix} a_1 & b_1 & c_1 \\ a_2 & b_2 & c_2 \\ a_3 & b_3 & c_3 \end{vmatrix}} = \dfrac{-\begin{vmatrix} a_1 & b_1 & d_1 \\ a_2 & b_2 & d_2 \\ a_3 & b_3 & d_3 \end{vmatrix}}{\begin{vmatrix} a_1 & b_1 & c_1 \\ a_2 & b_2 & c_2 \\ a_3 & b_3 & c_3 \end{vmatrix}}$$

These results can be combined, showing that

$$\dfrac{x}{|D_1|} = \dfrac{-y}{|D_2|} = \dfrac{z}{|D_3|} = \dfrac{-1}{|D|}$$

where $|D_1|$ is the determinant remaining when the x-column is 'covered up',

$|D_2|$ is the determinant remaining when the y-column is 'covered up',

$|D_3|$ is the determinant remaining when the z-column is 'covered up',

and $|D|$ is the determinant remaining when the constants-column is 'covered up'.

Thus, for the simultaneous equations:

$$a_1x + b_1y + c_1z + d_1 = 0$$
$$a_2x + b_2y + c_2z + d_2 = 0$$
$$a_3x + b_3y + c_3z + d_3 = 0$$

the solution is given by:

$$\frac{x}{\begin{vmatrix} b_1 & c_1 & d_1 \\ b_2 & c_2 & d_2 \\ b_3 & c_3 & d_3 \end{vmatrix}} = \frac{-y}{\begin{vmatrix} a_1 & c_1 & d_1 \\ a_2 & c_2 & d_2 \\ a_3 & c_3 & d_3 \end{vmatrix}} = \frac{z}{\begin{vmatrix} a_1 & b_1 & d_1 \\ a_2 & b_2 & d_2 \\ a_3 & b_3 & d_3 \end{vmatrix}} = \frac{-1}{\begin{vmatrix} a_1 & b_1 & c_1 \\ a_2 & b_2 & c_2 \\ a_3 & b_3 & c_3 \end{vmatrix}}$$

This relationship is used to solve simultaneous equations having three unknown values, as shown in the worked problems following.

Worked problems on the solution of simultaneous equations having three unknowns

Problem 1. Use determinants to solve the simultaneous equations:

$$2x + y = 2 \tag{1}$$
$$-4y + z = 0 \tag{2}$$
$$4x + z = 6 \tag{3}$$

Writing the equations in the form $a_1x + b_1y + c_1z + d_1 = 0$, gives

$$2x + y + 0z - 2 = 0$$
$$0x - 4y + z + 0 = 0$$
$$4x + 0y + z - 6 = 0$$

The solution can be obtained from

$$\frac{x}{|D_1|} = \frac{-y}{|D_2|} = \frac{z}{|D_3|} = \frac{-1}{|D|}$$

$|D_1|$ is the third-order determinant obtained by covering up the x-column and writing down the remaining coefficients.

i.e. $|D_1| = \begin{vmatrix} 1 & 0 & -2 \\ -4 & 1 & 0 \\ 0 & 1 & -6 \end{vmatrix} = 1\begin{vmatrix} 1 & 0 \\ 1 & -6 \end{vmatrix} - 0\begin{vmatrix} -4 & 0 \\ 0 & -6 \end{vmatrix} + (-2)\begin{vmatrix} -4 & 1 \\ 0 & 1 \end{vmatrix}$

$$= 2$$

$|D_2|$ is the third-order determinant remaining when the y-column is covered up, i.e.,

$|D_2| = \begin{vmatrix} 2 & 0 & -2 \\ 0 & 1 & 0 \\ 4 & 1 & -6 \end{vmatrix} = 0 + 1\begin{vmatrix} 2 & -2 \\ 4 & -6 \end{vmatrix} + 0 = -4$

$|D_3|$ is the third-order determinant remaining when the z-column is covered up, i.e.

$$|D_3| = \begin{vmatrix} 2 & 1 & -2 \\ 0 & -4 & 0 \\ 4 & 0 & -6 \end{vmatrix} = -4 \begin{vmatrix} 2 & -2 \\ 4 & -6 \end{vmatrix} = 16$$

$|D|$ is the third-order determinant remaining when the constants-column is covered up, i.e.

$$|D| = \begin{vmatrix} 2 & 1 & 0 \\ 0 & -4 & 1 \\ 4 & 0 & 1 \end{vmatrix} = 2 \begin{vmatrix} -4 & 1 \\ 0 & 1 \end{vmatrix} - 1 \begin{vmatrix} 0 & 1 \\ 4 & 1 \end{vmatrix} + 0 = -4$$

Thus, $\dfrac{x}{2} = \dfrac{-y}{-4} = \dfrac{z}{16} = \dfrac{-1}{-4}$

i.e. $x = \frac{1}{2}, y = 1, z = 4$.

Problem 2. Solve the simultaneous equations:

$$\frac{3}{x} - \frac{4}{y} - \frac{2}{z} = 1 \tag{1}$$

$$\frac{2}{x} + \frac{5}{y} - \frac{2}{z} = 3 \tag{2}$$

$$\frac{1}{x} + \frac{2}{y} + \frac{1}{z} = 2 \tag{3}$$

Letting $p = \dfrac{1}{x}$, $q = \dfrac{1}{y}$ and $r = \dfrac{1}{z}$ and writing the equations in the form $a_1 x + b_1 y + c_1 z + d_1 = 0$, gives

$3p - 4q - 2r - 1 = 0$
$2p + 5q - 2r - 3 = 0$
$p + 2q + r - 2 = 0$

The solution can be obtained from

$$\frac{p}{|D_1|} = \frac{-q}{|D_2|} = \frac{r}{|D_3|} = \frac{-1}{|D|}$$

$$|D_1| = \begin{vmatrix} -4 & -2 & -1 \\ 5 & -2 & -3 \\ 2 & 1 & -2 \end{vmatrix}$$

The arithmetic of evaluating this determinant can be simplified by taking twice the second-column elements from the first.

Thus:
$$\begin{vmatrix} -4 & -2 & -1 \\ 5 & -2 & -3 \\ 2 & 1 & -2 \end{vmatrix} = \begin{vmatrix} 0 & -2 & -1 \\ 9 & -2 & -3 \\ 0 & 1 & -2 \end{vmatrix}$$

$$= -9 \begin{vmatrix} -2 & -1 \\ 1 & -2 \end{vmatrix} = -45$$

$$|D_2| = \begin{vmatrix} 3 & -2 & -1 \\ 2 & -2 & -3 \\ 1 & 1 & -2 \end{vmatrix}$$

Taking the second-row elements from the first and then twice the first-column elements from the third, gives:

$$|D_2| = \begin{vmatrix} 1 & 0 & 2 \\ 2 & -2 & -3 \\ 1 & 1 & -2 \end{vmatrix} = \begin{vmatrix} 1 & 0 & 0 \\ 2 & -2 & -7 \\ 1 & 1 & -4 \end{vmatrix} = 1 \begin{vmatrix} -2 & -7 \\ 1 & -4 \end{vmatrix} = 15$$

$$|D_3| = \begin{vmatrix} 3 & -4 & -1 \\ 2 & 5 & -3 \\ 1 & 2 & -2 \end{vmatrix}$$

Adding the third-column elements to the second, and twice the first-column elements to the third gives:

$$|D_3| = \begin{vmatrix} 3 & -5 & -1 \\ 2 & 2 & -3 \\ 1 & 0 & -2 \end{vmatrix} = \begin{vmatrix} 3 & -5 & 5 \\ 2 & 2 & 1 \\ 1 & 0 & 0 \end{vmatrix} = 1 \begin{vmatrix} -5 & 5 \\ 2 & 1 \end{vmatrix} = -15$$

$$|D| = \begin{vmatrix} 3 & -4 & -2 \\ 2 & 5 & -2 \\ 1 & 2 & 1 \end{vmatrix}$$

Adding twice the third-row elements to the first, gives:

$$|D| = \begin{vmatrix} 5 & 0 & 0 \\ 2 & 5 & -2 \\ 1 & 2 & 1 \end{vmatrix} = 5 \begin{vmatrix} 5 & -2 \\ 2 & 1 \end{vmatrix} = 45$$

Thus, $\dfrac{p}{-45} = \dfrac{-q}{15} = \dfrac{r}{-15} = \dfrac{-1}{45}$

i.e. $p = 1, q = \dfrac{1}{3}, r = \dfrac{1}{3}$.

Thus $x = 1, y = 3$ and $z = 3$.

Further problems on the solution of simultaneous equations having three unknowns may be found in the following Section (3) (Problems 8 to 18), page 142.

3   Further problems

The properties of third-order determinants

In Problems 1 to 7, use the properties of determinants to simplify the determinants given and then evaluate them.

1. $\begin{vmatrix} 1 & 3 & 5 \\ 2 & 4 & 6 \\ 3 & 5 & 7 \end{vmatrix}$ [0]

2. $\begin{vmatrix} 13 & 2 & 23 \\ 30 & 7 & 53 \\ 39 & 9 & 70 \end{vmatrix}$ [34]

3. $\begin{vmatrix} 14 & 9 & 33 \\ 13 & 11 & 36 \\ 17 & 2 & 22 \end{vmatrix}$ [1]

4. $\begin{vmatrix} 1 & 11 & 16 \\ 23 & 37 & 58 \\ 16 & 9 & 1 \end{vmatrix}$ [3 310]

5. $\begin{vmatrix} 6 & 10 & 14 \\ -2 & 1 & 1 \\ -9 & -15 & -12 \end{vmatrix}$ [234]

6. $\begin{vmatrix} 22 & 25 & 28 \\ 26 & 28 & 31 \\ 24 & 27 & 29 \end{vmatrix}$ [40]

7. $\begin{vmatrix} 5.2 & 7.5 & 8.6 \\ 5.4 & 7.2 & 8.7 \\ 5.6 & 6.9 & 8.3 \end{vmatrix}$ [1.53]

The solution of simultaneous equations

In Problems 8 to 11, solve the simultaneous equations given, using determinants.

8. $11p + 7q + 2r = 31$
  $p + q + r = 4$
  $31p + 15q + 13r = 90$ $[p = 2, q = 1, r = 1]$

9. $4l + 9m + 2n = 21$
  $13l + 5m + 7n = 1$
  $17l + 19m + 8n = 26$ $[l = -7, m = 3, n = 11]$

10. $\dfrac{1}{x} + \dfrac{2}{y} + \dfrac{2}{z} = 4$

$\dfrac{3}{x} - \dfrac{1}{y} + \dfrac{4}{z} = 25$

$\dfrac{3}{x} + \dfrac{2}{y} - \dfrac{1}{z} = -4$ $[x = \dfrac{1}{2}, y = -\dfrac{1}{3}, z = \dfrac{1}{4}]$

11. $\dfrac{2x}{3} - y + \dfrac{2z}{3} = 2$

$x + 8y + 3z = -31$

$\dfrac{6x}{5} - \dfrac{4y}{5} + \dfrac{2z}{5} = -2$ $[x = -5, y = -4, z = 2]$

12. Kirchhoff's laws are applied to an electrical network and the following simultaneous equations for the current flowing in amperes in various closed circuits are obtained.

$$2i_1 \quad + i_2 \quad + i_3 \quad = 1.67$$
$$3i_1 \quad + 4.5i_2 - 1.5i_3 \quad = 0$$
$$2.25i_1 + 1.5i_2 + 5.25i_3 = 0$$

Use determinants to find the values of i_1, i_2 and i_3, correct to three significant figures. $[i_1 = 3.96A, i_2 = -3.54A, i_3 = -2.71A]$

13. The rate of working, $\dot{W}$, is given by:

$$\dot{W} = \Sigma F \cdot \dot{r},$$

where F is the force in newtons, r is the distance in metres from some reference point, and $\dot{r}$ is the velocity in metres per second. The following simultaneous equations arise from experiments carried out on a system.

$$\dot{r}_1 + 3\dot{r}_2 + 2\dot{r}_3 = -13$$
$$2\dot{r}_1 - 6\dot{r}_2 + 3\dot{r}_3 = 32$$
$$3\dot{r}_1 - 4\dot{r}_2 - \dot{r}_3 = 12$$

Use determinants to find the values of $\dot{r}_1$, $\dot{r}_2$ and $\dot{r}_3$.
$[\dot{r}_1 = -2 \text{ m s}^{-1}, \dot{r}_2 = -5 \text{ m s}^{-1}, \dot{r}_3 = 2 \text{ m s}^{-1}]$

14. The simultaneous equations representing the currents flowing in an unbalanced, three-phase, star-connected, electrical network are as follows:

$$3.6i_1 + 2.4i_2 + 4.8i_3 = 1.2$$
$$1.3i_1 - 3.9i_2 - 6.5i_3 = 2.6$$
$$11.9i_1 + 1.7i_2 + 8.5i_3 = 0$$

Use determinants to find the values of i_1, i_2 and i_3.
$[i_1 = 1, i_2 = 3, i_3 = -2]$

15. In a mass-spring-damper system, the acceleration $\ddot{x}$ m s^{-2}, velocity $\dot{x}$ m s^{-1} and displacement x m are related by the following simultaneous equations:

$$5.8\ddot{x} + 8.7\dot{x} + 14.5x = 23.2$$
$$8.1\ddot{x} + 5.4\dot{x} + 5.4x = 8.1$$
$$14.8\ddot{x} + 3.7\dot{x} - 14.8x = -22.2$$

By using determinants, determine the acceleration, velocity and displacement for the system, correct to two decimal places.
$[\ddot{x} = -0.33 \text{ m s}^{-2}, \dot{x} = 0.67 \text{ m s}^{-1}, x = 1.33 \text{ m}]$

16. The currents in a network are represented by I_1, I_2 and I_3. Application of Kirchhoff's laws to the circuit leads to the following three equations:

$$14I_1 - 5I_2 - 6I_3 = 10$$
$$-5I_1 + 14I_2 - 2I_3 = 3$$
$$-6I_1 - 2I_2 + 18I_3 = 5$$

Determine I_1, I_2 and I_3 correct to four significant figures.

$$[1.358, 0.816\ 5, 0.821\ 1]$$

17. The tensions in a simple framework, T_1, T_2 and T_3, are given by the equations:

$$6T_1 + 6T_2 + 6T_3 = 8.4$$
$$T_1 + 2T_2 + 3T_3 = 2.2$$
$$4T_1 + 2T_2 \qquad = 4.0$$

Determine T_1, T_2 and T_3. $\qquad\qquad$ $[0.8, 0.4, 0.2]$

18. When a number of mass/spring systems are connected together and have a mode of oscillation, and all masses oscillate with a frequency $\dfrac{n}{2\pi}$ but having different amplitudes, n can be given in terms of eigenvalues λ (where $\lambda = n^2$) by the determinant:

$$\begin{vmatrix} 1 - \lambda & -\dfrac{1}{2} & 0 \\[2mm] -\dfrac{3}{4} & \dfrac{6}{4} - \lambda & -\dfrac{3}{4} \\[2mm] 0 & -\dfrac{3}{4} & 1 - \lambda \end{vmatrix} = 0$$

Show that $16\lambda^3 - 56\lambda^2 + 49\lambda - 9 = 0$ and verify that $\lambda = 1$, $\dfrac{9}{4}$ and $\dfrac{1}{4}$.

Chapter 11

Differentiation of implicit functions (CA 15)

1 Implicit functions

In all previous work involving differentiation, the starting point has been equations of the form $y = f(x)$. When an equation is expressed in this form, it is said to be an **explicit** function. Thus:

$$y = 3x^3 + 4x - 5$$

$$y = e^x \sin x$$

and $\quad y = 3(x^2 - 1) \ln \dfrac{x}{2}$

are all examples of explicit functions. In many equations it is difficult, or indeed impossible, to make y the subject of the formula. The equation is then called an **implicit** function, there being an implied relationship between x and y. Thus:

$$x^2 - 4xy + 3y^3 = 8$$

$$e^x \sin y + 4xy^2 = 0$$

and $\quad 2 \sin \dfrac{x}{2} \cos y^2 = x^2 - y$

are all examples of implicit functions.

2 Differentiating implicit functions

When differentiating an explicit function, the function of a variable, say x, is differentiated with respect to x. However, when differentiating an implicit function, it becomes necessary to differentiate a function of one variable, say y, with respect to another variable, say x. This is achieved by using the 'chain' rule of differentiation, i.e.

$$\frac{du}{dx} = \frac{du}{dy} \times \frac{dy}{dx} .$$

Thus, to differentiate, say, y^2 with respect to x, let $u = y^2$.

Then $\frac{du}{dy} = 2y$ and $\frac{d(y^2)}{dx} = 2y \frac{dy}{dx}$ by using the 'chain' rule.

The rule when differentiating a function of, say y, with respect to, say x, is:

(i) differentiate the function of y with respect to y, and

(ii) always multiply by $\frac{dy}{dx}$.

This rule can be expressed mathematically as

$$\frac{d}{dx} [f(y)] = \frac{d}{dy} [f(y)] \times \frac{dy}{dx} \tag{1}$$

When a term is a product or a quotient of two functions of, say, x, the product or quotient rules of differentiation must be applied to find its differential coefficient. These same rules apply to products and quotients of terms having two variables (say $f(x, y)$). Thus if $z = x^2 y^3$, the product rule of differentiation must be used to find $\frac{dz}{dx}$, i.e. $\frac{dz}{dx} = u \frac{dv}{dx} + v \frac{du}{dx}$ where

$u = x^2$ and $v = y^3$. Since $u = x^2$, $\frac{du}{dx} = 2x$ and since $v = y^3$, $\frac{dv}{dx} = 3y^2 \frac{dy}{dx}$ from equation (1) above.

Hence $\frac{dz}{dx} = x^2 \cdot 3y^2 \frac{dy}{dx} + y^3 \cdot 2x$

$$= xy^2 \left(3x \frac{dy}{dx} + 2y \right)$$

For a quotient, say $z = \frac{3y^2}{\cos x}$, the quotient rule of differentiation is used,

i.e. $\frac{dz}{dx} = \frac{v \dfrac{du}{dx} - u \dfrac{dv}{dx}}{v^2}$, where $u = 3y^2$ and $v = \cos x$.

Since $u = 3y^2$, $\frac{du}{dx} = 6y \frac{dy}{dx}$ from equation (1) above and since $v = \cos x$,

$\frac{dv}{dx} = - \sin x,$

thus, $\frac{dz}{dx} = \dfrac{(\cos x) (6y \frac{dy}{dx}) - (3y^2)(- \sin x)}{\cos^2 x}$

$$= \dfrac{3y (2 \cos x \frac{dy}{dx} + y \sin x)}{\cos^2 x}$$

When differentiating an implicit function having several terms, each term is differentiated with respect to the variable. Thus, to differentiate $x^2 + 2xy + 3y^2 = 0$ with respect to x, it becomes:

$$\frac{d(x^2)}{dx} + \frac{d(2xy)}{dx} + \frac{d(3y^2)}{dx} = \frac{d(0)}{dx}$$

i.e. $2x + \left(2x\,\frac{dy}{dx}\right) + y \cdot 2 + 6y\,\frac{dy}{dx} = 0$

To express $\frac{dy}{dx}$ in terms of x and y, terms containing $\frac{dy}{dx}$ are grouped on the left hand side of the equation, and those not containing $\frac{dy}{dx}$ are grouped on the right-hand side. Thus,

$$2x\,\frac{dy}{dx} + 6y\,\frac{dy}{dx} = -2x - 2y$$

$$\frac{dy}{dx}\,(2x + 6y) = -2x - 2y$$

$$\frac{dy}{dx} = \frac{-2x - 2y}{2x + 6y} = \frac{-(x + y)}{x + 3y}$$

Worked problems on differentiating implicit functions

Problem 1. Differentiate the following functions with respect to x:

(a) y^5 (b) $\cos 2\theta$ (c) $3 \ln \frac{p}{2}$ (d) $\frac{1}{4}\,e^{2q+3}$

(a) Let $u = y^5$, then

$$\frac{du}{dx} = \frac{du}{dy} \times \frac{dy}{dx} = \frac{d}{dy}\,(y^5) \times \frac{dy}{dx} = 5y^4\,\frac{dy}{dx}$$

(b) Let $u = \cos 2\theta$.

$$\frac{du}{dx} = \frac{du}{d\theta} \times \frac{d\theta}{dx} = \frac{d}{d\theta}\,(\cos 2\theta) \times \frac{d\theta}{dx}$$

$$= -2 \sin 2\theta\,\frac{d\theta}{dx}$$

(c) Let $u = 3 \ln \frac{p}{2}$

$$\frac{du}{dx} = \frac{du}{dp} \times \frac{dp}{dx} = \frac{d}{dp}\left(3 \ln \frac{p}{2}\right) \times \frac{dp}{dx}$$

$$= \frac{3}{p}\,\frac{dp}{dx}$$

(d) Let $u = \dfrac{1}{4} e^{2q+3}$

$$\frac{du}{dx} = \frac{du}{dq} \times \frac{dq}{dx} = \frac{d}{dq}\left(\frac{1}{4} e^{2q+3}\right) \times \frac{dq}{dx}$$

$$= \frac{1}{2} e^{2q+3} \frac{dq}{dx}$$

Problem 2. If $y = f(m, p)$, differentiate with respect to p:

(a) $y = m^2 \cos 2p$ and (b) $y = \dfrac{(1-3p)^{\frac{1}{2}}}{2 \tan 2m}$

(a) Since y is a function of two variables, m and p, the product rule of differentiation is used.

Let $u = m^2$. Now $\dfrac{du}{dp} = \dfrac{du}{dm} \times \dfrac{dm}{dp} = \dfrac{d}{dm}(m^2) \times \dfrac{dm}{dp}$

$$= 2m \frac{dm}{dp}$$

Let $v = \cos 2p$, $\dfrac{dv}{dp} = -2 \sin 2p$.

Applying the product rule, $\dfrac{dy}{dp} = u\dfrac{dv}{dp} + v\dfrac{du}{dp}$ gives

$$\frac{dy}{dp} = m^2(-2\sin 2p) + \cos 2p\left(2m\frac{dm}{dp}\right)$$

$$= 2m\left(\cos 2p\frac{dm}{dp} - m\sin 2p\right)$$

(b) Since y is a function of both m and p, the quotient rule of differentiation is used.

Let $u = (1-3p)^{\frac{1}{2}}$, $\dfrac{du}{dp} = -\dfrac{3}{2}(1-3p)^{-\frac{1}{2}}$

Let $v = 2 \tan 2m$, $\dfrac{dv}{dp} = \dfrac{dv}{dm} \times \dfrac{dm}{dp}$

$$= 4\sec^2 2m \frac{dm}{dp}$$

Applying the quotient rule, $\dfrac{dy}{dp} = \dfrac{v\dfrac{du}{dp} - u\dfrac{dv}{dp}}{v^2}$, gives

$$\frac{dy}{dp} = \frac{2\tan 2m\left(-\dfrac{3}{2}(1-3p)^{-\frac{1}{2}}\right) - (1-3p)^{\frac{1}{2}}\left(4\sec^2 2m\dfrac{dm}{dp}\right)}{4\tan^2 2m}$$

$$= -\frac{3}{4(\tan 2m)(1-3p)^{\frac{1}{2}}} - \frac{(1-3p)^{\frac{1}{2}} \quad \sec^2 2m \dfrac{dm}{dp}}{\tan^2 2m}$$

Problem 3. Given that $x^2 \sin \theta - 3x^3 = \sec \theta$, determine the value of $\dfrac{dx}{d\theta}$ when $\theta = \pi$.

$$\frac{d}{d\theta}(x^2 \sin \theta) = x^2 \cos \theta + \sin \theta \left(2x \frac{dx}{d\theta} \right)$$

$$\frac{d}{d\theta}(3x^3) = 9x^2 \frac{dx}{d\theta}$$

$$\frac{d}{d\theta}(\sec \theta) = \sec \theta \tan \theta$$

Hence, $x^2 \cos \theta + 2x \sin \theta \dfrac{dx}{d\theta} - 9x^2 \dfrac{dx}{d\theta} = \sec \theta \tan \theta$

Grouping only terms containing $\dfrac{dx}{d\theta}$ on the left-hand side gives:

$$2x \sin \theta \frac{dx}{d\theta} - 9x^2 \frac{dx}{d\theta} = \sec \theta \tan \theta - x^2 \cos \theta$$

$$\frac{dx}{d\theta}(2x \sin \theta - 9x^2) = \sec \theta \tan \theta - x^2 \cos \theta$$

$$\frac{dx}{d\theta} = \frac{\sec \theta \tan \theta - x^2 \cos \theta}{2x \sin \theta - 9x^2}$$

To determine the value of this expression, the value of x when $\theta = \pi$ is needed.

Substituting $\theta = \pi$ in the original equation, gives:

$$x^2 \sin \pi - 3x^3 = \sec \pi.$$

Now, $\sin \pi = 0$ and $\sec \pi = -1$. Hence, $0 - 3x^3 = -1$, i.e. $x^3 = \dfrac{1}{3}$

i.e. $x = \sqrt[3]{\tfrac{1}{3}}$

Thus $\dfrac{dx}{d\theta} = \dfrac{\sec \pi \tan \pi - \sqrt[3]{(\tfrac{1}{3})^2} \cos \pi}{2\sqrt[3]{(\tfrac{1}{3})} \sin \pi - 9\sqrt[3]{(\tfrac{1}{3})^2}}$

$$= \frac{(-1)(0) - \sqrt[3]{(\tfrac{1}{3})^2} \cdot (-1)}{2\sqrt[3]{(\tfrac{1}{3})}(0) - 9\sqrt[3]{(\tfrac{1}{3})^2}}$$

$$= \frac{\sqrt[3]{(\tfrac{1}{3})^2}}{-9\sqrt[3]{(\tfrac{1}{3})^2}} = -\frac{1}{9}$$

Problem 4. Determine the values of the gradients of the tangents drawn to the circle $x^2 + y^2 - 3x + 4y + 1 = 0$ at $x = 1$, correct to four significant figures.

The equation of the gradient of the tangent is given by $\dfrac{dy}{dx}$.

Differentiating each term with respect to x, gives:

$$2x + 2y\,\frac{dy}{dx} - 3 + 4\,\frac{dy}{dx} + 0 = 0$$

$$\text{i.e.}\ \ \frac{dy}{dx}(2y + 4) = 3 - 2x$$

$$\frac{dy}{dx} = \frac{3 - 2x}{2y + 4}$$

The value of y when $x = 1$ is obtained by substituting $x = 1$ in the original equation. Thus:

$$(1)^2 + y^2 - 3(1) + 4y + 1 = 0$$
$$\text{i.e.}\ y^2 + 4y - 1 = 0$$
$$y = \frac{-4 \pm \sqrt{[4^2 - 4 \times 1 \times (-1)]}}{2}$$

$$y = 0.236\,07 \text{ or } -4.236\,07.$$

Substituting $x = 1$, $y = 0.236\,07$ in the equation for $\dfrac{dy}{dx}$ gives:

$$\frac{dy}{dx} = \frac{3 - 2(1)}{2 \times 0.236\,07 + 4} = 0.223\,6, \text{ correct to four significant figures.}$$

Substituting $x = 1$ and $y = -4.236\,07$ in the equation for $\dfrac{dy}{dx}$ gives:

$$\frac{dy}{dx} = \frac{3 - 2(1)}{2 \times (-4.236\,07) + 4} = -0.223\,6, \text{ correct to four significant figures.}$$

That is, the gradients to the tangents are $\pm\,0.223\,6$, correct to four significant figures.

Further problems on differentiating implicit functions may be found in the following Section (3) (Problems 1 to 20).

3 Further problems

Differentiating implicit functions

1. Differentiate with respect to x:

    (a) y^6 (b) $3z^{\frac{1}{2}}$ (c) $\sin\dfrac{p}{3}$.

    (a) $\left[\, 6y^5\,\dfrac{dy}{dx} \,\right]$ (b) $\left[\, \dfrac{3}{2}\,z^{-\frac{1}{2}}\dfrac{dz}{dx} \,\right]$ (c) $\left[\, \dfrac{1}{3}\,\cos\dfrac{p}{3}\,\dfrac{dp}{dx} \,\right]$

2. Differentiate with respect to y:

(a) $\sin 2p$ (b) $\dfrac{1}{2} \cos 3x$ (c) $2e^{2m+4}$.

(a) $\left[2 \cos 2p \dfrac{dp}{dy} \right]$ (b) $\left[-\dfrac{3}{2} \sin 3x \dfrac{dx}{dy} \right]$ (c) $\left[4e^{2m+4} \dfrac{dm}{dy} \right]$

3. Differentiate with respect to u:

(a) $(3 - 4x)^{\frac{1}{2}}$ (b) $2 \ln(3y + 2)$ (c) $\dfrac{1}{4} \tan(4u + \pi)$.

(a) $\left[-2(3 - 4x)^{-\frac{1}{2}} \dfrac{dx}{du} \right]$ (b) $\left[\dfrac{6}{3y + 2} \dfrac{dy}{du} \right]$

(c) $[\sec^2 (4u + \pi)]$

4. Differentiate with respect to w:

(a) $-\dfrac{1}{2} \sec 2\theta$ (b) $\dfrac{6}{3y + 2}$ (c) $\dfrac{3}{2} \cot \left(\dfrac{\pi}{2} - \alpha \right)$.

(a) $\left[-\sec 2\theta \tan 2\theta \dfrac{d\theta}{dw} \right]$ (b) $\left[\dfrac{-18}{(3y + 2)^2} \dfrac{dy}{dw} \right]$

(c) $\left[\dfrac{3}{2} \text{cosec}^2 \left(\dfrac{\pi}{2} - \alpha \right) \dfrac{d\alpha}{dw} \right]$

In Problems 5 to 8, find $\dfrac{dz}{dx}$.

5. $z = x^2 y^3$. $\left[3x^2 y^2 \dfrac{dy}{dx} + 2x y^3 \right]$

6. $z = 3y^2 \cos x$. $\left[-3y^2 \sin x + 6y \cos x \dfrac{dy}{dx} \right]$

7. $z = x^2 \ln y$. $\left[\dfrac{x^2}{y} \dfrac{dy}{dx} + 2x \ln y \right]$

8. $z = \dfrac{2x^{\frac{1}{3}}}{\tan y}$. $\left[\dfrac{\dfrac{2}{3} (\tan y) x^{-\frac{2}{3}} - 2x^{\frac{1}{3}} \sec^2 y \dfrac{dy}{dx}}{\tan^2 y} \right]$

In Problems 9 to 12, differentiate with respect to y.

9. $x^2 + \sin x \cos y$. $\left[2x \dfrac{dx}{dy} - \sin x \sin y + \cos x \cos y \dfrac{dx}{dy} \right]$

10. $e^{2x} \ln y - 4x^2$. $\left[\dfrac{e^{2x}}{y} + 2e^{2x} \ln y \dfrac{dx}{dy} - 8x \dfrac{dx}{dy} \right]$

11. $x^2 + 3xy - y^2$. $\left[2x \dfrac{dx}{dy} + 3x + 3y \dfrac{dx}{dy} - 2y \right]$

12. $y^3 - 4x^2 \cos y + \dfrac{\ln x}{\sec y}$.

$$\left[\dfrac{3y^2 + 4x^2 \sin y - 8x \cos y \dfrac{dx}{dy} + \dfrac{\frac{1}{x} \dfrac{dx}{dy} - \ln x \tan y}{\sec y}}{}\right]$$

In Problems 13 to 16, find $\dfrac{dy}{dx}$ in terms of x and y.

13. $x^2 + y^2 + 6x + 7y + 3 = 0$. $\left[\dfrac{dy}{dx} = \dfrac{-2(x+3)}{2y+7}\right]$

14. $2x^3 - 4x^2 y = \cos y$. $\left[\dfrac{dy}{dx} = \dfrac{2x(3x-4y)}{4x^2 \sin y}\right]$

15. $\sin 2x \cos 3y = x^{\frac{1}{2}} y^{\frac{1}{3}}$. $\left[\dfrac{dy}{dx} = \dfrac{2 \cos 2x \cos 3y - y^{\frac{1}{3}}/(2x^{\frac{1}{2}})}{3 \sin 2x \sin 3y + x^{\frac{1}{2}}/(3y^{\frac{2}{3}})}\right]$

16. $2y - \dfrac{x^3}{\sec y} = y^3$. $\left[\dfrac{dy}{dx} = \dfrac{3x^2}{x^3 \tan y + \sec y (2 - 3y^2)}\right]$

17. Determine the gradients of the tangents drawn to the circle $x^2 + y^2 = 4$ at the point $x = 1$.
$$\left[\pm \dfrac{1}{\sqrt{3}} = \pm 0.5774\right]$$

18. Find the gradients of the tangents drawn to the ellipse $2x^2 + y^2 = 9$ at the point $x = 2$ and to the hyperbola $x^2 - y^2 = 8$ at the point $x = 3$.
$$[\pm 4; \pm 3.]$$

19. If the distance moved by a body is given by $x = 3 \tan \theta$, the angular velocity, w, is $\dfrac{d\theta}{dt}$ and the velocity v is $\dfrac{dx}{dt}$, show that $w = \dfrac{v}{3} \cos^2 \theta$.

20. The pressure, p, and volume, V, of a gas are related by the law $p V^n = C$, where n and C are constants. Show that the rate of change of pressure, $\dfrac{dp}{dt} = -n \dfrac{p}{V} \dfrac{dV}{dt}$.

Chapter 12

Differentiation of functions defined parametrically (CA 16)

1 Parametric representation of points

Certain mathematical relationships can be expressed more simply by stating both x and y in terms of a third variable, say, θ. By doing this, subsequent work is frequently simplified. When both x and y are expressed in terms of the same variable, this variable is called a **parameter**. The equation of any point on a circle, centre at $x = 0$ and $y = 0$, and of radius r, is given by $x^2 + y^2 = r^2$.

Such a circle is shown in Fig. 1. Any point on the circumference of the circle may be expressed in terms of the radius and the angle θ and when this is done, θ is the parameter used.

With reference to Fig. 1, from triangle OAB, $x = r \cos \theta$ and $y = r \sin \theta$. These two equations are called the **parametric equations** for a circle. A check may be made when parametric equations have been formed by substituting for x and y in the given equation. Thus, substituting $x = r \cos \theta$ and $y = r \sin \theta$ in the left-hand side of the equation $x^2 + y^2 = r^2$ gives:

$$\text{L.H.S.} = r^2 \cos^2\theta + r^2 \sin^2 \theta \; = r^2 (\cos^2 \theta + \sin^2 \theta)$$
$$= r^2,$$

which is equal to the right-hand side of the original equation. Hence $x = r \cos \theta$, $y = r \sin \theta$ are suitable parametric equations of circle $x^2 + y^2 = r^2$.

When the parametric equations of a curve are given, the curve may be plotted as shown below. The parametric equations for a parabola are of the form $x = at^2$, $y = 2at$. Checking these parametric equations in the equation of a parabola $y^2 = 4ax$ gives:

L.H.S. $y^2 = (2at)^2 = 4a^2t^2$.
R.H.S. $4ax = 4a(at^2) = 4a^2t^2$.

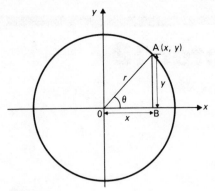

Figure 1

Since L.H.S. = R.H.S., $x = at^2$, $y = 2\,at$ are suitable parametric equations. To plot the graph for values of t between, say, -3 and 3, the values of x and y are calculated from the parametric equations. Thus, when

$t =$	-3	-2	-1	0	1	2	3
$x = at^2 =$	$9a$	$4a$	a	0	a	$4a$	$9a$
$y = 2\,at =$	$-6a$	$-4a$	$-2a$	0	$2a$	$4a$	$6a$

The parabola represented by $x = at^2$, $t = 2\,at$, over a range $-3 \leqslant t \leqslant 3$ is shown in Fig. 2.

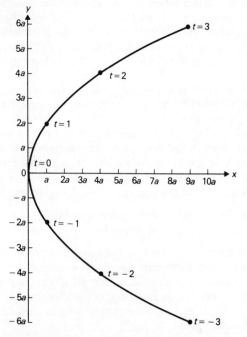

Figure 2

There can be more than one parametric form of the co-ordinates used to represent a particular curve. For example, the circle having parameter θ and parametric equations $x = a \cos \theta$, $y = a \sin \theta$ may be expressed in terms of a different parameter t, as

$$x = \frac{a(1-t^2)}{1+t^2} , \quad y = \frac{2at}{1+t^2} ,$$

where the parameter t is given by $t = \tan \dfrac{\theta}{2}$.

The list below shows some of the more common parametric equations used in mathematics.

Parabola	$x = at^2$,	$y = 2at$
Ellipse	$x = a \cos \theta$,	$y = b \sin \theta$
Hyperbola	$x = a \sec \theta$,	$y = b \tan \theta$
(also	$x = \pm a \cosh u$,	$y = b \sinh u$)
Rectangular Hyperbola	$x = ct$,	$y = \dfrac{c}{t}$
Astroid	$x = a \cos^3 \theta$,	$y = a \sin^3 \theta$
Cardioid	$x = a(2\cos\theta - \cos 2\theta)$,	$y = a(2\sin\theta - \sin 2\theta)$
Cycloid	$x = a(\theta - \sin\theta)$,	$y = a(1 - \cos\theta)$

Sketches showing the approximate shapes of these curves are shown in Fig. 3.

2 Differentiation in parameters

When x and y are given in terms of a parameter, say, t, then by the chain rule of differentiation:

$$\frac{dy}{dx} = \frac{dy}{dt} \times \frac{dt}{dx}$$

Also, if y is a function of x, then x must be a function of y. The differential coefficient, $\dfrac{dx}{dy}$, is defined as

$$\frac{dx}{dy} = \lim_{\delta y \to 0} \frac{\delta x}{\delta y}$$

Since δx and δy are finite, measurable quantities, the basic rules of fractions apply, and $\dfrac{dx}{dy}$ may be written as

$$\frac{dx}{dy} = \lim_{\delta y \to 0} \frac{1}{\frac{\delta y}{\delta x}}$$

As δy approaches zero, δx also approaches zero, hence

$$\frac{dx}{dy} = \lim_{\delta x \to 0} \frac{1}{\frac{\delta x}{\delta y}} = \frac{1}{\frac{dy}{dx}} , \text{ i.e. } \frac{dy}{dx} = \frac{1}{\frac{dx}{dy}}$$

156

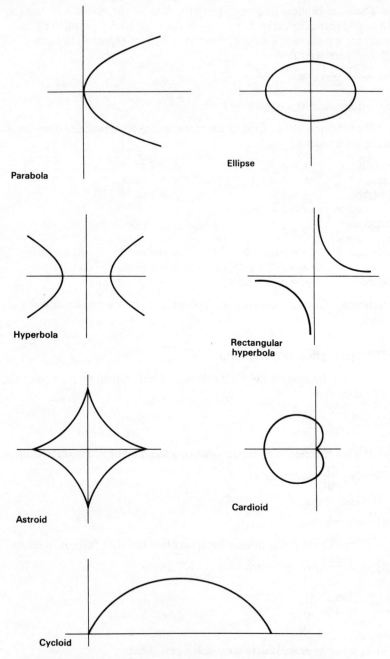

Figure 3 Parabola, ellipse, hyperbola, rectangular hyperbola, astroid, cardioid, cycloid.

Thus the chain rule can be written as:

$$\frac{dy}{dx} = \frac{\dfrac{dy}{dt}}{\dfrac{dx}{dt}}$$ (1)

The second derivative of y with respect to x can also be expressed in terms of a parameter t as follows:

$$\frac{d^2y}{dx^2} = \frac{d}{dx}\left(\frac{dy}{dx}\right)$$

Let $u = \dfrac{dy}{dx}$. Then by the chain rule of differentiation and equation (1)

above, $\dfrac{du}{dx} = \dfrac{du}{dt} \times \dfrac{dt}{dx} = \dfrac{\dfrac{du}{dt}}{\dfrac{dx}{dt}}$

Since $u = \dfrac{dy}{dx}$, then $\dfrac{d}{dx}\dfrac{dy}{dx} = \dfrac{d^2y}{dx^2} = \dfrac{\dfrac{d}{dt}\left(\dfrac{dy}{dx}\right)}{\dfrac{dx}{dt}}$ (2)

Worked problems on differentiation in parameters

Problem 1. If $x = 3t + 1$ and $y = 3t(t-2)$, find $\dfrac{dy}{dx}$ in terms of parameter t.

$x = 3t + 1$, hence $\dfrac{dx}{dt} = 3$

$y = 3t(t-2)$, hence $\dfrac{dy}{dt} = 6(t-1)$

From equation (1), $\dfrac{dy}{dx} = \dfrac{\dfrac{dy}{dt}}{\dfrac{dx}{dt}} = \dfrac{6(t-1)}{3} = 2(t-1)$

Problem 2. The equation of a tangent drawn to a curve at point (x_1, y_1) is given by:

$$y - y_1 = \frac{dy_1}{dx_1}(x - x_1)$$

Determine the equation of the tangent drawn to the parabola $x = 3t^2$, $y = 6t$, at the point t.

At point t, $x_1 = 3t^2$ and $y_1 = 6t$.

$$\frac{dx_1}{dt} = 6t \quad \frac{dy_1}{dt} = 6$$

From equation (1), $\dfrac{dy_1}{dx_1} = \dfrac{\dfrac{dy_1}{dt}}{\dfrac{dx_1}{dt}} = \dfrac{6}{6t} = \dfrac{1}{t}$

Hence, the equation of the tangent is: $y - 6t = \dfrac{1}{t} (x - 3t^2)$

Problem 3. The equation of the normal drawn to a curve at point (x_1, y_1) is given by:

$$y - y_1 = \frac{-1}{\dfrac{dy_1}{dx_1}} (x - x_1)$$

Determine the equation of the normal drawn to the astroid $x = a \cos^3 \theta$, $y = a \sin^3 \theta$ at the point $\theta = \dfrac{\pi}{4}$.

$x = a \cos^3 \theta, \dfrac{dx}{d\theta} = -3 a \cos^2 \theta \sin \theta$

$y = a \sin^3 \theta, \dfrac{dy}{d\theta} = 3 a \sin^2 \theta \cos \theta$

From equation (1), $\dfrac{dy}{dx} = \dfrac{\dfrac{dy}{d\theta}}{\dfrac{dx}{d\theta}} = \dfrac{3 a \sin^2 \theta \cos \theta}{-3 a \cos^2 \theta \sin \theta}$

$$= - \tan \theta$$

When $\theta = \dfrac{\pi}{4}$, $\dfrac{dy_1}{dx_1} = -1$,

$$x_1 = a \cos^3 \frac{\pi}{4} = a \left(\frac{1}{\sqrt{2}}\right)^3 = \frac{a}{2\sqrt{2}}$$

$$y_1 = a \sin^3 \frac{\pi}{4} = a \left(\frac{1}{\sqrt{2}}\right)^3 = \frac{a}{2\sqrt{2}}$$

Hence the equation of the normal is:

$$y - \frac{a}{2\sqrt{2}} = \frac{-1}{-1} \left(x - \frac{a}{2\sqrt{2}} \right)$$

i.e. $\quad x \quad = y$

Problem 4. Determine the values of $\dfrac{dy}{dx}$ and $\dfrac{d^2 y}{dx^2}$ for the cycloid

$x = 3(\theta - \sin \theta), y = 3(1 - \cos \theta)$ at the point $\theta = \dfrac{\pi}{3}$.

$$x = 3(\theta - \sin\theta), \quad \frac{dx}{d\theta} = 3(1 - \cos\theta)$$

$$y = 3(1 - \cos\theta), \quad \frac{dy}{d\theta} = 3\sin\theta$$

From equation (1) $\dfrac{dy}{dx} = \dfrac{\dfrac{dy}{d\theta}}{\dfrac{dx}{d\theta}} = \dfrac{3\sin\theta}{3(1-\cos\theta)} = \dfrac{\sin\theta}{1-\cos\theta}$

When $\theta = \dfrac{\pi}{3}$, $\dfrac{dy}{dx} = \dfrac{\sin\dfrac{\pi}{3}}{1 - \cos\dfrac{\pi}{3}} = \dfrac{\dfrac{\sqrt{3}}{2}}{\left(1 - \dfrac{1}{2}\right)} = \sqrt{3}$

From equation (2) $\dfrac{d^2y}{dx^2} = \dfrac{\dfrac{d}{d\theta}\left(\dfrac{dy}{dx}\right)}{\dfrac{dx}{d\theta}} = \dfrac{\dfrac{d}{d\theta}\left(\dfrac{\sin\theta}{1-\cos\theta}\right)}{3(1-\cos\theta)}$

$$= \frac{(1-\cos\theta)\cos\theta - \sin\theta\,(\sin\theta)}{3(1-\cos\theta)^3}$$

$$= \frac{\cos\theta - (\cos^2\theta + \sin^2\theta)}{3(1-\cos\theta)^3} = -\frac{4}{3}$$

Problem 5. When determining the surface tension of a liquid, the radius of curvature, ρ, of part of the surface is given by

$$\rho = \frac{\left[1 + \left(\dfrac{dy}{dx}\right)^2\right]^{\frac{3}{2}}}{\dfrac{d^2y}{dx^2}}$$

Determine the radius of curvature of the part of the surface having the parametric equations $x = 2t^2$, $y = 4t$ at the point $t = 1$.

$$x = 2t^2, \quad \frac{dx}{dt} = 4t$$

$$y = 4t, \quad \frac{dy}{dt} = 4$$

From equation (1), $\dfrac{dy}{dx} = \dfrac{\dfrac{dy}{dt}}{\dfrac{dx}{dt}} = \dfrac{4}{4t} = \dfrac{1}{t}$

From equation (2), $\dfrac{d^2y}{dx^2} = \dfrac{\dfrac{d}{dt}\left(\dfrac{dy}{dx}\right)}{\dfrac{dx}{dt}} = \dfrac{\dfrac{d}{dt}\left(\dfrac{1}{t}\right)}{\dfrac{dx}{dt}}$

$$= \dfrac{-\dfrac{1}{t^2}}{4t} = \dfrac{1}{4t^3}$$

$$\rho = \dfrac{\left[1 + \left(\dfrac{dy}{dx}\right)^2\right]^{\frac{3}{2}}}{\dfrac{d^2 y}{dx^2}} = \dfrac{\left[1 + \left(\dfrac{1}{t}\right)^2\right]^{\frac{3}{2}}}{-\dfrac{1}{4t^3}}$$

When $t = 1$, $\rho = \dfrac{\sqrt{8}}{-\dfrac{1}{4}} = -4\sqrt{8} = -8\sqrt{2}$

Problem 6. Determine the radius of curvature, ρ, of the cardioid $x = 3(2 \cos \theta - \cos 2\theta), y = 3(2 \sin \theta - \sin 2\theta)$ at the point $\theta = \dfrac{\pi}{6}$ radians. Express the result correct to four decimal places.

$$\left(\rho = \left[1 + \left(\dfrac{dy}{dx}\right)^2\right]^{\frac{3}{2}} \bigg/ \dfrac{d^2 y}{dx^2}\right)$$

$x = 3(2 \cos \theta - \cos 2\theta),$ $y = 3(2 \sin \theta - \sin 2\theta)$

$\dfrac{dx}{d\theta} = -6 \sin \theta + 6 \sin 2\theta,$ $\dfrac{dy}{d\theta} = 6 \cos \theta - 6 \cos 2\theta$

From equation (1), $\dfrac{dy}{dx} = \dfrac{\dfrac{dy}{d\theta}}{\dfrac{dx}{d\theta}} = \dfrac{6 \cos \theta - 6 \cos 2\theta}{-6 \sin \theta + 6 \sin 2\theta}$

$$= \dfrac{\cos \theta - \cos 2\theta}{\sin 2\theta - \sin \theta}$$

From equation (2), $\dfrac{d^2 y}{dx^2} = \dfrac{\dfrac{d}{d\theta}\left(\dfrac{dy}{dx}\right)}{\dfrac{dx}{d\theta}}$

$$\dfrac{d}{d\theta}\left(\dfrac{dy}{dx}\right) = \dfrac{(\sin 2\theta - \sin \theta)(-\sin \theta + 2 \sin 2\theta) - (\cos \theta - \cos 2\theta)(2\cos 2\theta - \cos \theta)}{(\sin 2\theta - \sin \theta)^2}$$

$$= \dfrac{\left[\begin{array}{c}(-\sin 2\theta \sin \theta + 2 \sin^2 2\theta + \sin^2 \theta - 2 \sin \theta \sin 2\theta) \\ - (2 \cos \theta \cos 2\theta - \cos^2 \theta - 2 \cos^2 2\theta + \cos \theta \cos 2\theta)\end{array}\right]}{(\sin 2\theta - \sin \theta)^2}$$

Since $\sin^2 \theta + \cos^2 \theta = 1$ and $\sin^2 2\theta + \cos^2 2\theta = 1$

$$\dfrac{d^2 y}{dx^2} = \dfrac{3 - 3 \sin \theta \sin 2\theta - 3 \cos \theta \cos 2\theta}{(\sin 2\theta - \sin \theta)^2 \, (6) \, (\sin 2\theta - \sin \theta)}$$

$$= \frac{1 - \sin\theta \sin 2\theta - \cos\theta \cos 2\theta}{2(\sin 2\theta - \sin\theta)^3}$$

Radius of curvature, $\rho = \left[1 + \left(\dfrac{dy}{dx}\right)^2\right]^{3/2} \Big/ \dfrac{d^2y}{dx^2}$

$$= \frac{\left[1 + \left(\dfrac{\cos\theta - \cos 2\theta}{\sin 2\theta - \sin\theta}\right)^2\right]^{3/2}}{(1 - \sin\theta \sin 2\theta - \cos\theta \cos 2\theta)/2(\sin 2\theta - \sin\theta)^3}$$

When $\theta = \dfrac{\pi}{6}$, $\sin\theta = \dfrac{1}{2}$, $\cos\theta = \dfrac{\sqrt{3}}{2}$, $\sin 2\theta = \dfrac{\sqrt{3}}{2}$ and $\cos 2\theta = \dfrac{1}{2}$

Hence $\rho = \dfrac{\left[1 + \left(\dfrac{\frac{\sqrt{3}}{2} - \frac{1}{2}}{\frac{\sqrt{3}}{2} - \frac{1}{2}}\right)^2\right]^{3/2} \left[2\left(\frac{\sqrt{3}}{2} - \frac{1}{2}\right)^3\right]}{\left(1 - \dfrac{1}{2} \cdot \dfrac{\sqrt{3}}{2} - \dfrac{\sqrt{3}}{2} \cdot \dfrac{1}{2}\right)}$

$$= \frac{(\sqrt{8})(0.098\,08)}{0.133\,97} = \textbf{2.070\,7, correct to four decimal places.}$$

Further problems on differentiating in parameters may be found in the following Section (3) (Problems 1 to 20).

3 Further problems

Differentiation in parameters

In Problems 1 to 5, determine $\dfrac{dy}{dx}$ in terms of the parameter given and hence find the value of $\dfrac{dy}{dx}$ at the point stated.

1. The ellipse $x = 2\cos\theta$, $y = 3\sin\theta$ at $\theta = \dfrac{\pi}{6}$. $\left[-\dfrac{3\sqrt{3}}{2}\right]$

2. The parabola $x = \dfrac{1}{2}t^2$, $y = t$ at $t = \ln 1.3$, correct to four significant figures.   [3.811]

3. The rectangular hyperbola $x = 4t$, $y = \dfrac{4}{t}$ at $t = 0.25$.   [−16]

4. The hyperbola $x = 2\sec\theta$, $y = 5\tan\theta$ at $\theta = 1.4$ radians, correct to four decimal places.   [2.536 9]

5. The cycloid $x = 3(\theta - \cos\theta)$, $y = 3(1 - \cos\theta)$ at $\theta = 0.57$ radians, correct to four decimal places.   [0.350 5]

In problems 6 to 10, given that the equation of the tangent drawn to a curve at point (x_1, y_1) is $y - y_1 = \dfrac{dy_1}{dx_1} (x - x_1)$ and that the equation of the normal drawn to a curve at point (x_1, y_1) is $y - y_1 = \dfrac{-1}{\dfrac{dy_1}{dx_1}} (x - x_1)$, determine the equation stated, expressing it in the form $y = mx + c$.

6. The tangent drawn to the ellipse $x = 2 \cos \theta$, $y = 3 \sin \theta$ at $\theta = \dfrac{\pi}{3}$.

$$\left[y = - \frac{\sqrt{3}}{2} x + 2\sqrt{3} \right]$$

7. The normal to the parabola $x = \dfrac{1}{2} t^2$, $y = t$ at $t = \dfrac{1}{4}$.

$$\left[y = - \frac{1}{4} x + \frac{33}{128} \right]$$

8. The tangent to the rectangular hyperbola $x = 4t$, $y = \dfrac{4}{t}$ at $t = 3$.

$$\left[y = - \frac{1}{9} x + 2 \frac{2}{3} \right]$$

9. The normal to the hyperbola $x = 2 \sec \theta$, $y = 5 \tan \theta$ at $\theta = \dfrac{7\pi}{6}$ radians, correct to four decimal places. $\qquad [y = 0.2000 x + 3.3486]$

10. The equations of the tangent and normal to the cycloid $x = 3 (\theta - \sin\theta)$, $y = 3 (1 - \cos \theta)$ at $\theta = 1.3$ radians, correct to four decimal places.

$$\begin{bmatrix} \text{Tangent } y = 1.3154 x + 0.8698, \\ \text{normal } y = 0.7602 x + 2.9648 \end{bmatrix}$$

In Problems 11 to 15, find the values of $\dfrac{dy}{dx}$ and $\dfrac{d^2 y}{dx^2}$ at the points stated, for the curves represented by the parametric equations given.

11. The hyperbola $x = 2.3 \sec \theta$, $y = 3.4 \tan \theta$ at $\theta = 1$ radian, correct to four decimal places. $\qquad [1.7568, -0.3913]$

12. The parabola $x = \sqrt{5}t^2$, $y = 2\sqrt{5}t$ at $t = 2.83$, correct to four significant figures. $\qquad [0.3534, \pm 0.009866]$

13. The ellipse $x = 3.7 \cos \theta$, $y = 4.4 \sin \theta$ at $\theta = 1.43$ radians, correct to four decimal places. $\qquad [-0.1685, -0.3311]$

14. The rectangular hyperbola $x = \sqrt{3}t$, $y = \dfrac{\sqrt{3}}{t}$ at $t = 0.18$, correct to four significant figures. $\qquad [-30.86, \pm 198.0]$

15. The cardioid $x = 4 (2 \cos \theta - \cos 2\theta)$, $y = 4 (2 \sin \theta - \sin 2\theta)$ at $\theta = 3.5$ radians, correct to four decimal places. $\qquad [-1.6773, 7.2282]$

16. If $x = \sqrt{2} (1 - \cos \theta)$ and $y = \sqrt{2} \sin \theta$, show that $\dfrac{dy}{dx} = \cot \theta$ and that

$$\frac{d^2 y}{dx^2} = \frac{- \operatorname{cosec}^3 \theta}{\sqrt{2}}.$$

17. A point on a curve is given by $x = 7 \cot t + 3.5 \cos 2t$, $y = 7 \sin t - 3.5$ $\sin 2t$. Express $\dfrac{d^2y}{dx^2}$ in terms of t. $\qquad \left[\dfrac{1}{28} \cos^3 \left(\dfrac{'t}{2} \right) \sin \dfrac{3t}{2} \right]$

18. For the curve described by $x = 2\sqrt{2}t^2$, $y = 2\sqrt{2}t^3$, find $\dfrac{d^2y}{dx^2}$ in terms of t. $\qquad \left[\dfrac{3}{8\sqrt{2t}} \quad \text{or} \quad \dfrac{3\sqrt{2}}{16t} \right]$

In Problems 19 and 20, determine the radius of curvature, ρ, for the curves stated in terms of their parameters, given that

$$\rho = \frac{\left[1 + \left(\dfrac{dy}{dx} \right)^2 \right]^{\frac{3}{2}}}{\dfrac{d^2y}{dx^2}}$$

19. The astroid, $x = 5 \cos^3 t$, $y = 5 \sin^3 t$. $\qquad$ [15 $\sin t \cos t$ or 7.5 $\sin 2t$]

20. The curve, $x = 3 \sin 2\theta \, (1 + \cos 2\theta)$, $y = 3 \cos 2\theta \, (1 - \cos 2\theta)$.
$$[12 \cos 3\theta]$$

Chapter 13

Logarithmic differentiation (CA 17)

1 The laws of logarithms applied to functions

The laws of logarithms may be used to change problems involving products to additions, quotients to subtractions and powers to products. They are stated mathematically as:

$$\log (A \cdot B) = \log A + \log B$$
$$\log \left(\frac{A}{B}\right) = \log A - \log B$$
$$\text{and} \quad \log A^n = n \log A,$$

where 'log' means the logarithm to any base. In calculus, logarithms to the base of 'e' are invariably used. Also, the constants A and B can be changed to functions of x, and the laws stated above are still true. Thus for two functions of x, $f(x)$ and $g(x)$, the laws may be expressed as:

$$\ln (f(x) \cdot g(x)) = \ln f(x) + \ln g(x) \tag{1}$$
$$\ln \left(\frac{f(x)}{g(x)}\right) = \ln f(x) - \ln g(x) \tag{2}$$
$$\text{and} \quad \ln [f(x)]^n = n \ln f(x) \tag{3}$$

If $y = (x^3 + 4) \sin 2x$, let $(x^3 + 4) = f(x)$ and $\sin 2x = g(x)$. Taking logarithms to the base of e of each side of the equation gives:

$\ln y = \ln [(x^3 + 4) \cdot \sin 2x]$, and by applying law (1) above:

$\ln y = \ln (x^3 + 4) + \ln (\sin 2x)$.

Functions may be made up of a mixture of products and quotients.

When logarithms are applied to numbers,

$$\log\left(\frac{A \cdot B}{C}\right) = \log(A \cdot B) - \log C$$
$$= \log A + \log B - \log C$$

Applying this principle to functions of x, say $f(x), g(x)$ and $F(x)$ gives:

$$y = \frac{f(x) \cdot g(x)}{F(x)}$$

Then, $\ln y = \ln f(x) + \ln g(x) - \ln F(x)$

For a more complicated expression such as

$$y = \frac{f(x) \cdot [g(x)]^n}{F(x) \cdot [G(x)]^m}$$

The three laws of logarithms may be applied as follows:

$$\ln y = \ln\left\{\frac{f(x) \cdot [g(x)]^n}{F(x) \cdot [G(x)]^m}\right\}$$

Applying law (2), $\ln y = \ln\{f(x) \cdot [g(x)]^n\} - \ln\{F(x) \cdot [G(x)]^m\}$

Applying law (1), $\ln y = \{\ln f(x) + \ln[g(x)]^n\} - \{\ln F(x) + \ln[G(x)]^m\}$
$$= \ln f(x) + \ln[g(x)]^n - \ln F(x) - \ln[G(x)]^m$$

Applying law (3), $\ln y = \ln f(x) + n\ln g(x) - \ln F(x) - m\ln G(x).$

These principles are used in the worked problems following, and also form the basis of logarithmic differentiation introduced in section 2.

Worked problems on the laws of logarithms applied to functions
In worked problems 1 and 2 below, by taking Naperian logarithms of each side of the equations given, convert the functions from 'product, quotient, power' form to 'addition, subtraction, product' form.

Problem 1.   (a) $y = (x^2 + 3)\cos 2x$

   (b) $p = \dfrac{3\tan 2q}{4e^{\left(\frac{q}{2} - 3\right)}}$

   (c) $\alpha = 4\sin^3\dfrac{\theta}{2}$

(a) $y = (x^2 + 3)\cos 2x$

   Taking logarithms to the base of e of each side of the equation, gives:

   $\ln y = \ln[(x^2 + 3) \cdot \cos 2x]$

   Applying law (1), **$\ln y = \ln(x^2 + 3) + \ln\cos 2x$**

(b) $p = \dfrac{3 \tan 2q}{4 \, e^{\left(\frac{q}{2} - 3\right)}}$

Taking Naperian logarithms of each side of the equation gives:

$\ln p = \ln \left\{ \dfrac{3 \tan 2q}{4 \, e^{\left(\frac{q}{2} - 3\right)}} \right\}$

Applying law (2), $\ln p = \ln (3 \tan 2q) - \ln (4e^{\left(\frac{q}{2} - 3\right)})$

Applying law (1), $\ln p = \ln 3 + \ln \tan 2q - \ln 4 - \ln e^{\left(\frac{q}{2} - 3\right)}$

Applying law (3), $\ln p = \ln 3 + \ln \tan 2q - \ln 4 - \left(\dfrac{q}{2} - 3 \right) \ln e$. From the definition of a logarithm to any base, a, if

$$\begin{aligned} y &= a^x && (1)\\ \text{then} \quad x &= \log_a y && (2) \end{aligned}$$

When $a = y$, giving $\log_a a$ in equation (2), then from equation (1), $y = y^x$, i.e. $x = 1$.

It follows that $\log_a a = 1$, where a is any value, thus $\ln e = \log_e e = 1$.

Hence, **$\ln p = \ln 3 + \ln \tan 2q - \ln 4 - \left(\dfrac{q}{2} - 3 \right)$.**

(c) $\alpha = 4 \sin^3 \dfrac{\theta}{2}$

$\ln \alpha = \ln \left(4 \sin^3 \dfrac{\theta}{2} \right)$

$\qquad = \ln \left[4 \left(\sin \dfrac{\theta}{2} \right)^3 \right]$

Applying law (1), $\ln \alpha = \ln 4 + \ln \left(\sin \dfrac{\theta}{2} \right)^3$

Applying law (3), **$\ln \alpha = \ln 4 + 3 \ln \sin \dfrac{\theta}{2}$**

Problem 2. (a) $(2x^2 - 3x)^{\frac{1}{2}} \sec^3 x$

(b) $\dfrac{x^2 \sin^2 x}{e^{2x} (3 - 4x^{\frac{1}{2}})^5}$

(c) $\dfrac{e^{3x} \ln (x^2)}{\ln (x^3) \operatorname{cosec}^3 x}$

(a) Let $y = (2x^2 - 3x)^{\frac{1}{2}} \sec^3 x$

Taking logarithms to the base of e of each side of the equation gives:

$$\ln y = \ln \left[(2x^2 - 3x)^{\frac{1}{2}} \sec^3 x\right]$$

Applying law (1), $\ln y = \ln (2x^2 - 3x)^{\frac{1}{2}} + \ln \sec^3 x$

Applying law (3), $\ln y = \dfrac{1}{2} \ln (2x^2 - 3x) + 3 \ln \sec x$

i.e. $\ln y = \dfrac{1}{2} \ln x (2x - 3) + 3 \ln \sec x$

Applying law (1) to the first term,

$$\ln y = \frac{1}{2} \ln x + \frac{1}{2} \ln (2x - 3) + 3 \ln \sec x$$

(b) Let $y = \dfrac{x^2 \sin^2 x}{e^x (3 - 4x^{\frac{1}{2}})^5}$

Taking logarithms to the base of e of each side of the equation, gives:

$$\ln y = \ln \left[\frac{x^2 \sin^2 x}{e^x (3 - 4x^{\frac{1}{2}})^5}\right]$$

Applying law (2), $\ln y = \ln (x^2 \sin^2 x) - \ln [e^x (3 - 4x^{\frac{1}{2}})^5$

Applying law (1), $\ln y = (\ln x^2 + \ln \sin^2 x) - [\ln e^x + \ln (3 - 4x^{\frac{1}{2}})^5]$

Applying law (3), $\ln y = 2 \ln x + 2 \ln \sin x - x \ln e - 5 \ln (3 - 4x^{\frac{1}{2}})$

But $\ln e = 1$, hence $\ln y = 2 (\ln x + \ln \sin x) - x - 5 \ln (3 - 4x^{\frac{1}{2}})$

(c) Let $y = \dfrac{e^{3x} \ln (x^2)}{\ln (x^3) \operatorname{cosec}^3 x}$

Taking logarithms to the base of e of each side of the equation gives:

$$\ln y = \ln \left[\frac{e^{3x} \ln (x^2)}{\ln (x^3) \operatorname{cosec}^3 x}\right]$$

Applying law (3), $\ln y = \ln \left[\dfrac{e^{3x} \cdot 2\cancel{\ln x}}{3 \cancel{\ln x} \cdot \operatorname{cosec}^3 x}\right]$

$$= \ln \left[\frac{2 e^{3x}}{3 \operatorname{cosec}^3 x}\right]$$

Applying law (2), $\ln y = \ln 2 + \ln e^{3x} - \ln 3 - \ln \operatorname{cosec}^3 x$

Applying law (3), $\ln y = \ln 2 + 3x \ln e - \ln 3 - 3 \ln \operatorname{cosec} x$

But $\ln e = 1$,

Hence, $\qquad \ln y = \ln 2 + 3x - \ln 3 - 3 \ln \operatorname{cosec} x$

Applying law (2), $\ln y = \ln \dfrac{2}{3} + 3x - 3 \ln \operatorname{cosec} x$

Further problems on the laws of logarithms applied to functions may be found in Section 3 (Problems 1 to 5), page 171.

2 Logarithmic differentiation

To differentiate a function which contains products and/or quotients and/or powers, it is often easier to firstly take logarithms to a base of e and to express the function in a different form. For example, a simple function such as, say, $y = x^2 \sin x$ may be differentiated using the product rule of differentiation, giving:

$$\frac{dy}{dx} = x^2 \cos x + 2x \sin x$$

An alternative method of differentiating this function is by logarithmic differentiation as follows:

Taking Naperian logarithms of each side of the equation gives:

$\ln y = \ln (x^2 \sin x)$
i.e. $\ln y = 2 \ln x + \ln \sin x$

from the laws of logarithms introduced in Section 1. In Chapter 11, dealing with implicit functions, it is shown that

$$\frac{d}{dx} (f(y)) = \frac{d}{dy} (f(y)) \times \frac{dy}{dx}$$

Thus the left-hand side, i.e. $\ln y$, becomes $\frac{1}{y} \frac{dy}{dx}$ when differentiated with respect to x. With logarithmic differentiation, the right-hand side of the equation is of the form $\ln f(x) \pm \ln g(x) \pm \dots$. To differentiate, say, $\ln f(x)$ with respect to x, a substitution method is used.

Thus, let $u = f(x)$, then $\frac{du}{dx} = f'(x)$

Hence, $\frac{d}{dx} (\ln f(x)) = \frac{d}{du} (\ln u) \times \frac{du}{dx}$

$$= \frac{1}{u} \frac{du}{dx}$$

i.e. $\frac{d}{dx} (\ln f(x)) = \frac{f'(x)}{f(x)}$ \hfill (4)

Applying equation (4) to the right-hand side, i.e. $2 \ln x + \ln \sin x$, gives:

$\frac{2}{x} + \frac{\cos x}{\sin x}$ when it is differentiated with respect to x. Thus, applying logarithmic differentiation to the equation $y = x^2 \sin x$, gives:

$$\frac{1}{y}\frac{dy}{dx} = \frac{2}{x} + \frac{\cos x}{\sin x}$$

i.e. $\quad \dfrac{dy}{dx} = y\left(\dfrac{2}{x} + \cos x\right)$

But $\quad y \quad = x^2 \sin x,$ hence

$$\frac{dy}{dx} = x^2 \sin x \left(\frac{2}{x} + \frac{\cos x}{\sin x}\right)$$

$$= 2x \sin x + x^2 \cos x,$$

as obtained by applying the product rule of differentiation.

In this case, using a comparatively simple function, the process of logarithmic differentiation is longer than that of applying the product rule of differentiation. However, for more complicated functions, logarithmic differentiation is usually the simplest method of differentiating the function.

Procedure for logarithmic differentiation

The procedure for differentiating a function of the form, say,
$y = \dfrac{f(x)\,[g(x)]^n}{h(x)}$ is shown below.

(i) Take Naperian logarithms of each side of the equation, giving

$$\ln y = \left\{\frac{f(x)\,[g(x)]^n}{h(x)}\right\}.$$

(ii) Apply the laws of logarithms to change products to addition, etc. This gives:

$$\ln y = \ln f(x) + n \ln g(x) - \ln h(x)$$

(iii) Differentiate each side with respect to the variable, giving:

$$\frac{1}{y}\frac{dy}{dx} = \frac{f'(x)}{f(x)} + \frac{n g'(x)}{g(x)} - \frac{h'(x)}{h(x)}$$

(iv) Multiply throughout by y, giving

$$\frac{dy}{dx} = y\left\{\frac{f'(x)}{f(x)} + \frac{n g'(x)}{g(x)} - \frac{h'(x)}{h(x)}\right\}.$$

(v) Substitute for y in terms of x, giving:

$$\frac{dy}{dx} = \frac{f(x)\,[g(x)]^n}{h(x)}\left\{\frac{f'(x)}{f(x)} + \frac{n g'(x)}{g(x)} - \frac{h'(x)}{h(x)}\right\}$$

(vi) Simplify the expression obtained in (v) where possible.

This procedure is used to determine the differential coefficients in the worked problems following.

Worked problems on logarithmic differentiation

Problem 1. Use logarithmic differentiation to differentiate

$$y = \frac{3(4-x^2)}{\tan x}$$

With reference to the procedure for logarithmic differentiation given above:

(i) $\ln y \quad = \quad \ln \left\{ \dfrac{3(4-x^2)}{\tan x} \right\}$

(ii) $\ln y \quad = \quad \ln 3 + \ln(4-x^2) - \ln \tan x$

(iii) $\dfrac{1}{y}\dfrac{dy}{dx} \quad = \quad 0 + \dfrac{(-2x)}{4-x^2} - \dfrac{\sec^2 x}{\tan x}$

(iv) $\dfrac{dy}{dx} \quad = \quad y\left\{ \dfrac{-2x}{4-x^2} - \dfrac{\sec^2 x}{\tan x} \right\}$

(v) $\dfrac{dy}{dx} \quad = \quad \dfrac{3(4-x^2)}{\tan x}\left\{ -\dfrac{2x}{4-x^2} - \dfrac{\sec^2 x}{\tan x} \right\}$

(vi) $\dfrac{dy}{dx} \quad = -\dfrac{6x}{\tan x} - \dfrac{3(4-x^2)\sec^2 x}{\tan^2 x}$

$$= -\frac{3}{\tan x}\left\{ 2x + \frac{(4-x^2)\sec^2 x}{\tan x} \right\}$$

$$= -3\cot x \left\{ 2x + \frac{4-x^2}{\sin x \cos x} \right\}$$

Problem 2. Use logarithmic differentiation to differentiate

(a) $y \quad = \quad \dfrac{4e^{-2x}\sec x}{\left(x^2 + \dfrac{1}{2}\right)^{3/2}}$

(b) $p \quad = \quad \dfrac{q^3 \ln 2q}{(2-q)^{1/3}\operatorname{cosec} 2q}$

(a) $y \quad = \quad \dfrac{4e^{-2x}\sec x}{\left(x^2 + \dfrac{1}{2}\right)^{3/2}}$

Applying the laws of logarithms:

$$\ln y = \ln 4 + (-2x)\ln e + \ln \sec x - \frac{3}{2}\ln\left(x^2 + \frac{1}{2}\right)$$

Differentiating each term:

$$\frac{1}{y}\frac{dy}{dx} = -2 + \frac{\sec x \tan x}{\sec x} - \frac{3(2x)}{2\left(x^2 + \dfrac{1}{2}\right)}$$

$$\frac{dy}{dx} = \frac{4e^{-2x} \sec x}{(x^2 + \frac{1}{2})^{3/2}} \left[-2 + \tan x - \frac{3x}{(x^2 + \frac{1}{2})} \right]$$

(b) $$p = \frac{q^3 \ln 2q}{(2-q)^{\frac{1}{3}} \operatorname{cosec} 2q}$$

$$\ln p = 3 \ln q + \ln (\ln 2q) - \frac{1}{3} \ln (2-q) - \ln \operatorname{cosec} 2q$$

$$\frac{1}{p} \frac{dp}{dq} = \frac{3}{q} + \frac{\frac{2}{2q}}{\ln 2q} - \frac{(-1)}{3(2-q)} - \frac{-2 \operatorname{cosec} 2q \cot 2q}{\operatorname{cosec} 2q}$$

$$\frac{dp}{dq} = \frac{q^3 \ln 2q}{(2-q)^{\frac{1}{3}} \operatorname{cosec} 2q} \left\{ \frac{3}{q} + \frac{1}{q \ln 2q} + \frac{1}{3(2-q)} + 2 \cot 2q \right\}$$

Further problems on logarithmic differentiation may be found in the following Section (3) (Problems 6 to 17).

3 Further Problems

Laws of logarithms applied to functions

In Problems 1 to 5, by taking Napierian logarithms of each side of the equation, convert the functions from 'product, quotient, power' form to 'addition, subtraction, product' form.

1. $y = (3x - 4) \tan 2x$ $[\ln y = \ln (3x - 4) + \ln (\tan 2x)]$

2. $p = \dfrac{3 \cot 2q}{e^{3-q}}$. $[\ln p = \ln 3 + \ln \cot 2q - (3 - q)]$

3. $m = \dfrac{n^3 e^{2n}}{\operatorname{cosec} \dfrac{n}{2}}$. $[\ln m = 3 \ln n + 2n - \ln \operatorname{cosec} \dfrac{n}{2}]$

4. $y = \dfrac{2(1 + x^{3/2})^{1/2} \cos (x - 2)}{3 e^{(3x+4)} \ln 2x}$.

$[\ln y = \ln 2 + \dfrac{1}{2} \ln (1 + x^{3/2}) + \ln \cos (x - 2) - \ln 3 - (3x + 4) - \ln (\ln 2x)]$

5. $\alpha = \dfrac{3\theta^{3/2} \sin^3 \theta}{(2 - \theta^{1/2})^{1/3} \sec^3 \theta}$.

$[\ln \alpha = \ln 3 + \dfrac{3}{2} \ln \theta + 3 \ln \sin \theta - \dfrac{1}{3} \ln (2 - \theta^{1/2}) - 3 \ln \sec \theta]$

Logarithmic Differentiation

In Problems 6 to 11, use logarithmic differentiation to differentiate the functions given.

6. $y = 3\sqrt{x} \sin 2x.$ $[\dfrac{dy}{dx} = 3\sqrt{x} \left(\dfrac{\sin 2x}{2x} + 2\cos 2x\right)]$

7. $l = 4 e^{(2-3m)} \operatorname{cosec} \dfrac{m}{2}.$

$$[\dfrac{dl}{dm} = -4e^{(2-3m)} \operatorname{cosec} \dfrac{m}{2} \left(3 + \dfrac{1}{2} \cot \dfrac{m}{2}\right)]$$

8. $p = \dfrac{2(1-q^2)}{3 \sec (\pi - q)}.$

$$\left[\dfrac{dp}{dq} = \dfrac{2}{3 \sec (\pi - q)} \left\{ (1-q^2) \tan (\pi - q) - 2q \right\}\right]$$

9. $r = \dfrac{2 \ln (4 - \frac{s}{3})}{3 \cot (s^2)}.$ $\left[\dfrac{dr}{ds} = \dfrac{2}{3 \cot (s^2)} \left\{ -\dfrac{1}{12-s} + \dfrac{2 s \ln (4 - \frac{s}{3})}{\sin s^2 \cos s^2} \right\}\right]$

10. $u = 1.8\sqrt{\sin^3 v}.$ $[\dfrac{du}{dv} = 2.7\sqrt{(\sin v)} \cos v \ ]$

11. $\alpha = \dfrac{4}{9}\left[3 - (2\theta)^{1/3}\right]^{1/2}.$

$$\left[\dfrac{d\alpha}{d\theta} = -\dfrac{4}{27}\left\{ \dfrac{1}{(2\theta)^{2/3}(3-(2\theta)^{1/3})^{1/2}} \right\}\right]$$

12. If $\dfrac{y}{e^{2x}} = \sin^3 (4 - 2x)$, find $\dfrac{dy}{dx}$.

$$\left[\dfrac{dy}{dx} = 2e^{2x} \sin^3 (4-2x) \left\{ 1 - 3 \cot (4-2x) \right\}\right]$$

13. If $\dfrac{r}{\sec^2 \frac{x}{2}} = \ln 3x \tan 3x$, show that

$$\dfrac{dr}{dx} = \ln 3x \cdot \sec^2 \dfrac{x}{2} \tan 3x \left\{ \dfrac{1}{x \ln 3x} + \tan \dfrac{x}{2} + \dfrac{3 \sec^2 3x}{\tan 3x} \right\}$$

14. If $\dfrac{p}{4(3 - \frac{q}{4})^{1/4}} = e^{(2q-4)}\sqrt{(\cos (3 - 2q))}$, find $\dfrac{dp}{dq}$

$$\left[\dfrac{dp}{dq} = 4(3 - \dfrac{q}{4})^{1/4} e^{(2q-4)} \sqrt{(\cos (3-2q))} \left\{ 2 + \tan (3-2q) - \dfrac{1}{4(12-q)} \right\}\right.$$

15. If $y \left(\ln \dfrac{x}{4} \right)^2 = (x \sin x)^3$, show that

$$\frac{dy}{dx} = \frac{(x \sin x)^3}{(\ln \frac{x}{4})^2} \left\{ \cdot \frac{3}{x} + 3 \cot x - \frac{2}{x \ln \frac{x}{4}} \right\}.$$

16. If $u(v-1)^{1/2} \cot^3 v = v^3 \sec^2 v$, find $\dfrac{du}{dv}$.

$$\left[\frac{du}{dv} = \frac{v^3 \sec^2 v}{(v-1)^{1/2} \cot^3 v} \left\{ \frac{3}{v} + 2 \tan v - \frac{1}{2(v-1)} + \frac{3}{\sin v \cos v} \right\} \right]$$

17. If $pe^{q/3} \sqrt{(\frac{q}{2})} = 16 \cot^2 q$, show that

$$\frac{dp}{dq} = - \frac{16 \cot^2 q}{\sqrt{(\frac{q}{2})} e^{q/3}} \left\{ \frac{2}{\sin q \cos q} + \frac{1}{2q} + \frac{1}{3} \right\},$$

Chapter 14

Differentiation of inverse trigonometric and inverse hyperbolic functions (CA 19)

1 Inverse functions

If y is a function of x, i.e. $y = f(x)$, it is often possible to transpose the equation to find x in terms of y.

For example, if $y = 2x - 1$ then $x = \dfrac{y + 1}{2}$

The two functions, i.e. $y = 2x - 1$ and $x = \dfrac{y + 1}{2}$, are called **inverse functions**.

Further examples of inverse functions are:

(i)   if $y = x^2$ then $x = \sqrt{y}$,
(ii)   if $y = 10^x$ then $x = \log_{10} y$,
(iii)   if $y = \sin x$ then x is the inverse sine of y, which is written as $x = \arcsin y$,
(iv)   if $y = \cosh x$ then x is the inverse hyperbolic cosine of y, which is written as $x = \operatorname{arcosh} y$.

The inverse circular functions are denoted by prefixing the function with 'arc'. For example, $\arcsin x$, $\operatorname{arcsec} x$, etc. The former notation $\sin^{-1} x$, $\tan^{-1} x$, etc., is discouraged because of possible confusion with $(\sin x)^{-1}$, $(\tan x)^{-1}$, etc.

The inverse hyperbolic functions are denoted by prefixing the function with 'ar'. For example, $\operatorname{arsinh} x$, $\operatorname{arcosech} x$, etc. The former notation $\sinh^{-1} x$, $\coth^{-1} x$, etc., is discouraged because of possible confusion with $(\sinh x)^{-1}$, $(\coth x)^{-1}$, etc.

2 Differentiation of inverse trigonometric functions

(a) $y = \arcsin x$

If $y = \arcsin x$ then $x = \sin y$.

Differentiating x with respect to y gives $\dfrac{\mathrm{d}x}{\mathrm{d}y} = \cos y$.

Now $\cos^2 y + \sin^2 y = 1$. Hence $\cos y = \sqrt{(1 - \sin^2 y)} = \sqrt{(1 - x^2)}$

$$\text{Hence } \quad \frac{\mathrm{d}x}{\mathrm{d}y} = \sqrt{(1 - x^2)}$$

It may be shown that $\dfrac{\mathrm{d}y}{\mathrm{d}x} = \dfrac{1}{\mathrm{d}x/\mathrm{d}y}$ (see Chapter 12, page 155).

Thus when $y = \arcsin x$, $\dfrac{\mathrm{d}y}{\mathrm{d}x} = \dfrac{1}{\sqrt{(1 - x^2)}}$ and there are two possible values, one positive and one negative.

A sketch of part of the graph $y = \arcsin x$ is shown in Fig. 1(a) and it is seen that there are an infinite number of values for y for a given value of x. When y has more than one value for a given value of x, such a is the relationship $y = \arcsin x$, the numerically least of these values is called the the **principal value** of y. If there are two numerically equal least values, the positive one is called the principal value.

The principal value of $\arcsin x$ is defined as the value between $-\dfrac{\pi}{2}$ and $+\dfrac{\pi}{2}$, shown between points P and Q in Fig. 1(a). This range of values covers every possible value of x that can occur and when adopting principal values, if given a particular value of x, only one value of y is possible. The gradient of the curve between P and Q is positive for all values of x, and hence if $\arcsin x$ is understood to mean the principal value of $\arcsin x$ then

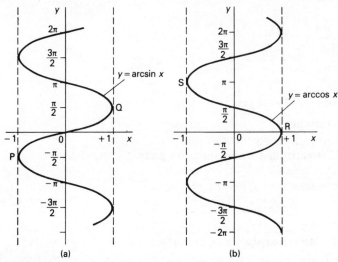

(a)   (b)

Figure 1

the differential coefficient will be positive and only the positive value is taken when evaluating $\sqrt{(1-x^2)}$.

Hence if $y = \arcsin x$ **then** $\dfrac{dy}{dx} = \dfrac{1}{\sqrt{(1-x^2)}}$

If $y = \arcsin \dfrac{x}{a}$ then $\dfrac{x}{a} = \sin y$

i.e. $x = a \sin y$

$$\frac{dx}{dy} = a \cos y = a\sqrt{(1 - \sin^2 y)} = a\sqrt{\left[1 - \left(\frac{x}{a}\right)^2\right]}$$

i.e. $\dfrac{dx}{dy} = a\sqrt{\left(\dfrac{a^2 - x^2}{a^2}\right)}$

$$= a\,\frac{\sqrt{(a^2 - x^2)}}{a} = \sqrt{(a^2 - x^2)}$$

Hence $\dfrac{dy}{dx} = \dfrac{1}{dx/dy} = \dfrac{1}{\sqrt{(a^2 - x^2)}}$

This latter form is important when integrating functions of the form

$\dfrac{1}{\sqrt{(a^2 - x^2)}}$, i.e. $\displaystyle\int \dfrac{1}{\sqrt{(a^2 - x^2)}}\ dx = \arcsin \dfrac{x}{a} + c.$

(This is discussed in Chapter 18, page 238.)

The most general form of the differential coefficient of $\arcsin f(x)$ is obtained by using the 'chain rule' of differentiation for 'functions of a function'. That is, given $y = \arcsin f(x)$, let $u = f(x)$

then $\dfrac{du}{dx} = f'(x)$

$\dfrac{dy}{du} = \dfrac{1}{\sqrt{(1 - u^2)}}$

Thus $\dfrac{dy}{dx} = \dfrac{dy}{du} \cdot \dfrac{du}{dx} = \dfrac{1}{\sqrt{(1 - u^2)}}\,f'(x),$

i.e. $\dfrac{dy}{dx} = \dfrac{1}{\sqrt{\{1 - [f(x)]^2\}}}\,f'(x)$

(b) $y = \arccos x$

If $y = \arccos x$ then $x = \cos y$.

Differentiating x with respect to y gives:

$\dfrac{dx}{dy} = -\sin y = -\sqrt{(1 - \cos^2 y)}$

$= -\sqrt{(1 - x^2)}$

Hence $\dfrac{dy}{dx} = \dfrac{1}{dx/dy} - -\dfrac{1}{\sqrt{(1 - x^2)}}$

A sketch of part of the graph $y = \arccos x$ is shown in Fig. 1(b) and $\arccos x$ is defined as that angle which lies between 0 and π shown between points R and S, i.e. the principal value of $y = \arccos x$ is between 0 and π.

(If the range $-\dfrac{\pi}{2}$ to $+\dfrac{\pi}{2}$ were used, as in the case of arcsin x, then the cosine of an angle would always be positive. Thus $\arccos\left(-\dfrac{1}{3}\right)$, for example, would have no meaning.) The gradient of the curve between R and S is negative for all values of x and hence the differential coefficient is negative as shown above.

It may be shown that if $y = \arccos\dfrac{x}{a}$ then $\dfrac{dy}{dx} = -\dfrac{1}{\sqrt{(a^2 - x^2)}}$

The most general form of the differential coefficient of arccos $f(x)$ is:

$$\frac{d}{dx}\ [\ \arccos f(x)] = \frac{1}{\sqrt{\{1 - [f(x)]^2\}}}\ [f'(x)]$$

(c) $y = \arctan x$

A sketch of part of the graph of $y = \arctan x$ is shown in Fig. 2(a). By definition the principal value of $y = \arctan x$ is defined as that angle which lies between $-\dfrac{\pi}{2}$ and $+\dfrac{\pi}{2}$. Hence the gradient (i.e. $\dfrac{dy}{dx}$) is always positive.

If $y = \arctan x$, then $x = \tan y$

$$\frac{dx}{dy} = \sec^2 y$$

Now $1 + \tan^2 y = \sec^2 y$

Hence $\dfrac{dx}{dy} = 1 + \tan^2 y = 1 + x^2$

$$\frac{dy}{dx} = \frac{1}{dx/dy} = \frac{1}{1+x^2}$$

If $y = \arctan\dfrac{x}{a}$, then $\dfrac{x}{a} = \tan y$, i.e. $x = a \tan y$

$$\frac{dx}{dy} = a \sec^2 y = a(1 + \tan^2 y) = a\left[1 + \left(\frac{x}{a}\right)^2\right]$$

$$= a\left(\frac{a^2 + x^2}{a^2}\right) = \frac{a^2 + x^2}{a}$$

Hence $\dfrac{dy}{dx} = \dfrac{a}{a^2 + x^2}$

This latter form is important when integrating functions of the form $\dfrac{a}{a^2 + x^2}$,

i.e. $\displaystyle\int \frac{a}{a^2 + x^2}\ dx = \arctan\frac{x}{a} + c$ or $\displaystyle\int \frac{1}{a^2 + x^2}\ dx = \frac{1}{a}\arctan\frac{x}{a} + c.$

(This is discussed in Chapter 18, page 240.)

The most general form of the differential coefficient of arctan $f(x)$ is:

$$\frac{d}{dx}\ [\arctan f(x)] = \frac{1}{\{1 + [f(x)]^2\}}\ [f'(x)]$$

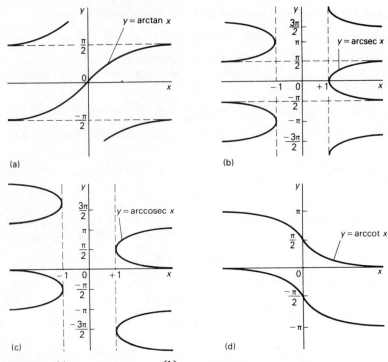

Figure 2   (*a*) $y = \arctan x$ (*b*) $y = \operatorname{arcsec} x$
                (*c*) $y = \operatorname{arccosec} x$ (*d*) $y = \operatorname{arccot} x$

(d) $y = \operatorname{arcsec} x$

A sketch of part of the graph of $y = \operatorname{arcsec} x$ is shown in Fig. 2(b). It is a discontinuous curve with no part of it between $x = -1$ and $x = +1$. By definition, the principal value of $y = \operatorname{arcsec} x$ is defined as that angle which lies between 0 and π. In this range the gradient of the curve is always positive.

    If $y = \operatorname{arcsec} x$ then $x = \sec y$

$$\frac{\mathrm{d}x}{\mathrm{d}y} = \sec y \tan y$$

Now $1 + \tan^2 y = \sec^2 y$. Hence $\tan y = \sqrt{(\sec^2 y - 1)} = \sqrt{(x^2 - 1)}$

Thus $\dfrac{\mathrm{d}x}{\mathrm{d}y} = \sec y \tan y = x\sqrt{(x^2 - 1)}$

$$\frac{\mathrm{d}y}{\mathrm{d}x} = \frac{1}{\mathrm{d}x/\mathrm{d}y} = \frac{1}{x\sqrt{(x^2 - 1)}}$$

By a similar method it may be shown that if $y = \operatorname{arcsec} \dfrac{x}{a}$, then

$$\frac{\mathrm{d}y}{\mathrm{d}x} = \frac{a}{x\sqrt{(x^2 - a^2)}}.$$

The most general form of the differential coefficient of arcsec $f(x)$ is:

$$\frac{d}{dx}\left[\,\text{arcsec}\,f(x)\,\right] = \frac{1}{f(x)\sqrt{\{[f(x)]^2 - 1\}}}\,[f'(x)]$$

(e) $y = \text{arccosec}\,x$

A sketch of part of the graph of $y = \text{arccosec}\,x$ is shown in Fig. 2(c). It is a discontinuous curve with no part of it between $x = -1$ and $x = +1$. By definition, the principal value of $y = \text{arccosec}\,x$ is defined as that angle which lies between $-\frac{\pi}{2}$ and $+\frac{\pi}{2}$. In this range the gradient of the curve is always negative.

If $y = \text{arccosec}\,x$ then $x = \text{cosec}\,y$

$$\frac{dx}{dy} = -\text{cosec}\,y \cot y.$$

Now $\cot^2 y + 1 = \text{cosec}^2 y$. Hence $\cot y = \sqrt{(\text{cosec}^2 y - 1)} = \sqrt{(x^2 - 1)}$

Thus $\quad \dfrac{dx}{dy} = -\text{cosec}\,y \cot y = -x\,\sqrt{(x^2 - 1)}$

$$\frac{dy}{dx} = \frac{1}{dx/dy} = -\frac{1}{x\,\sqrt{(x^2 - 1)}}$$

By a similar method it may be shown that if $y = \text{arccosec}\,\dfrac{x}{a}$, then

$$\frac{dy}{dx} = \frac{-a}{x\,\sqrt{(x^2 - a^2)}}\,.$$

The most general form of the differential coefficient of arccosec $f(x)$ is:

$$\frac{d}{dx}\left[\,\text{arccosec}\,f(x)\,\right] = -\frac{1}{f(x)\sqrt{\{[f(x)]^2 - 1\}}}\,[f'(x)]$$

(f) $y = \text{arccot}\,x$

A sketch of part of the graph of $y = \text{arccot}\,x$ is shown in Fig. 2(d). By definition, the principal value of $y = \text{arccot}\,x$ is defined as that angle which lies between $-\frac{\pi}{2}$ and $+\frac{\pi}{2}$.

In this range the gradient of the curve is always negative.

If $y = \text{arccot}\,x$ then $x = \cot y$

$$\frac{dx}{dy} = -\text{cosec}^2 y$$

Now $\cot^2 y + 1 = \text{cosec}^2 y$. Hence $\dfrac{dx}{dy} = -(\cot^2 y + 1) = -(x^2 + 1)$.

Thus $\quad \dfrac{dy}{dx} = \dfrac{1}{dx/dy} = -\dfrac{1}{(1 + x^2)}$

By a similar method it may be shown that if $y = \text{arccot}\,\dfrac{x}{a}$, then

$$\frac{dy}{dx} = \frac{-a}{(a^2 + x^2)}$$

The most general form of the differential coefficient of arccot $f(x)$ is:

$$\frac{d}{dx}\,[\text{arccot}\,f(x)] = -\frac{1}{\{1 + [f(x)]^2\}}\,[f'(x)]$$

Summary of differential coefficients of inverse trigonometrical functions

	y or $f(x)$	$\dfrac{dy}{dx}$ or $f'(x)$
(a)	$\text{arcsin}\,\dfrac{x}{a}$	$\dfrac{1}{\sqrt{(a^2 - x^2)}}$
	$\text{arcsin}\,f(x)$	$\dfrac{1}{\sqrt{\{1 - [f(x)]^2\}}}[f'(x)]$
(b)	$\text{arccos}\,\dfrac{x}{a}$	$\dfrac{-1}{\sqrt{(a^2 - x^2)}}$
	$\text{arccos}\,f(x)$	$\dfrac{-1}{\sqrt{\{1 - [f(x)]^2\}}}[f'(x)]$
(c)	$\text{arctan}\,\dfrac{x}{a}$	$\dfrac{a}{a^2 + x^2}$
	$\text{arctan}\,f(x)$	$\dfrac{1}{\{1 + [f(x)]^2\}}\,[f'(x)]$
(d)	$\text{arcsec}\,\dfrac{x}{a}$	$\dfrac{a}{x\sqrt{(x^2 - a^2)}}$
	$\text{arcsec}\,f(x)$	$\dfrac{1}{f(x)\sqrt{\{[f(x)]^2 - 1\}}}\,[f'(x)]$
(e)	$\text{arccosec}\,\dfrac{x}{a}$	$\dfrac{-a}{x\sqrt{(x^2 - a^2)}}$
	$\text{arccosec}\,f(x)$	$\dfrac{-1}{f(x)\sqrt{\{[f(x)]^2 - 1\}}}\,[f'(x)]$
(f)	$\text{arccot}\,\dfrac{x}{a}$	$\dfrac{-a}{a^2 + x^2}$
	$\text{arccot}\,f(x)$	$\dfrac{-1}{\{1 + [f(x)]^2\}}\,[f'(x)]$

Worked problems on differentiating inverse trigonometric functions

Problem 1. Find the differential coefficient of $y = \arcsin 3x^2$

$$\frac{dy}{dx} = \frac{1}{\sqrt{\{1 - [f(x)]^2\}}} \; [f'(x)] = \frac{1}{\sqrt{\{1 - [3x^2]^2\}}} \; [6x] = \frac{6x}{\sqrt{(1 - 9x^4)}}$$

Problem 2. Differentiate $y = \ln (\arccos 2t)$.

Let $\quad u = \arccos 2t$.
then $\quad y = \ln u$

$$\frac{dy}{dt} = \frac{dy}{du} \cdot \frac{du}{dt} = \frac{1}{u} \cdot \frac{d}{dt} (\arccos 2t)$$

$$= \left(\frac{1}{\arccos 2t}\right) \left(-\frac{2}{\sqrt{\{1 - (2t)^2\}}}\right)$$

Hence $\dfrac{d}{dt} (\ln \arccos 2t) = \dfrac{-2}{\sqrt{(1 - 4t^2)} \arccos 2t}$

Problem 3. Find $\dfrac{d}{d\theta} \left(\arctan \dfrac{2}{\theta^2}\right)$.

$$\frac{d}{d\theta} \left(\arctan \frac{2}{\theta^2}\right) = \left(\frac{1}{1 + \left(\frac{2}{\theta^2}\right)^2}\right) \left(\frac{-4}{\theta^3}\right) = \left(\frac{1}{\frac{\theta^4 + 4}{\theta^4}}\right) \left(\frac{-4}{\theta^3}\right) = \frac{-4\theta}{\theta^4 + 4}$$

Problem 4. Find the differential coefficient of $f(x) = x \operatorname{arcsec} x$.

$$f'(x) = (x)\left(\frac{1}{x\sqrt{(x^2 - 1)}}\right) + (\operatorname{arcsec} x)(1), \text{ by the product rule of differentiation}$$

$$= \frac{1}{\sqrt{(x^2 - 1)}} + \operatorname{arcsec} x.$$

Problem 5. Differentiate $y = \dfrac{\operatorname{arccot} x}{(1 + x^2)}$

$$\frac{dy}{dx} = \frac{(1 + x^2)\left(\frac{-1}{1 + x^2}\right) - (\operatorname{arccot} x)(2x)}{(1 + x^2)^2}, \text{ by the quotient rule of differentiation}$$

$$= \frac{-(1 + 2x \operatorname{arccot} x)}{(1 + x^2)^2}$$

Problem 6. Show that if $y = \operatorname{arccot} \left(\dfrac{\cos \theta}{1 - \sin \theta}\right)$ then $\dfrac{dy}{d\theta} = -\dfrac{1}{2}$

$$\frac{dy}{d\theta} = \frac{-1}{\{1 + [f(x)]^2\}} \; [f'(x)]$$

$$= -\frac{1}{\left\{1 + \dfrac{\cos\theta}{1-\sin\theta}\right\}^2} \left\{\frac{(1-\sin\theta)(-\sin\theta)-(\cos\theta)(-\cos\theta)}{[(1-\sin\theta)^2}\right\}$$

$$= \left\{\frac{-1}{\dfrac{(1-\sin\theta)^2 + \cos^2\theta}{(1-\sin\theta)^2}}\right\} \left\{\frac{-\sin\theta + \sin^2\theta + \cos^2\theta}{(1-\sin\theta)^2}\right\}$$

$$= \left\{\frac{-(1-\sin\theta)^2}{(1-\sin\theta)^2 + \cos^2\theta}\right\} \left\{\frac{1-\sin\theta}{(1-\sin\theta)^2}\right\}$$

$$= \left\{\frac{-1}{1-2\sin\theta + \sin^2\theta + \cos^2\theta}\right\} (1-\sin\theta)$$

$$= \frac{-1}{(2-2\sin\theta)} (1-\sin\theta) = \frac{-(1-\sin\theta)}{2(1-\sin\theta)} = -\frac{1}{2}$$

Further problems on differentiating inverse trigonometric functions may be found in Section 5, (Problems 1 to 16), page 193.

3 Logarithmic forms of the inverse hyperbolic functions

(i) arsinh $\dfrac{x}{a}$

If $y = \text{arsinh } \dfrac{x}{a}$ then $\dfrac{x}{a} = \sinh y$.

From Chapter 4, $e^y = \cosh y + \sinh y$. Also $\cosh^2 y - \sinh^2 y = 1$, from which $\cosh y = \sqrt{(1 + \sinh^2 y)}$, which is positive since $\cosh y$ is always positive (see Fig. 1, Chapter 4).

Hence $e^y = \sqrt{(1 + \sinh^2 y)} + \sinh y$

i.e. $\qquad e^y = \sqrt{\left[1 + \left(\dfrac{x}{a}\right)^2\right]} + \dfrac{x}{a} = \dfrac{\sqrt{(a^2 + x^2)}}{a} + \dfrac{x}{a}$

Taking Napierian logarithms of both sides gives:

$$y = \ln\left\{\frac{x + \sqrt{(a^2 + x^2)}}{a}\right\}$$

Hence arsinh $\dfrac{x}{a} = \ln\left\{\dfrac{x + \sqrt{(a^2 + x^2)}}{a}\right\}$

(ii) arcosh $\dfrac{x}{a}$

If $y = \text{arcosh } \dfrac{x}{a}$ then $\dfrac{x}{a} = \cosh y$.

$e^y = \cosh y + \sinh y$

$\quad = \cosh y \pm \sqrt{(\cosh^2 y - 1)}$, since $\sinh y$ may be positive or negative

(see Fig. 2, Chapter 4).

i.e. $e^y = \dfrac{x}{a} \pm \sqrt{\left[\left(\dfrac{x}{a}\right)^2 - 1\right]} = \dfrac{x}{a} \pm \dfrac{\sqrt{(x^2 - a^2)}}{a}$

Taking Napierian logarithms of both sides gives:

$$y = \ln\left\{\frac{x \pm \sqrt{(x^2 - a^2)}}{a}\right\}$$

The two values obtained are $\quad y = \ln\left\{\dfrac{x + \sqrt{(x^2 - a^2)}}{a}\right\}$ and

$$y = \ln\left\{\frac{x - \sqrt{(x^2 - a^2)}}{a}\right\}$$

Adding these two values of y gives:

$$\ln\left\{\frac{x + \sqrt{(x^2 - a^2)}}{a}\right\} + \ln\left\{\frac{x - \sqrt{(x^2 - a^2)}}{a}\right\}$$

$$= \ln\left\{\frac{x + \sqrt{(x^2 - a^2)}}{a}\right\}\left\{\frac{x - \sqrt{(x^2 - a^2)}}{a}\right\} \quad \text{from the laws of logarithms}$$

$$= \ln\left\{\frac{x^2 - (x^2 - a^2)}{a^2}\right\} = \ln 1 = 0.$$

Hence the two values of y are equal but opposite in sign.

Assuming the principal value, $\mathbf{arccosh}\ \dfrac{x}{a} = \ln\left\{\dfrac{x + \sqrt{(x^2 - a^2)}}{a}\right\}$

(iii) artanh $\dfrac{x}{a}$

If $y = \text{artanh}\ \dfrac{x}{a}$ then $\dfrac{x}{a} = \tanh y$

Now $\tanh y = \dfrac{\sinh y}{\cosh y} = \dfrac{\frac{1}{2}(e^y - e^{-y})}{\frac{1}{2}(e^y + e^{-y})} = \dfrac{e^{2y} - 1}{e^{2y} + 1}$

Then $\dfrac{x}{a} = \dfrac{e^{2y} - 1}{e^{2y} + 1}$ and $x(e^{2y} + 1) = a(e^{2y} - 1)$

Hence $\qquad\qquad\qquad x + a = ae^{2y} - xe^{2y}$

$$e^{2y} = \frac{a + x}{a - x}$$

Taking Napierian logarithms of both sides gives:

$$2y = \ln\left(\frac{a + x}{a - x}\right)$$

$$y = \tfrac{1}{2}\ln\left(\frac{a + x}{a - x}\right)$$

Hence artanh $\dfrac{x}{a} = \dfrac{1}{2}\ln\left(\dfrac{a + x}{a - x}\right)$

Worked problem on evaluating inverse hyperbolic functions

Problem 1. Evaluate, correct to four decimal places:

(a) $\operatorname{arsinh} \dfrac{4}{3}$ (b) $\operatorname{arcosh} 3$ (c) $\operatorname{artanh} 0.3$

(a) In logarithmic form, $\operatorname{arsinh} \dfrac{x}{a} = \ln \left\{ \dfrac{x + \sqrt{(a^2 + x^2)}}{a} \right\}$

    If $x = 4$ and $a = 3$ then $\operatorname{arsinh} \dfrac{4}{3} = \ln \left\{ \dfrac{4 + \sqrt{(3^2 + 4^2)}}{3} \right\}$

$$= \ln \left(\frac{4 + 5}{3} \right) = \ln 3 \text{ or } \mathbf{1.098\ 6}$$

(b) In logarithmic form, $\operatorname{arcosh} \dfrac{x}{a} = \ln \left\{ \dfrac{x + \sqrt{(x^2 - a^2)}}{a} \right\}$

    If $x = 3$ and $a = 1$ then $\operatorname{arcosh} \dfrac{3}{1} = \ln \left\{ \dfrac{3 + \sqrt{(3^2 - 1^2)}}{1} \right\}$

$$= \ln (3 + \sqrt{8}) = \ln 5.828\ 4 = \mathbf{1.762\ 7}$$

(c) In logarithmic form, $\operatorname{artanh} \dfrac{x}{a} = \tfrac{1}{2} \ln \left(\dfrac{a + x}{a - x} \right)$

    $\operatorname{artanh} 0.3 = \operatorname{artanh} \dfrac{3}{10}$

    If $x = 3$ and $a = 10$ then $\operatorname{artanh} \dfrac{3}{10} = \dfrac{1}{2} \ln \left(\dfrac{10 + 3}{10 - 3} \right) = \dfrac{1}{2} \ln \dfrac{13}{7}$

$$= \mathbf{0.309\ 5}$$

Further problems on evaluating inverse hyperbolic functions may be found in Section 5 (Problems 17 to 19), page 195.

4 Differentiation of inverse hyperbolic functions

(a) $y = \operatorname{arsinh} \dfrac{x}{a}$

If $y = \operatorname{arsinh} \dfrac{x}{a}$ then $\dfrac{x}{a} = \sinh y$ and $x = a \sinh y$.

$\dfrac{dx}{dy} = a \cosh y.$ (For differential coefficients of hyperbolic functions, see Chapter 4.)

Now $\cosh^2 y - \sinh^2 y = 1$. Thus $\cosh y = \sqrt{(1 + \sinh^2 y)} = \sqrt{\left\{ 1 + \left(\dfrac{x}{a} \right)^2 \right\}}$

Hence $\dfrac{dx}{dy} = a \cosh y = (a) \left[\dfrac{\sqrt{(a^2 + x^2)}}{a} \right] = \sqrt{(a^2 + x^2)}$

$\dfrac{dy}{dx} = \dfrac{1}{dx/dy} = \dfrac{1}{\sqrt{(a^2 + x^2)}}$, and there are two possible values, one positive and one negative, due to the square root sign. A sketch of part of the graph $y = \text{arsinh}\, x$ is shown in Fig. 3(a) where it is seen that the gradient (i.e. $\frac{dy}{dx}$) is always positive.

Hence if $y = \text{arsinh}\, \dfrac{x}{a}$ **then** $\dfrac{dy}{dx} = \dfrac{1}{\sqrt{(a^2 + x^2)}}$

Alternatively, since $y = \text{arsinh}\, \dfrac{x}{a} = \ln\left\{\dfrac{x + \sqrt{(a^2 + x^2)}}{a}\right\}$ then

$$\frac{d}{dx}\left(\text{arsinh}\, \frac{x}{a}\right) = \frac{d}{dx}\left[\ln\left\{\frac{x + \sqrt{(a^2 + x^2)}}{a}\right\}\right]$$

$$= \left(\frac{1}{\dfrac{x + \sqrt{(a^2 + x^2)}}{a}}\right)\left(\frac{1}{a}\right)\left[1 + \tfrac{1}{2}(a^2 + x^2)^{-1/2}\, 2x\right]$$

$$= \left(\frac{a}{x + \sqrt{(a^2 + x^2)}}\right)\left(\frac{1}{a}\right)\left(1 + \frac{x}{\sqrt{(a^2 + x^2)}}\right)$$

$$= \left(\frac{1}{x + \sqrt{(a^2 + x^2)}}\right)\left(\frac{\sqrt{(a^2 + x^2)} + x}{\sqrt{(a^2 + x^2)}}\right)$$

$$= \frac{1}{\sqrt{(a^2 + x^2)}} \text{ , as above}$$

When $a = 1$, $\dfrac{d}{dx}(\text{arsinh}\, x) = \dfrac{1}{\sqrt{(1 + x^2)}}$

The most general form of the differential coefficient of arsinh $f(x)$ is:

$$\frac{d}{dx}[\text{arsinh}\, f(x)] = \frac{1}{\sqrt{\{1 + [f(x)]^2\}}}\, [f'(x)]$$

(b) $y = \text{arcosh}\, \dfrac{x}{a}$

If $y = \text{arcosh}\, \dfrac{x}{a}$ then $\dfrac{x}{a} = \cosh y$ and $x = a \cosh y$.

$\dfrac{dx}{dy} = a \sinh y = (a)\,[\sqrt{(\cosh^2 y - 1)}]$, since $\cosh^2 y - \sinh^2 y = 1$.

$\dfrac{dx}{dy} = (a)\sqrt{\left[\left(\frac{x}{a}\right)^2 - 1\right]} = (a)\left(\frac{\sqrt{(x^2 - a^2)}}{a}\right) = \sqrt{(x^2 - a^2)}$

$\dfrac{dy}{dx} = \dfrac{1}{dx/dy} = \dfrac{1}{\sqrt{(x^2 - a^2)}}$ and there are two possible values, one positive and one negative, due to the square root sign. A sketch of part of the graph $y = \text{arcosh}\, x$ is shown in Fig. 3(b), where it is seen that when x is greater than $+1$ there are two values of y which correspond to a particular value of x. The positive value of y is defined as the principal value of arcosh $\dfrac{x}{a}$.

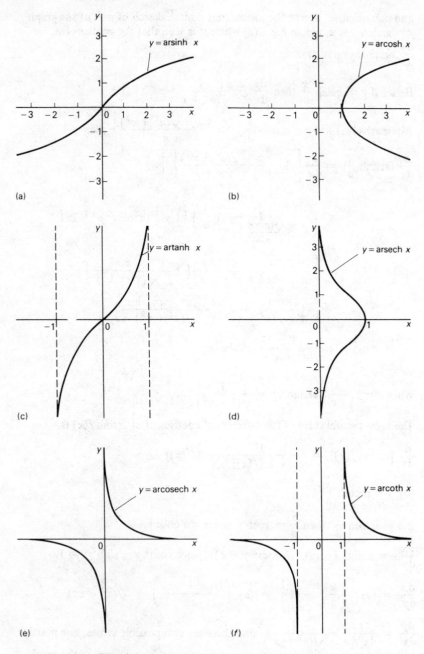

Figure 3 (a) $y = \text{arsinh}\, x$ (b) $y = \text{arcosh}\, x$
                (c) $y = \text{artanh}\, x$ (d) $y = \text{arsech}\, x$
                (e) $y = \text{arcosech}\, x$ (f) $y = \text{arcoth}\, x$

Thus if $y = \text{arcosh } \dfrac{x}{a}$ **then** $\dfrac{dy}{dx} = \dfrac{1}{\sqrt{(x^2 - a^2)}}$

Alternatively, the logarithmic equivalent of arcosh $\dfrac{x}{a}$, that is

$\ln \left\{ \dfrac{x + \sqrt{(x^2 - a^2)}}{a} \right\}$, may be differentiated with the same result. When

$a = 1$, $\dfrac{d}{dx} (\text{arcosh } x) = \dfrac{1}{\sqrt{(x^2 - 1)}}$

The most general form of the differential coefficient of arcosh $f(x)$ is:

$\dfrac{d}{dx} [\text{arcosh } f(x)] = \dfrac{1}{\sqrt{\{[f(x)]^2 - 1\}}} [f'(x)]$

(c) $y = \text{artanh } \dfrac{x}{a}$

If $y = \text{artanh } \dfrac{x}{a}$, then $\dfrac{x}{a} = \tanh y$ and $x = a \tanh y$

$\dfrac{dx}{dy} = a \text{ sech}^2 y$

Now $1 - \text{sech}^2 y = \tanh^2 y$ thus $\text{sech}^2 y = 1 - \tanh^2 y$

Hence $\dfrac{dx}{dy} = a(1 - \tanh^2 y) = a \left[1 - \left(\dfrac{x}{a} \right)^2 \right] = a \left(\dfrac{a^2 - x^2}{a^2} \right)$

i.e. $\dfrac{dx}{dy} = \dfrac{a^2 - x^2}{a}$

Hence $\dfrac{dy}{dx} = \dfrac{1}{dx/dy} = \dfrac{a}{a^2 - x^2}$

Alternatively, the logarithmic equivalent of artanh $\dfrac{x}{a}$, that is $\frac{1}{2} \ln \left(\dfrac{a + x}{a - x} \right)$,

may be differentiated with the same result. When $a = 1$,

$\dfrac{d}{dx} (\text{artanh } x) = \dfrac{1}{1 - x^2}$

The most general form of the differential coefficient of artanh $f(x)$ is:

$\dfrac{d}{dx} [\text{artanh } f(x)] = \dfrac{1}{\{1 - [f(x)]^2\}} [f'(x)]$

A sketch of part of the graph $y = \text{artanh } x$ is shown in Fig. 3(c), where the gradient (i.e. $\dfrac{dy}{dx}$) is seen to be always positive.

(d) $y = \text{arsech } \dfrac{x}{a}$

If $y = \text{arsech } \dfrac{x}{a}$, then $\dfrac{x}{a} = \text{sech } y$ and $x = a \text{ sech } y$.

$$\frac{dx}{dy} = -a \operatorname{sech} y \tanh y$$

Now $1 - \tanh^2 y = \operatorname{sech}^2 y$, thus $\tanh y = \sqrt{(1 - \operatorname{sech}^2 y)}$

$$= \sqrt{\left[1 - \left(\frac{x}{a}\right)^2\right]} = \frac{\sqrt{(a^2 - x^2)}}{a},$$

the positive value being taken, since the positive value is defined as the principal value.

$$\frac{dx}{dy} = -a \operatorname{sech} y \tanh y = -a\left(\frac{x}{a}\right)\left[\frac{\sqrt{(a^2 - x^2)}}{a}\right]$$

Hence $\dfrac{dy}{dx} = \dfrac{1}{dx/dy} = \dfrac{-a}{x\sqrt{(a^2 - x^2)}}$

When $a = 1$, $\dfrac{d}{dx} (\operatorname{arsech} x) = \dfrac{-1}{x\sqrt{(1 - x^2)}}$

A sketch of part of the graph $y = \operatorname{arsech} x$ is shown in Fig. 3(d), where it is seen that the gradient is negative for principal values of y.

The most general form of the differential coefficient of $\operatorname{arsech} f(x)$ is:

$$\frac{d}{dx} [\operatorname{arsech} f(x)] = \frac{-1}{f(x)\sqrt{\{1 - [f(x)]^2\}}} [f'(x)]$$

(e) $y = \operatorname{arcosech} \dfrac{x}{a}$

If $y = \operatorname{arcosech} \dfrac{x}{a}$, then $\dfrac{x}{a} = \operatorname{cosech} y$ and $x = a \operatorname{cosech} y$.

$$\frac{dx}{dy} = -a \operatorname{cosech} y \coth y.$$

Now $\coth^2 y - 1 = \operatorname{cosech}^2 y$, thus $\coth y = \sqrt{(\operatorname{cosech}^2 y + 1)} = \sqrt{\left[\left(\frac{x}{a}\right)^2 + 1\right]}$

$$= \frac{\sqrt{(x^2 + a^2)}}{a},$$

the positive value being taken since the positive value is defined as the principal value.

Hence $\dfrac{dx}{dy} = -a \operatorname{cosech} y \coth y = -a\left(\dfrac{x}{a}\right)\left[\dfrac{\sqrt{(x^2 + a^2)}}{a}\right] = -\dfrac{x}{a}\sqrt{(x^2 + a^2)}$

$$\frac{dy}{dx} = \frac{1}{dx/dy} = \frac{-a}{x\sqrt{(x^2 + a^2)}}$$

When $a = 1$, $\dfrac{d}{dx} (\operatorname{arcosech} x) = \dfrac{-1}{x\sqrt{(x^2 + 1)}}$

a sketch of part of the graph $y = \operatorname{arcosech} x$ is shown in Fig. 3(e), where it is seen that the gradient is negative for principal values of y.

The most general form of the differential coefficient of arcosech $f(x)$ is:

$$\frac{d}{dx}\left[\text{arcosech}\, f(x)\right] = \frac{-1}{f(x)\sqrt{\{[f(x)]^2 + 1\}}}\left[f'(x)\right]$$

(f) $y = \text{arcoth}\,\dfrac{x}{a}$

If $y = \text{arcoth}\,\dfrac{x}{a}$ then $\dfrac{x}{a} = \coth y$ and $x = a\coth y$.

$$\frac{dx}{dy} = -a\,\text{cosech}^2\, y$$

Now $\coth^2 y - 1 = \text{cosech}^2\, y$. Hence $\dfrac{dx}{dy} = -a\,(\coth^2 y - 1)$

$$= -a\left[\left(\frac{x}{a}\right)^2 - 1\right]$$

$$= -\frac{(x^2 - a^2)}{a}$$

$$\frac{dy}{dx} = \frac{1}{dx/dy} = \frac{-a}{x^2 - a^2} = \frac{a}{a^2 - x^2}$$

(i.e. the same result as obtained for $\text{artanh}\,\dfrac{x}{a}$)

When $a = 1$, $\dfrac{d}{dx}(\text{arcoth}\, x) = \dfrac{1}{1-x^2}$

A sketch of part of the graph $y = \text{arcoth}\, x$ is shown in Fig. 3(f), where it is seen that the gradient is negative for positive values of y.

The most general form of the differential coefficient of arcoth $f(x)$ is:

$$\frac{d}{dx}\left[\text{arcoth}\, f(x)\right] = \frac{1}{\{1-[f(x)]^2\}}\left[f'(x)\right]$$

Summary of differential coefficients of inverse hyperbolic functions

y or $f(x)$	$\dfrac{dy}{dx}$ or $f'(x)$
(a) $\text{arsinh}\,\dfrac{x}{a}$	$\dfrac{1}{\sqrt{(x^2 + a^2)}}$
$\text{arsinh}\, f(x)$	$\dfrac{1}{\sqrt{\{[f(x)]^2 + 1\}}}\,[f'(x)]$
(b) $\text{arcosh}\,\dfrac{x}{a}$	$\dfrac{1}{\sqrt{(x^2 - a^2)}}$
$\text{arcosh}\, f(x)$	$\dfrac{1}{\sqrt{\{[f(x)]^2 - 1\}}}\,[f'(x)]$

y or $f(x)$	$\dfrac{dy}{dx}$ or $f'(x)$
(c) $\operatorname{artanh} \dfrac{x}{a}$	$\dfrac{a}{(a^2 - x^2)}$
$\operatorname{artanh} f(x)$	$\dfrac{1}{\{1 - [f(x)]^2\}} \, [f'(x)]$
(d) $\operatorname{arsech} \dfrac{x}{a}$	$\dfrac{-a}{x\sqrt{(a^2 - x^2)}}$
$\operatorname{arsech} f(x)$	$\dfrac{-1}{f(x)\sqrt{\{1 - [f(x)]^2\}}} \, [f'(x)]$
(e) $\operatorname{arcosech} \dfrac{x}{a}$	$\dfrac{-a}{x\sqrt{(x^2 + a^2)}}$
$\operatorname{arcosech} f(x)$	$\dfrac{-1}{f(x)\sqrt{\{[f(x)]^2 + 1\}}} \, [f'(x)]$
(f) $\operatorname{arcoth} \dfrac{x}{a}$	$\dfrac{a}{a^2 - x^2}$
$\operatorname{arcoth} f(x)$	$\dfrac{1}{\{1 - [f(x)]^2\}} \, [f'(x)]$

From the above results it may be seen that:

$$\int \frac{1}{\sqrt{(x^2 + a^2)}}\, dx = \operatorname{arsinh} \frac{x}{a} + c \ \text{ or } \ \ln\left\{\frac{x + \sqrt{(x^2 + a^2)}}{a}\right\} + c,$$

$$\int \frac{1}{\sqrt{(x^2 - a^2)}}\, dx = \operatorname{arcosh} \frac{x}{a} + c \ \text{ or } \ \ln\left\{\frac{x + \sqrt{(x^2 - a^2)}}{a}\right\} + c,$$

and

$$\int \frac{1}{(a^2 - x^2)}\, dx = \frac{1}{a} \operatorname{artanh} \frac{x}{a} + c \ \text{ or } \ \frac{1}{2a} \ln\left(\frac{a + x}{a - x}\right) + c.$$

The method of determining such integrals is discussed in Chapter 18, Section 3.

Worked problems on differentiating inverse hyperbolic functions

Problem 1. Find the differential coefficient of (a) arsinh $3x$ and (b) $\operatorname{arcosh} \sqrt{(1 + t^2)}$.

(a) $\dfrac{d}{dx} [\operatorname{arsinh} f(x)] = \dfrac{1}{\sqrt{\{[f(x)]^2 + 1\}}} \, [f'(x)]$

Hence $\dfrac{d}{dx} [\operatorname{arsinh} 3x] = \dfrac{1}{\sqrt{\{(3x)^2 + 1\}}} \, [3] = \dfrac{3}{\sqrt{(9x^2 + 1)}}$

(b) $\dfrac{d}{dt}$ [arcosh $f(t)$] $= \dfrac{1}{\sqrt{\{[f(t)]^2-1\}}}$ [$f'(t)$]

Hence $\dfrac{d}{dt}$ [arcosh $\sqrt{(1+t^2)}$] $= \dfrac{1}{\sqrt{\{[\sqrt{(1+t^2)}]^2-1\}}}$ [$\tfrac{1}{2}(1+t^2)^{-1/2}2t$]

$$= \frac{1}{\sqrt{(1+t^2-1)}}\left[\frac{t}{\sqrt{(1+t^2)}}\right]$$

$$= \frac{1}{t}\left[\frac{t}{\sqrt{(1+t^2)}}\right] = \frac{1}{\sqrt{(1+t^2)}}$$

Problem 2. Differentiate (a) artanh $\dfrac{3x}{4}$, (b) arcosech (sinh x).

(a) $\dfrac{d}{dx}$ [artanh $f(x)$] $= \dfrac{1}{\{1-[f(x)]^2\}}$ [$f'(x)$]

Hence $\dfrac{d}{dx}$ [artanh $\dfrac{3x}{4}$] $= \dfrac{1}{\left\{1-\left(\dfrac{3x}{4}\right)^2\right\}}\left[\dfrac{3}{4}\right] = \dfrac{1}{\left(1-\dfrac{9x^2}{16}\right)}\left[\dfrac{3}{4}\right]$

$$= \frac{1}{\left(\dfrac{16-9x^2}{16}\right)}\left[\frac{3}{4}\right] = \frac{16}{(16-9x^2)}\left[\frac{3}{4}\right]$$

$$= \frac{12}{16-9x^2}$$

(b) $\dfrac{d}{dx}$ [arcosech $f(x)$] $= \dfrac{-1}{f(x)\sqrt{\{[f(x)]^2+1\}}}$ [$f'(x)$]

Hence $\dfrac{d}{dx}$ [arcosech (sinh x)] $= \dfrac{-1}{\sinh x\sqrt{\{\sinh^2 x+1\}}}$ [cosh x]

Now $\cosh^2 x - \sinh^2 x = 1$, hence $\sinh^2 x + 1 = \cosh^2 x$.

Therefore $\dfrac{d}{dx}$ [arcosech (sinh x)] $= \dfrac{-1}{\sinh x\sqrt{(\cosh^2 x)}}$ [cosh x]

$$= \frac{-1}{\sinh x} = -\text{cosech } x.$$

Problem 3. Find (a) $\dfrac{d}{dx}$ [arsech $(3x-1)$] and (b) $\dfrac{d}{d\theta}$ [arcoth $\sqrt{(1-2\theta^2)}$].

(a) $\dfrac{d}{dx}$ [arsech $f(x)$] $= \dfrac{-1}{f(x)\sqrt{\{1-[f(x)]^2\}}}$ [$f'(x)$]

Hence $\dfrac{d}{dx}$ [arsech $(3x-1)$] $= \dfrac{-1}{(3x-1)\sqrt{\{1-(3x-1)^2\}}}$ [3]

$$= \frac{-3}{(3x-1)\sqrt{\{1-(9x^2-6x+1)\}}}$$

$$= \frac{-3}{(3x-1)\sqrt{(6x-9x^2)}}$$

$$= \frac{-3}{(3x-1)\sqrt{[3x(2-3x)]}}$$

(b) $\dfrac{d}{d\theta}$ [arcoth $f(\theta)$] $= \dfrac{1}{\{1-[f(\theta)]^2\}}$ $[f'(\theta)]$

Hence $\dfrac{d}{d\theta}$ [arcoth $\sqrt{(1-2\theta^2)}$]

$$= \frac{1}{\{1-[\sqrt{(1-2\theta^2)}]^2\}} \left[\tfrac{1}{2}(1-2\theta^2)^{-1/2}(-4\theta)\right]$$

$$= \frac{1}{\{1-(1-2\theta^2)\}} \left[\frac{-2\theta}{\sqrt{(1-2\theta^2)}}\right]$$

$$= \frac{1}{2\theta^2} \left[\frac{-2\theta}{\sqrt{(1-2\theta^2)}}\right] = \frac{-1}{\theta\sqrt{(1-2\theta^2)}}$$

Problem 4. Differentiate: (a) $(1-t^2)$ arcoth t, (b) $\dfrac{\text{arcosh (sec }x)}{x}$

(a) $\quad f(t) = (1-t^2)$ arcoth t

Hence $f'(t) = (1-t^2)\dfrac{1}{(1-t^2)}$ $+$ (arcoth $t)(-2t)$, by the product rule of differentiation

$$= 1 - 2t \text{ arcoth } t.$$

(b) $\quad f(x) = \dfrac{\text{arcosh (sec }x)}{x}$

Hence $f'(x) = \dfrac{(x)\left[\dfrac{1}{\sqrt{(\sec^2 x - 1)}}(\sec x \tan x)\right] - [\text{arcosh (sec }x)](1)}{x^2}$,

by the quotient rule of differentiation

$$= \frac{(x)\left(\dfrac{\sec x \tan x}{\tan x}\right) - \text{arcosh (sec }x)}{x^2}, \text{ since } \sqrt{(\sec^2 x - 1)} = \tan x.$$

$$= \frac{x \sec x - \text{arcosh (sec }x)}{x^2}$$

Further problems on differentiating inverse hyperbolic functions may be found in the following Section (5) (Problems 20 to 35), page 195,

5 Further problems

Differentiation of inverse trigonometrical functions
In Problems 1 to 8, differentiate with respect to the variable.

1. (a) $\arcsin 3x$ (b) $\arcsin \dfrac{x}{5}$

$$(a)\left[\frac{3}{\sqrt{(1-9x^2)}}\right] \quad (b)\left[\frac{1}{\sqrt{(25-x^2)}}\right]$$

2. (a) $2 \arccos 4x$ (b) $3 \arccos \dfrac{2x}{3}$

$$(a)\left[\frac{-8}{\sqrt{(1-16x^2)}}\right] \quad (b)\left[\frac{-6}{\sqrt{(9-4x^2)}}\right]$$

3. (a) $2 \arctan 2x$ (b) $\arctan \sqrt{(x-1)}$

$$(a)\left[\frac{4}{1+4x^2}\right] \quad (b)\left[\frac{1}{2x\sqrt{(x-1)}}\right]$$

4. (a) $\text{arcsec } 4\theta$ (b) $2 \text{ arcsec } \sqrt{x}$.

$$(a)\left[\frac{1}{\theta\sqrt{(16\theta^2-1)}}\right] \quad (b)\left[\frac{1}{x\sqrt{(x-1)}}\right]$$

5. (a) $\text{arccosec } \dfrac{x}{3}$ (b) $\text{arccosec } (x^2+1)$.

$$(a)\left[\frac{-3}{x\sqrt{(x^2-9)}}\right] \quad (b)\left[\frac{-2}{(x^2+1)\sqrt{(x^2+2)}}\right]$$

6. (a) $\text{arccot } x^2$ (b) $3 \text{ arccot } \sqrt{(x^2-1)}$.

$$(a)\left[\frac{-2x}{1+x^4}\right] \quad (b)\left[\frac{-3}{x\sqrt{(x^2-1)}}\right]$$

7. (a) $\ln(\arcsin x)$ (b) $3e^{\text{arccot } x}$

$$(a)\left[\frac{1}{\sqrt{(1-x^2)}\arcsin x}\right] \quad (b)\left[\frac{-3e^{\text{arccot } x}}{1+x^2}\right]$$

8. (a) $\arctan\left(\dfrac{2t}{1-t^2}\right)$ (b) $\dfrac{1}{2} \text{ arcsec}\left(\dfrac{x^2+1}{x^2-1}\right)$

$$(a)\left[\frac{2}{1+t^2}\right] \quad (b)\left[\frac{-1}{(x^2+1)}\right]$$

9. Show that (a) $\dfrac{d}{dx}\left[\arcsin\left(\dfrac{2x^2+3}{5}\right)\right] = \dfrac{2x}{\sqrt{(4-3x^2-x^4)}}$

(b) $\dfrac{d}{dt}\left[\text{arcsec}\sqrt{\left(\dfrac{t}{2}\right)}\right] = \dfrac{1}{t\sqrt{[2(t-2)]}}$.

10. Show that (a) if $y = \arctan\left(\dfrac{\sin\theta}{1-\cos\theta}\right)$ then $\dfrac{dy}{d\theta} = -\dfrac{1}{2}$,

(b) if $y = 2 \arcsec \sqrt{(\cos t)}$ then $\dfrac{dy}{dt} = \dfrac{-\tan t}{\sqrt{(\cos t - 1)}}$

In Problems 11 to 16 differentiate with respect to the variable.

11. (a) $x \arcsin \dfrac{x}{2}$ (b) $x^2 \text{arccosec} \, 2x$

(a) $\left[\dfrac{x}{\sqrt{(4 - x^2)}} + \arcsin \dfrac{x}{2} \right]$

(b) $\left[x \left(2 \, \text{arccosec} \, 2x - \dfrac{1}{\sqrt{(4x^2 - 1)}} \right) \right]$

12. (a) $t \arccos (2t^2 - 1)$ (b) $3t^3 \arccot t$.

(a) $\left[\dfrac{-2t}{\sqrt{(1 - t^2)}} + \arccos (2t^2 - 1) \right]$

(b) $\left[3t^2 \left(3 \arccot t - \dfrac{t}{1 + t^2} \right) \right]$

13. (a) $(1 + \theta^2) \arctan \theta$ (b) $\sqrt{(1 - x^2)} \arcsin x$.

(a) $[1 + 2\theta \arctan \theta]$ (b) $\left[1 - \dfrac{x \arcsin x}{\sqrt{(1 - x^2)}} \right]$

14. (a) $\sin \left(\arcsec \dfrac{1}{x} \right)$ (b) $2t^2 \arccos (t - 1)$.

(a) $\left[\dfrac{-\cos (\arcsec 1/x)}{\sqrt{(1 - x^2)}} \right]$

(b) $\left[4t \arccos (t - 1) - \dfrac{2t^2}{\sqrt{[t(2 - t)]}} \right]$

15. (a) $\dfrac{\arcsin 2x}{x}$ (b) $\dfrac{\arccos x}{\sqrt{(1 - x^2)}}$.

(a) $\left[\dfrac{1}{x^2} \left(\dfrac{2x}{\sqrt{(1 - 4x^2)}} - \arcsin 2x \right) \right]$

(b) $\left[\dfrac{-1}{(1 - x^2)} + \dfrac{x \arccos x}{\sqrt{(1 - x^2)^3}} \right]$

16. (a) $\dfrac{\arcsec \sqrt{(x^2 + 1)}}{\sqrt{(x^2 + 1)}}$ (b) $\dfrac{\arccot x}{(1 + x^2)}$.

(a) $\left[\dfrac{1 - x \arcsec \sqrt{(x^2 + 1)}}{\sqrt{(x^2 + 1)^3}} \right]$

(b) $\left[\dfrac{-(1 + 2x \arccot x)}{(1 + x^2)^2} \right]$

Evaluation of inverse hyperbolic functions

In Problems 17 to 19 use logarithmic equivalents of inverse hyperbolic functions functions to evaluate correct to four decimal places.

17. (a) arsinh $\frac{2}{3}$ (b) arsinh $\frac{3}{4}$ (c) arsinh 0.5

           (a) [0.625 1]   (b) [0.693 1]   (c) [0.481 2]

18. (a) arcosh $\frac{5}{3}$ (b) arcosh 2   (c) arcosh $\frac{13}{12}$.

           (a) [1.098 6]   (b) [1.317 0]   (c) [0.405 5]

19. (a) artanh $\frac{1}{2}$ (b) artanh $\frac{3}{5}$ (c) artanh 0.3.

           (a) [0.549 3]   (b) [0.693 1]   (c) [0.309 5]

Differentiation of inverse hyperbolic functions

In Problems 20 to 34 differentiate with respect to the variable.

20. (a) arsinh $\frac{x}{2}$ (b) arsinh $5x$.

       (a) $\left[\dfrac{1}{\sqrt{(x^2 + 4)}} \right]$ (b) $\left[\dfrac{5}{\sqrt{(25x^2 + 1)}} \right]$

21. (a) arcosh $\frac{x}{3}$ (b) arcosh $4x$.

       (a) $\left[\dfrac{1}{\sqrt{(x^2 - 9)}} \right]$ (b) $\left[\dfrac{4}{\sqrt{(16x^2 - 1)}} \right]$

22. (a) artanh $\frac{2x}{3}$ (b) artanh $3x$.

       (a) $\left[\dfrac{6}{9 - 4x^2} \right]$ (b) $\left[\dfrac{3}{1 - 9x^2} \right]$

23. (a) arsech $\frac{3t}{4}$ (b) arsech $3t$.

       (a) $\left[\dfrac{-4}{t\sqrt{(16 - 9t^2)}} \right]$ (b) $\left[\dfrac{-1}{t\sqrt{(1 - 9t^2)}} \right]$

24. (a) arcosech $\frac{x}{5}$ (b) arcosech $2x$.

       (a) $\left[\dfrac{-5}{x\sqrt{(x^2 + 25)}} \right]$ (b) $\left[\dfrac{-1}{x\sqrt{(4x^2 + 1)}} \right]$

25. (a) arcoth $\frac{2\theta}{5}$ (b) arcoth 4θ.

       (a) $\left[\dfrac{10}{25 - 4\theta^2} \right]$ (b) $\left[\dfrac{4}{1 - 16\theta^2} \right]$

26. (a) arsinh $\sqrt{(t^2 - 1)}$ (b) arsech $(2x - 1)$.

       (a) $\left[\dfrac{1}{\sqrt{(t^2 - 1)}} \right]$ (b) $\left[\dfrac{-1}{(2x - 1)\sqrt{x(1 - x)}} \right]$

27. (a) arcosh $\sqrt{(1 + x^2)}$ (b) arcosh $(\cosh x)$

(a) $\left[\dfrac{1}{\sqrt{(1+x^2)}}\right]$ (b) $[1]$

28. (a) $\operatorname{artanh}\sqrt{(2t^2+1)}$ (b) $\operatorname{arcoth}(\sin x)$.

(a) $\left[\dfrac{-1}{t\sqrt{(2t^2+1)}}\right]$ (b) $[\sec x]$

29. (a) $\operatorname{arcosech}\sqrt{(x^2-1)}$ (b) $\operatorname{arcoth}\sqrt{(1-3t^2)}$

(a) $\left[\dfrac{-1}{(x^2-1)}\right]$ (b) $\left[\dfrac{-1}{t\sqrt{(1-3t^2)}}\right]$

30. (a) $x\operatorname{arsinh}x$ (b) $(1-x^2)\operatorname{artanh}x$.

(a) $\left[\dfrac{x}{\sqrt{(x^2+1)}}+\operatorname{arsinh}x\right]$ (b) $[1-2x\operatorname{artanh}x]$

31. (a) $3x\operatorname{arsech}2x$ (b) $x^2\operatorname{arcoth}\sqrt{(1-x^2)}$.

(a) $\left[\dfrac{-3}{\sqrt{(1-4x^2)}}+3\operatorname{arsech}2x\right]$

(b) $\left[2x\operatorname{arcoth}\sqrt{(1-x^2)}\dfrac{-x}{\sqrt{(1-x^2)}}\right]$

32. (a) $\sqrt{(x^2-1)}\operatorname{arcosh}x$ (b) $\sqrt{(x^2-1)}\operatorname{arcosech}\sqrt{(x^2-1)}$.

(a) $\left[1+\dfrac{x\operatorname{arcosh}x}{\sqrt{(x^2-1)}}\right]$

(b) $\left[\dfrac{1}{\sqrt{(x^2-1)}}\ (x\operatorname{arcosech}\sqrt{(x^2-1)}-1)\right]$

33. (a) $\dfrac{\operatorname{arsinh}\sqrt{(x^2-1)}}{x}$ (b) $\dfrac{\operatorname{artanh}(2x-1)}{2x}$

(a) $\left[\dfrac{1}{x^2}\left\{\dfrac{x}{\sqrt{(x^2-1)}}-\operatorname{arsinh}\sqrt{(x^2-1)}\right\}\right]$

(b) $\left[\dfrac{1}{4x^2}\left\{\dfrac{1}{(1-x)}-2\operatorname{artanh}(2x-1)\right\}\right]$

34. (a) $\dfrac{\operatorname{arsech}x}{\sqrt{(1-x^2)}}$ (b) $\dfrac{\operatorname{arcoth}x/2}{4-x^2}$

(a) $\left[\dfrac{1}{(1-x^2)}\left\{-\dfrac{1}{x}+\dfrac{x\operatorname{arsech}x}{\sqrt{(1-x^2)}}\right\}\right]$

(b) $\left[\dfrac{2(1+x\operatorname{arcoth}x/2)}{(4-x^2)^2}\right]$

35. Show that $\dfrac{d}{dx}[x\operatorname{arcosh}(\sec x)]=x\sec x+\operatorname{arcosh}(\sec x)$.

Chapter 15

Partial differentiation (CC 17)

1 Differentiating a function having two variables

When differentiating functions previously, problems have always been associated with one variable only. For a given function of x, say $y = f(x)$, the problem has been to find $\dfrac{\mathrm{d}y}{\mathrm{d}x}$, i.e. $f'(x)$. When applying the product or quotient rules, $y = uv$ or $y = \dfrac{u}{v}$, y, u, and v are all functions of x only.

However, there are many formulae and occurrences in mathematics, engineering and science, in which the variation of one function may depend on changes taking place in two or more variables. Some examples include:

(i) The volume of a cylinder, $V = \pi r^2 h$. The volume of the cylinder will change if either radius r or height h is changed.

(ii) Resonant frequency, $f_0 = \dfrac{1}{2\pi\sqrt{(LC)}}$. The resonant frequency will change if either inductance L or capacitance C is changed.

(iii) The pressure of an ideal gas, $p = \dfrac{mRT}{V}$. The pressure will change if either the thermodynamic temperature T or the volume V is changed.

(iv) Torque, $T = I\alpha$. The torque on a shaft will change if either the moment of inertia, I, or the angular acceleration, α, is changed, and so on.

For the formula giving the volume of a right circular cylinder, the value of V depends on the values of r and h and this is expressed mathematically as $V = f(r, h)$, called 'V is some function of r and h'. A statement such as '$V = f(r, h)$ and $V = \pi r^2 h$' indicates that both r and h are variable quantities.

When differentiating a function having two variables, one variable is kept constant and the differential coefficient is found of the other variable, with respect to that variable. The differential coefficient obtained is called a **partial derivative** of the function. Thus, for a right circular cylinder, the increase in volume with respect to radius $\dfrac{dV}{dr}$, when h remains constant, is the partial derivative of V with respect to r, and is written $\dfrac{\partial V}{\partial r}$. The 'curly' dee, ∂, is used to denote a differential coefficient in an expression containing more than one variable. Alternatively,

$$\left[\frac{dV}{dh}\right]_{r \text{ constant}} = \frac{\partial V}{\partial h},$$

the partial derivative of V with respect to h, with r kept constant.

2 First order partial derivatives

By applying the principles given in section 1 to the volume of a right circular cylinder, the partial derivatives $\dfrac{\partial V}{\partial r}$ and $\dfrac{\partial V}{\partial h}$ may be found. These are called **first order partial derivatives** since $n = 1$ when they are written in the form $\dfrac{\partial^n V}{\partial r^n}$. Second order partial derivatives contain at least one term in which $n = 2$, i.e. a term of the form $\dfrac{\partial^2 V}{\partial r^2}$. The partial derivative of V with respect to r is found by keeping h constant, i.e. since $V = \pi r^2 h$, then (πh) becomes the constant term. Thus $\dfrac{\partial V}{\partial r} = (\pi h) \dfrac{d(r^2)}{dr} = 2\pi rh$. The partial derivative of V with respect to h is found by keeping r constant, i.e. since $V = \pi r^2 h$ then (πr^2) becomes the constant term. Thus $\dfrac{\partial V}{\partial h} = (\pi r^2) \dfrac{d(h)}{dh} = \pi r^2$.

Worked problems on first order partial derivatives

Problem 1. The pressure p of a given mass of gas is given by $pV = kT$, where k is a constant, where V is the volume and T is the thermodynamic temperature. Determine equations for $\dfrac{\partial p}{\partial V}$ and $\dfrac{\partial p}{\partial T}$.

Since $pV = kT$, $p = \dfrac{kT}{V}$

To find $\dfrac{\partial p}{\partial V}$, T is kept constant and $p = (kT) V^{-1}$.

Hence, $\dfrac{\partial p}{\partial V} = (kT) \dfrac{d(V^{-1})}{dV} = kT(-1) V^{-2} = -\dfrac{kT}{V^2}$

To find $\dfrac{\partial V}{\partial T}$, V is kept constant and $p = \left(\dfrac{k}{V}\right) \cdot T$.

Hence, $\dfrac{\partial V}{\partial T} = \left(\dfrac{k}{V}\right) \cdot \dfrac{d\,(T)}{dT} = \left(\dfrac{k}{V}\right) \cdot 1 = \dfrac{k}{V}$

Problem 2. If $z = (x + y)\ln\dfrac{x}{y}$, prove that

$$x\,\dfrac{\partial z}{\partial x} + y\,\dfrac{\partial z}{\partial y} = z.$$

As there is a product of two functions having x in each, $\dfrac{\partial z}{\partial x}$ is found by keeping y constant and applying the product rule of differentiation. The partial differential coefficient of the first function with respect to x is:

$$\dfrac{\partial(x + y)}{\partial x} = 1,$$

since y is a constant. The partial differential coefficient of the second function with respect to x is:

$$\dfrac{\partial(\ln x/y)}{\partial x} = \dfrac{1}{x/y} \times \dfrac{1}{y} = \dfrac{1}{x}$$

Alternatively, $\dfrac{\partial(\ln x/y)}{\partial x} = \dfrac{\partial(\ln x - \ln y)}{\partial x} = \dfrac{1}{x}$

since $\ln y$ is a constant.

Hence, by the product rule, $\dfrac{\partial z}{\partial x} = (x + y) \cdot \dfrac{1}{x} + \left(\ln\dfrac{x}{y}\right) \cdot 1$

$\dfrac{\partial z}{\partial y}$ can be found by keeping x constant.

As there are two functions of y in each, the product rule of differentiation is again used. The partial differential coefficient of the first function with respect to y is: $\dfrac{\partial(x + y)}{\partial y} = 1$, since x is a constant. The partial differential coefficient of the second function with respect to y is:

$$\dfrac{\partial(\ln x/y)}{\partial y} = \dfrac{\partial(\ln x - \ln y)}{\partial y} = -\dfrac{1}{y}\,.$$

Thus, $\dfrac{\partial z}{\partial y} = (x + y)\left(-\dfrac{1}{y}\right) + \left(\ln\dfrac{x}{y}\right) \cdot 1.$

Taking the L.H.S. of the identity and substituting for $\dfrac{\partial z}{\partial x}$ and $\dfrac{\partial z}{\partial y}$ gives:

$$x\left[\dfrac{1}{x}(x + y) + \ln\dfrac{x}{y}\right] + y\left[-\dfrac{1}{y}(x + y) + \ln\dfrac{x}{y}\right]$$

i.e. L.H.S. $= (x + y) + x\ln\dfrac{x}{y} - (x + y) + y\ln\dfrac{x}{y}$

$$= (x + y) \ln \frac{x}{y} = z = \text{R.H.S.}$$

Problem 3. If $z = f(\theta, \phi)$, find $\dfrac{\partial z}{\partial \theta}$ and $\dfrac{\partial z}{\partial \phi}$, if $z = 3 \sin 2\theta \cos 3\phi$.

$\dfrac{\partial z}{\partial \theta}$ may be found by keeping ϕ constant, i.e. (3 cos 3ϕ) is the constant term.

Thus, $\dfrac{\partial z}{\partial \theta} = 3 \cos 3\phi. \dfrac{d (\sin 2\theta)}{d\theta}$,

i.e. $\dfrac{\partial z}{\partial \theta} = 3 \cos 3\phi. 2 \cos 2\theta$.

$= \mathbf{6 \cos 2\theta \cos 3\phi}$

$\dfrac{\partial z}{\partial \phi}$ may be found by keeping θ constant, i.e. (3 sin 2θ) is the constant term.

Thus, $\dfrac{\partial z}{\partial \phi} = 3 \sin 2\theta \cdot \dfrac{d (\cos 3\phi)}{d\phi}$

$= 3 \sin 2\theta \cdot (-3 \sin 3\phi)$

$= \mathbf{-9 \sin 2\theta \sin 3\phi.}$

Problem 4. If $z = f(x^2 + y^2)$, show that $x \dfrac{\partial z}{\partial y} - y \dfrac{\partial z}{\partial x} = 0$.

The actual function of z is not known and z could be equal to, say, $\sin (x^2 + y^2)$ or $\ln (x^2 + y^2)$ or $(x^2 + y^2)^{3/2}$ or any other function of $(x^2 + y^2)$. However, proofs based on functions are often possible, even though the actual function is not known. Such techniques are sometimes used in solving partial differential equations. Making a substitution such as let $u = (x^2 + y^2)$ usually helps in proving the identity. If such a substitution is made in this case, then $z = f(u)$.

Also, since $u = x^2 + y^2$, $\dfrac{\partial u}{\partial x} = 2x$ (y is kept constant) and $\dfrac{\partial u}{\partial y} = 2y$ (x is kept constant). The chain rule of differentiation applies equally to partial derivatives, hence

$$\frac{\partial z}{\partial x} = \frac{\partial z}{\partial u} \times \frac{\partial u}{\partial x} = f'(u) \times 2x$$

Similarly,

$$\frac{\partial z}{\partial y} = \frac{\partial z}{\partial u} \times \frac{\partial u}{\partial y} = f'(u) \times 2y.$$

Substituting for $\dfrac{\partial z}{\partial x}$ and $\dfrac{\partial z}{\partial y}$ in the L.H.S. of the equation, gives:

$$x \cdot 2y f'(u) - y \cdot 2x f'(u) = 0 = \text{R.H.S.}$$

Further problems on first order partial derivatives may be found in Section 4 (Problems 1 to 7), page 204.

3 Second order partial derivatives

It is shown in Section 2 that if $z = f(x, y)$, then it is possible to find $\dfrac{\partial z}{\partial x}$ and $\dfrac{\partial z}{\partial y}$. The partial derivatives $\dfrac{\partial z}{\partial x}$ and $\dfrac{\partial z}{\partial y}$ are themselves functions of x and y and hence can be differentiated again with respect to x or with respect to y. When this is done, four **second order partial derivatives** are possible for a function containing two variables. These are:

(i)   differentiating $\dfrac{\partial z}{\partial x}$ with respect to x, (keeping y constant), gives

$$\frac{\partial \left(\dfrac{\partial z}{\partial x} \right)}{\partial x} \text{, written as } \frac{\partial^2 z}{\partial x^2} \text{ ,}$$

(ii)   differentiating $\dfrac{\partial z}{\partial x}$ with respect to y (keeping x constant) gives

$$\frac{\partial \left(\dfrac{\partial z}{\partial x} \right)}{\partial y} \text{, written as } \frac{\partial^2 z}{\partial y \partial x} \text{ ,}$$

(iii)   differentiating $\dfrac{\partial z}{\partial y}$ with respect to x (keeping y constant) gives

$$\frac{\partial \left(\dfrac{\partial z}{\partial y} \right)}{\partial x} \text{, written as } \frac{\partial^2 z}{\partial x \partial y} \text{ , and}$$

(iv)   differentiating $\dfrac{\partial z}{\partial y}$ with respect to y (keeping x constant) gives

$$\frac{\partial \left(\dfrac{\partial z}{\partial y} \right)}{\partial y} \text{, written as } \frac{\partial^2 z}{\partial y^2} \text{ .}$$

To illustrate these four possibilities, consider the function:

$z = \sin xy.$

$\dfrac{\partial z}{\partial x} = y \cos xy$ and $\dfrac{\partial z}{\partial y} = x \cos xy.$ Then,

(i) $\dfrac{\partial^2 z}{\partial x^2} = \dfrac{\partial \left(\dfrac{\partial z}{\partial x} \right)}{\partial x} = \dfrac{\partial (y \cos xy)}{\partial x} = -y^2 \sin xy$

(ii) $\dfrac{\partial^2 z}{\partial y \partial x} = \dfrac{\partial \left(\dfrac{\partial z}{\partial x} \right)}{\partial y} = \dfrac{\partial (y \cos xy)}{\partial y}$

$$= y \left(-x \sin xy \right) + 1 \left(\cos xy \right)$$

$$= \cos xy - xy \sin xy$$

(iii) $\dfrac{\partial^2 z}{\partial x \partial y} = \dfrac{\partial \left(\dfrac{\partial z}{\partial y}\right)}{\partial x} \cdot = \dfrac{\partial (x \cos xy)}{\partial x}$

$$= x(-y \sin xy) + 1(\cos xy)$$

$$= \cos xy - xy \sin xy$$

(iv) $\dfrac{\partial^2 z}{\partial y^2} = \dfrac{\partial \left(\dfrac{\partial z}{\partial y}\right)}{\partial y} = \dfrac{\partial (x \cos xy)}{\partial y} = -x^2 \sin xy.$

It can be seen that the values of $\dfrac{\partial^2 z}{\partial y \partial x}$ and $\dfrac{\partial^2 z}{\partial x \partial y}$ are the same in this case. This result is true for all functions which are continuous. (A function which is continuous means that the graph of the function has no sudden jumps or breaks.)

Since $\dfrac{\partial^2 z}{\partial y \partial x}$ is always equal to $\dfrac{\partial^2 z}{\partial x \partial y}$ for functions being considered at this level, there are in fact only three second order partial derivatives for functions of two variables, namely:

$$\dfrac{\partial^2 z}{\partial x^2}, \quad \dfrac{\partial^2 z}{\partial y^2} \quad \text{and either} \quad \dfrac{\partial^2 z}{\partial x \partial y} \quad \text{or} \quad \dfrac{\partial^2 z}{\partial y \partial x}.$$

Second order partial derivatives will be met in such topic areas as entropy and the continuity theorem in thermodynamics, the waveguide theory in electrical engineering and in the solution of partial differential equations.

Worked problems on second order partial derivatives

Problem 1. If $u = f(x, y)$ and $u = \dfrac{x - y}{x + y}$, find

$$\dfrac{\partial^2 u}{\partial x^2}, \quad \dfrac{\partial^2 u}{\partial y^2}, \quad \dfrac{\partial^2 u}{\partial x \partial y} \quad \text{and} \quad \dfrac{\partial^2 u}{\partial y \partial x}$$

To find $\dfrac{\partial u}{\partial x}$, the quotient rule of differentiation is used, y being treated as a constant. Hence

$$\dfrac{\partial u}{\partial x} = \dfrac{(x + y)(1) - (x - y)(1)}{(x + y)^2} = \dfrac{2y}{(x + y)^2} = 2y(x + y)^{-2}$$

$$\dfrac{\partial^2 u}{\partial x^2} = \dfrac{\partial \left(\dfrac{\partial u}{\partial x}\right)}{\partial x} = \dfrac{\partial \left[2y(x + y)^{-2}\right]}{\partial x}$$

$$= (2y)(-2)(x+y)^{-3}(1) = \frac{-4y}{(x+y)^3}$$

To find $\frac{\partial u}{\partial y}$, the quotient rule of differentiation is used, x being treated as a constant. Hence,

$$\frac{\partial u}{\partial y} = \frac{(x+y)(-1)-(x-y)(1)}{(x+y)^2} = \frac{-2x}{(x+y)^2} = -2x(x+y)^{-2}$$

$$\frac{\partial^2 u}{\partial y^2} = \frac{\partial\left(\frac{\partial u}{\partial y}\right)}{\partial y} = \frac{\partial\left(-2x(x+y)^{-2}\right)}{\partial y}$$

$$= (-2x)(-2)(x+y)^{-3}(1) = \frac{4x}{(x+y)^3}$$

$$\frac{\partial^2 u}{\partial x \partial y} = \frac{\partial\left(\frac{\partial u}{\partial y}\right)}{\partial x} = \frac{\partial\left(\frac{-2x}{(x+y)^2}\right)}{\partial x}$$

$$= \frac{(x+y)^2(-2)-(-2x)(2)(x+y)(1)}{(x+y)^4} = \frac{2(x-y)}{(x+y)^3}$$

$$\frac{\partial^2 u}{\partial y \partial x} = \frac{\partial\left(\frac{\partial u}{\partial x}\right)}{\partial y} = \frac{\partial\left(\frac{-2y}{(x+y)^2}\right)}{\partial y}$$

$$= \frac{(x+y)^2(2)-2y(2)(x+y)(1)}{(x+y)^4} = \frac{2(x-y)}{(x+y)^3}$$

The results obtained for $\frac{\partial^2 u}{\partial x \partial y}$ and $\frac{\partial^2 u}{\partial y \partial x}$ verify that in this case the relationship stated in the text that $\frac{\partial^2 u}{\partial x \partial y} = \frac{\partial^2 u}{\partial y \partial x}$ is true.

Problem 2. If $z = f(x, y)$ and $z = x \cos(x+y)$, find $\frac{\partial^2 z}{\partial x^2}$ and $\frac{\partial^2 z}{\partial y^2}$. Show that $\frac{\partial^2 z}{\partial x \partial y} = \frac{\partial^2 z}{\partial y \partial x}$.

Applying the product rule, gives:

$$\frac{\partial z}{\partial x} = x(-\sin(x+y))+1 \cdot \cos(x+y)$$
$$= -x\sin(x+y)+\cos(x+y).$$

Differentiating $\frac{\partial z}{\partial x}$ with respect to x gives:

$$\frac{\partial^2 x}{\partial x^2} = -x \cos (x + y) + (-1) \sin (x + y) - \sin (x + y)$$

$$= -x \cos (x + y) - 2 \sin (x + y).$$

Differentiating $z = x \cos (x + y)$ with respect to y gives:

$$\frac{\partial z}{\partial y} = -x \sin (x + y).$$

Differentiating $\frac{\partial z}{\partial y}$ with respect to y gives:

$$\frac{\partial^2 z}{\partial y^2} = -x \cos (x + y))$$

$$\frac{\partial^2 z}{\partial x \partial y} = \frac{\partial \left(\frac{\partial z}{\partial y}\right)}{\partial x} = \frac{\partial (-x \sin (x + y))}{\partial x}$$

$$= -x \cos (x + y) + (-1) \sin (x + y)$$

$$= -x \cos (x + y) - \sin (x + y) \qquad (1)$$

$$\frac{\partial^2 z}{\partial y \partial x} = \frac{\partial \left(\frac{\partial z}{\partial x}\right)}{\partial y} = \frac{\partial (-x \sin (x + y) + \cos (x + y))}{\partial y}$$

$$= -x \cos (x + y) - \sin (x + y) \qquad (2)$$

Since equation (1) is equal to equation (2), then

$$\frac{\partial^2 z}{\partial x \partial y} = \frac{\partial^2 z}{\partial y \partial x}$$

Further problems on second order partial derivatives may be found in the following Section (4), (Problems 8 to 19), page 206.

4 Further problems

First order partial derivatives

1. If $z = f(x, y)$, find $\frac{\partial z}{\partial x}$ and $\frac{\partial z}{\partial y}$ for the following equations:

   (a) $z = x^2 + 3xy + y^3$ (b) $z = (3x - 2y)^2$ (c) $z = e^y \ln (x + y)$.

   (a) $\left[\frac{\partial z}{\partial x} = 2x + 3y, \frac{\partial z}{\partial y} = 3 (x + y^2) \right]$

   (b) $\left[\frac{\partial z}{\partial x} = 6 (3x - 2y), \frac{\partial z}{\partial y} = -4 (3x - 2y) \right]$

   (c) $\left[\frac{\partial z}{\partial x} = \frac{e^y}{x + y}, \frac{\partial z}{\partial y} = e^y \left(\frac{1}{x + y} + \ln (x + y) \right) \right]$

2. If $u = f(p, q)$, find $\dfrac{\partial u}{\partial p}$ and $\dfrac{\partial u}{\partial q}$ for the following equations:

(a) $u = \dfrac{1}{\sqrt{(p^2 + q^2)}}$ (b) $u = 3^{(4p + q/2)} \tan (3p + 2q)$

(c) $u = pe^q + \dfrac{1}{p} \ln q.$

(a) $\left[\dfrac{\partial u}{\partial p} = - \dfrac{p}{\sqrt{(p^2 + q^2)^3}} , \dfrac{\partial u}{\partial q} = - \dfrac{q}{\sqrt{(p^2 + q^2)^3}} \right]$

(b) $\left[\dfrac{\partial u}{\partial p} = e^{(4p + q/2)} \left\{ 3 \sec^2 (3p + 2q) + 4 \tan (3p + 2q) \right\} \right.$

$\left. \dfrac{\partial u}{\partial q} = e^{(4p + q/2)} \left\{ 2 \sec^2 (3p + 2q) + \dfrac{1}{2} \tan (3p + 2q) \right\} \right]$

(c) $\left[\dfrac{\partial u}{\partial p} = e^q - \dfrac{\ln q}{p^2} , \dfrac{\partial u}{\partial q} = p e^q + \dfrac{1}{pq} \right]$

3. If $m = f(u, v)$ and $m = \cos uv$, prove that

$$\dfrac{1}{v} \dfrac{\partial m}{\partial u} = \dfrac{1}{u} \dfrac{\partial m}{\partial v}.$$

4. The volume V of a right circular cone of height h and base radius r is given by:

$$V = \dfrac{1}{3} \pi r^2 h.$$

Find $\dfrac{\partial V}{\partial r}$ and $\dfrac{\partial V}{\partial h}$. $\left[\dfrac{\partial V}{\partial r} = \dfrac{2}{3} \pi rh, \dfrac{\partial V}{\partial h} = \dfrac{1}{3} \pi r^2 \right]$

5. The time of oscillation T of a simple pendulum is given by

$T = 2\pi \sqrt{\left(\dfrac{l}{g} \right)}$, where l is the length of the pendulum and g is the free fall acceleration due to gravity. Find $\dfrac{\partial T}{\partial l}$ and $\dfrac{\partial T}{\partial g}$.

$$\left[\dfrac{\partial T}{\partial l} = \dfrac{\pi}{\sqrt{(lg)}} , \dfrac{\partial T}{\partial g} = - \pi \sqrt{\left(\dfrac{l}{g^3} \right)} \right]$$

6. If $z = f(p - q)$, prove that

$$\dfrac{\partial z}{\partial p} + \dfrac{\partial z}{\partial q} = 0.$$

7. If $z = m^2 f(m - n)$, show that

$$\dfrac{\partial z}{\partial m} + \dfrac{\partial z}{\partial n} = \dfrac{2z}{m}.$$

Second order partial derivatives

In Problems 8 to 10, find (a) $\dfrac{\partial^2 z}{\partial x^2}$ (b) $\dfrac{\partial^2 z}{\partial y \partial x}$ (c) $\dfrac{\partial^2 z}{\partial x \partial y}$ and (d)

$\dfrac{\partial^2 z}{\partial y^2}$, given that $z = f(x, y)$.

8. $z = \ln xy$.   (a) $\left[-\dfrac{1}{x^2}\right]$ (b) $[0]$ (c) $[0]$ (d) $\left[-\dfrac{1}{y^2}\right]$

9. $z = x \cos y - y \cos x$.   (a) $[y \cos x]$ (b) $[\sin x - \sin y]$
   (c) $[\sin x - \sin y]$ (d) $[-x \cos y]$

10. $z = \dfrac{x}{x+y}$.   (a) $\left[-\dfrac{2y}{(x+y)^3}\right]$ (b) $\left[\dfrac{x-y}{(x+y)^3}\right]$

   (c) $\left[\dfrac{x-y}{(x+y)^3}\right]$ (d) $\left[\dfrac{2x}{(x+y)^3}\right]$

11. If $z = f(x, y)$ and $z = e^{-y} \sin x$, show that

$$\dfrac{\partial^2 z}{\partial x^2} + \dfrac{\partial^2 z}{\partial y^2} = 0.$$

12. If $z = f(x, t)$ and $z = \sqrt{\left(\dfrac{x}{4t}\right)}$, find $\dfrac{\partial^2 z}{\partial x^2}$ and show that $\dfrac{\partial^2 z}{\partial x \partial t} = \dfrac{\partial^2 z}{\partial t \partial x}$.

$$\left[-\dfrac{1}{8\sqrt{(t x^3)}}\right]$$

13. If $z = f(p, q)$ and $z = p^3 \tan \dfrac{p}{q}$, prove that

$$\dfrac{\partial^2 z}{\partial p \partial q} = \dfrac{\partial^2 z}{\partial q \partial p} = -\dfrac{2p^3}{q^2}\left(\sec^2 \dfrac{p}{q}\right)\left\{\dfrac{p}{q} \tan \left(\dfrac{p}{q}\right) + 2\right\}.$$

14. If $u = f(p, q)$ and $u = \dfrac{2p}{q^2} \ln q$, prove that $\dfrac{\partial^2 u}{\partial q \partial p} - \dfrac{\partial^2 u}{\partial p \partial q} = 0.$

15. If $\phi = f(r, \theta)$ and $\phi = (Ar^n + Br^{-n}) \sin (n \theta + \alpha)$ where A, B, n and α are constants, show that

$$\dfrac{\partial^2 \phi}{\partial r^2} + \dfrac{1}{r} \dfrac{\partial \phi}{\partial r} + \dfrac{1}{r^2} \dfrac{\partial^2 \phi}{\partial \theta^2} = 0.$$

16. An equation used in thermodynamics is the Benedict–Webb–Rubine equation of state for the expansion of a gas. The equtaion is:

$$p = \dfrac{RT}{V} + \left(B_0 RT - A_0 - \dfrac{C_0}{T^2}\right)\dfrac{1}{V^2} + (bRT - a)\dfrac{1}{V^3} + \dfrac{A\alpha}{V^6}$$

$$+ \dfrac{C(1 + \gamma/V^2)}{T^2}\left(\dfrac{1}{V^3}\right) e^{-\gamma/V^2}$$

Show that $\dfrac{\partial^2 P}{\partial T^2} = \dfrac{3}{V^2 T^4}\left\{\dfrac{1}{V}\left(C + \dfrac{\gamma}{V^2}\right) e^{-\frac{\gamma}{V^2}} - 2 C_0\right\}$

17. An equation resulting from plucking a string is

$$y = \sin \dfrac{n\pi x}{l}\left\{k \cos \dfrac{n\pi b}{l} t + c \sin \dfrac{n\pi b}{l} t\right\}$$

Determine $\dfrac{\partial y}{\partial t}$ and $\dfrac{\partial y}{\partial x}$.

$$\left[\begin{array}{l}
\dfrac{\partial y}{\partial t} = \dfrac{bn\pi}{l} \sin \dfrac{n\pi}{l} x \left\{ c \cos \dfrac{n\pi b}{l} t - k \sin \dfrac{n\pi b}{l} \right\} \\[3mm]
\dfrac{\partial y}{\partial x} = \dfrac{n\pi}{l} \cos \dfrac{n\pi}{l} x \left\{ k \cos \dfrac{n\pi b}{l} t + c \sin \dfrac{n\pi b}{l} \right\}
\end{array}\right]$$

18. The magnetic field vector H due to a steady current I flowing round a circular wire of radius a and at a distance z from its centre is given by:

$$H = \pm \frac{I}{2} \frac{\partial}{\partial z} \left(\frac{z}{\sqrt{(a^2 + z^2)}} \right)$$

Determine H.
$$\left[H = \pm \frac{a^2 I}{2\sqrt{(a^2 + z^2)^3}} \right]$$

19. In a thermodynamic system
$$k = Ae^{\left(\frac{T\Delta S - \Delta H}{RT} \right)},$$

where R, k and A are constants.
Find $\dfrac{\partial k}{\partial T}$, $\dfrac{\partial A}{\partial T}$, $\dfrac{\partial (\Delta S)}{\partial T}$ and $\dfrac{\partial (\Delta H)}{\partial T}$.

$$\left[\begin{array}{l}
\dfrac{\partial k}{\partial T} = \dfrac{A \Delta H}{RT^2} e^{\left(\frac{T\Delta S - \Delta H}{RT} \right)}; \\[3mm]
\dfrac{\partial A}{\partial T} = \dfrac{-k \Delta H}{RT^2} e^{\left(\frac{\Delta H - T\Delta S}{RT} \right)}; \\[3mm]
\dfrac{\partial (\Delta S)}{\partial T} = \dfrac{-\Delta H}{T^2}; \quad \dfrac{\partial (\Delta H)}{\partial T} = \Delta S - R \ln \left(\dfrac{k}{A} \right)
\end{array}\right]$$

Chapter 16

Total differentiation, rates of change and small changes (CC 18)

1 Total differential

In Chapter 15 the principles of partial differentiation are introduced for the case where only one variable changes at a time, the other variables being kept constant. In practice, variables may all be changing at the same time. Consider a right cylindrical block of metal being heated for a time δt. Let the initial volume of the block, V, be $\pi r^2 h$, where r is the radius of the base and h is the height of the cylinder. If the metal of the block has a positive temperature coefficient then both r and h will increase as the block is heated. Let the increase in r be δr, the increase in h be δh and the resultant increase in volume be δV. Then, after time δt:

the new volume, $V + \delta V \quad = \pi (r + \delta r)^2 (h + \delta h)$
The change in volume, $\delta V \quad = \pi (r + \delta r)^2 (h + \delta h) - V$
$= \pi (r^2 + 2r\delta r + \delta r^2)(h + \delta h) - \pi r^2 h$

i.e. $\delta V = \pi r^2 h + 2\pi rh\delta r + \pi h\delta r^2 + \pi r^2 \delta h + 2\pi r\delta r\delta h + \pi \delta r^2 \delta h - \pi r^2 h$.

i.e. $\delta V = 2\pi rh\delta r + \pi h\delta r^2 + \pi r^2 \delta h + 2\pi r\delta r\delta h + \pi \delta r^2 \delta h$ \quad (1)

Change of volume in unit time, $\dfrac{\delta V}{\delta t}$, is given by:

$$\frac{\delta V}{\delta t} = 2\pi rh \frac{\delta r}{\delta t} + \pi h\delta r \frac{\delta r}{\delta t} + \pi r^2 \frac{\delta h}{\delta t} + 2\pi r\delta r \frac{\delta h}{\delta t} + \pi \delta r^2 \frac{\delta h}{\delta t} \quad (2)$$

As $\delta t \to 0$, the ratios $\dfrac{\delta V}{\delta t}, \dfrac{\delta r}{\delta t}$ and $\dfrac{\delta h}{\delta t}$ become $\dfrac{dV}{dt}, \dfrac{dr}{dt}$ and $\dfrac{dh}{dt}$ respectively, and terms containing the factors δr and δr^2 become zero. Hence rate of volume, $\dfrac{dV}{dt} = 2\pi rh \dfrac{dr}{dt} + \pi r^2 \dfrac{dh}{dt}$. But it is shown in Chapter 15 that if $V = \pi r^2 h$, then $\dfrac{\partial V}{\partial r} = 2\pi rh$ and $\dfrac{\partial V}{\partial h} = \pi r^2$

Thus, $\dfrac{dV}{dt} = \dfrac{\partial V}{\partial r}\dfrac{dr}{dt} + \dfrac{\partial V}{\partial h}\dfrac{dh}{dt}$ \hfill (3)

Equation (3) may be made independant of time by multiplying throughout by dt, giving

$$\dfrac{dV}{\cancel{dt}}\,\cancel{dt} = \dfrac{\partial V}{\partial r}\dfrac{dr}{\cancel{dt}}\,\cancel{dt} + \dfrac{\partial V}{\partial h}\dfrac{dh}{dt}\,dt$$

i.e. $dV = \dfrac{\partial V}{\partial r}\,dr + \dfrac{\partial V}{\partial h}\,dh$ \hfill (4)

In this form, dV is called the **total differential** of V and $\dfrac{\partial V}{\partial r}\,dr$ and $\dfrac{\partial V}{\partial h}\,dh$ are called the **partial differentials** of V. Thus the total differential is the sum of the separate partial differentials. The total differential can be used as a basis for finding the change of volume with respect to some other quantity. For example, to find the change of volume with respect to, say, thermodynamic temperature, T, each term is divided by dT, giving

$$\dfrac{dV}{dT} = \dfrac{\partial V}{\partial r}\dfrac{dr}{dT} + \dfrac{\partial V}{\partial h}\dfrac{dh}{dT}$$

The total differential is also used as a basis for solving partial differential equations.

The result for the total differential of a right circular cylindrical block can be generalised. If $z = f(u, v, w,)$, then from equation (4),

$$dz = \dfrac{\partial z}{\partial u}\,du + \dfrac{\partial z}{\partial v}\,dv + \dfrac{\partial z}{\partial w}\,dw + \hspace{3em} (5)$$

Worked problems on total differentials

Problem 1. If $z = f(x, y)$ and $z = x^3 y^2 + \dfrac{y}{x}$, find the total differential, dz.

$\dfrac{\partial z}{\partial x} = y^2\,(3x^2) - \dfrac{y}{x^2}$ (y is kept constant)

$\dfrac{\partial z}{\partial y} = x^3\,(2y) + \dfrac{1}{x}$ (x is kept constant)

The total differential is the sum of the partial differentials,

i.e. $dz = \dfrac{\partial z}{\partial x}\,dx + \dfrac{\partial z}{\partial y}\,dy$

Thus, $dz = \left(3x^2 y^2 - \dfrac{y}{x^2}\right)dx + \left(2x^3 y + \dfrac{1}{x}\right)dy$

Problem 2. The pressure p, volume V and temperature T of unit mass of a gas are related by the formula $pV = RT$, where R is a constant.

Prove that $dp = \dfrac{p}{T} dT - \dfrac{p}{V} dV$, and also that

$$dT = \frac{T}{V} dV + \frac{T}{p} dp$$

Since $pV = RT$, $p = \dfrac{RT}{V}$, $\dfrac{\partial p}{\partial T} = \dfrac{R}{V}$ and $\dfrac{\partial p}{\partial V} = -\dfrac{RT}{V^2}$

From equation (5), $dp = \dfrac{\partial p}{\partial T} dT + \dfrac{\partial p}{\partial V} dV.$

$$= \frac{R}{V} dT + \left(-\frac{RT}{V^2}\right)dV$$

But from the initial equation, $\dfrac{R}{V} = \dfrac{p}{T}$, hence substituting for $\dfrac{R}{V}$ gives:

$$dp = \frac{p}{T} dT - \left(\frac{p}{T} \cdot \frac{T}{V}\right)dV$$

$$= \frac{p}{T} dT - \frac{p}{V} dV$$

Since $pV = RT$, $T = \dfrac{pV}{R}$, $\dfrac{\partial T}{\partial p} = \dfrac{V}{R}$ and $\dfrac{\partial T}{\partial V} = \dfrac{p}{R}$

From equation (5), $dT = \dfrac{\partial T}{\partial V} dV + \dfrac{\partial T}{\partial p} dp$

$$= \frac{p}{R} dV + \frac{V}{R} dp$$

But from the initial equation, $\dfrac{p}{R} = \dfrac{T}{V}$ and $\dfrac{V}{R} = \dfrac{T}{p}$,

hence, $dT = \dfrac{T}{V} dV + \dfrac{T}{p} dp$.

Further problems on total differentials may be found in Section 4 (Problems 1 to 5), page 215.

2 Rates of change

It is shown in Section 1, equation (3), that for a right circular cylindrical block being heated, the rate of change of volume is given by

$$\frac{dV}{dt} = \frac{\partial V}{\partial r} \frac{dr}{dt} + \frac{\partial V}{\partial h} \frac{dh}{dt}$$

This result may be generalised and if $z = f(u, v, w, \ldots)$ then

$$\frac{dz}{dt} = \frac{\partial z}{\partial u} \frac{du}{dt} + \frac{\partial z}{\partial v} \frac{dv}{dt} + \frac{\partial z}{\partial w} \frac{dw}{dt} + \cdots \qquad (6)$$

Equation (6) may be used to solve problems in which different quantities have different rates of change. The principles used are shown in the worked problems following.

Worked problems on rates of change

Problem 1. If $z = f(u, v)$ and $z = e^u \sin 2v$, find the rate of change of z, correct to four decimal places, when u is increasing at 3 units per second, v is decreasing at 5 units per second, u is 0.1 units and v is 0.5 units.

Since $z = e^u \sin 2v$, $\dfrac{\partial z}{\partial u} = e^u \sin 2v$ and $\dfrac{\partial z}{\partial v} = 2e^u \cos 2v$.

If u is increasing at 3 units/s, then $\dfrac{du}{dt} = 3$. Similarly, $\dfrac{dv}{dt} = -5$

From equation (6), $\dfrac{dz}{dt} = \dfrac{\partial z}{\partial u} \dfrac{du}{dt} + \dfrac{\partial z}{\partial v} \dfrac{dv}{dt}$

$$= (e^u \sin 2v)(3) + (2e^u \cos 2v)(-5)$$

But $u = 0.1$ and $v = 0.5$ units, hence

$$\frac{dz}{dt} = 3e^{0.1} \sin 1 - 10e^{0.1} \cos 1$$

Sin 1 means the sine of 1 radian, i.e. $\sin \left(\dfrac{180}{\pi}\right)^\circ$. Also cos 1 means $\cos \left(\dfrac{180}{\pi}\right)^\circ$.

Hence, $\dfrac{dz}{dt} = e^{0.1} \left[3 \sin \left(\dfrac{180}{\pi}\right)^\circ - 10 \cos \left(\dfrac{180}{\pi}\right)^\circ \right]$, i.e. $\dfrac{dz}{dt} = -3.181\,4$,
correct to four decimal places.

That is, *z* **is decreasing at 3.181 4 units per second**, correct to four decimal places.

Problem 2. The radius of a right circular cylinder is increasing at a rate of 2 cm s^{-1} and the height is decreasing at a rate of 3 cm s^{-1}. Find the rate at which the volume is changing when the radius is 8 cm and the height is 5 cm.

The volume of a right cylinder, $V = \pi r^2 h$.

From equation (6), $\dfrac{dV}{dt} = \dfrac{\partial V}{\partial r} \dfrac{dr}{dt} + \dfrac{\partial V}{\partial h} \dfrac{dh}{dt}$

$$\frac{\partial V}{\partial r} = 2\pi rh \text{ and } \frac{\partial V}{\partial h} = \pi r^2$$

Since the radius is increasing at 2 cm s^{-1}, $\dfrac{dr}{dt} = 2$.

Since the height is decreasing at 3 cm s^{-1}, $\dfrac{dh}{dt} = -3$.

Thus, rate of change of volume, $\dfrac{dV}{dt} = (2\,\pi\,rh)\,(2) + \pi\,r^2\,(-3)$.

When $r = 8$ and $h = 5$, $\dfrac{dV}{dt} = 4\,\pi\,(8)\,(5) - 3\,\pi\,(8)^2$

$$= 160\,\pi - 192\,\pi$$
$$= -32\,\pi$$

i.e. **the volume is decreasing at a rate of 32 π cm^3 s^{-1}.**

Problem 3. A rectangular box has sides of length x mm, y mm and z mm. Sides x and y are expanding at rates of 1.5 and 2.5 mm s^{-1} respectively and side z is contracting at a rate of 0.25 mm s^{-1}. Find the rate of change of volume when x is 15 mm, y is 12 mm and z is 10 mm.

Volume of box, $\quad V = x\,y\,z$.

From equation (6), $\quad \dfrac{dV}{dt} = \dfrac{\partial V}{\partial x}\dfrac{dx}{dt} + \dfrac{\partial V}{\partial y}\dfrac{dy}{dt} + \dfrac{\partial V}{\partial z}\dfrac{dz}{dt}$

$\dfrac{\partial V}{\partial x} = y\,z$, $\dfrac{\partial V}{\partial y} = x\,z$ and $\dfrac{\partial V}{\partial z} = x\,y$

Also, $\dfrac{dx}{dt} = 1.5$, $\dfrac{dy}{dt} = 2.5$ and $\dfrac{dz}{dt} = -0.25$

Hence, $\dfrac{dV}{dt} = y\,z\,(1.5) + x\,z\,(2.5) + x\,y\,(-0.25)$

When $x = 15, y = 12$ and $z = 10$,

$\dfrac{dV}{dt} = 12 \times 10 \times 1.5 + 15 \times 10 \times 2.5 - 15 \times 12 \times 0.25$

$= +510,$

i.e. **the rate of increase of volume is 510 mm^3 s^{-1}.**

Further problems on rates of change may be found in Section 4, (Problems 6 to 10), page 215.

3 Small changes

It is shown in Section 1, equation (1), that for a right circular cylinder being heated, the change of volume is given by

$$\delta V = 2\,\pi\,rh\delta r + \pi\,h\,\delta r^2 + \pi\,r^2\delta h + 2\,\pi\,r\delta r\delta h + \pi\,\delta r^2\,\delta h$$

If δr and δh are small compared with r and h, then products of these small terms such as δr^2, $\delta r\delta h$ and $\delta r^2\delta h$ will be very small and may be neglected. Thus the change of volume simplifies to:

$$\delta V \simeq 2\,\pi\,rh\delta r + \pi\,r^2\delta h$$

But, $V = \pi r^2 h$, so $\dfrac{\partial V}{\partial r} = 2\pi r h$ and $\dfrac{\partial V}{\partial h} = \pi r^2$.

Hence, $\delta V \simeq \dfrac{\partial V}{\partial r}\,\delta r + \dfrac{\partial V}{\partial h}\,\delta h$

This result may be generalised and if $z = f(u, v, w, \ldots)$, then the approximate change in z, δz, for small changes in u, v, and w, denoted by δu, δv and δw respectively, is given by

$$\delta z \simeq \frac{\partial z}{\partial u}\,\delta u + \frac{\partial z}{\partial v}\,\delta v + \frac{\partial z}{\partial w}\,\delta w + \cdots \tag{7}$$

Worked problems on small changes

Problem 1. The current in an electrical circuit is given by $i = \dfrac{v}{R}$ amperes.

Determine the approximate change in current if the voltage falls from 240 V to 238 V and the value of the resistance is increased from 100 Ω to 100.5 Ω.

The relationship is $i = f(v, R)$ and $i = \dfrac{v}{R}$.

From equation (7), the change in current $\delta i \simeq \dfrac{\partial i}{\partial v}\,\delta v + \dfrac{\partial i}{\partial R}\,\delta R$.

Since $i = \dfrac{v}{R}$, $\dfrac{\partial i}{\partial v} = \dfrac{1}{R}$ and $\dfrac{\partial i}{\partial R} = -\dfrac{v}{R^2}$.

The small change in voltage, δv is -2 volts and change in resistance δR is $+0.5$ ohms.

Thus, $\delta i = \dfrac{1}{R}(-2) + \left(-\dfrac{v}{R^2}\right)(0.5)$.

But $R = 100\ \Omega$ and $v = 240$ volts (the initial values),

hence, $\delta i = -\dfrac{2}{100} - \dfrac{240}{2 \times 100^2} = -0.032$.

That is, the approximate change in current is a decrease of 0.032 A.

Problem 2. The pressure p and volume V of a gas are related by the equation $pV^{1.4} = C$.

Find the approximate percentage change in C when the pressure is increased by 2.3 per cent and the volume is decreased by 0.84 per cent.

Let p, V and C refer to the initial values.

From equation (7), $\delta C = \dfrac{\partial C}{\partial p}\,\delta p + \dfrac{\partial C}{\partial V}\,\delta V$.

Since $C = f(p, V)$ and $C = p\,V^{1.4}$, then

$\dfrac{\partial C}{\partial p} = V^{1.4}$ and $\dfrac{\partial C}{\partial V} = 1.4\,p\,V^{0.4}$.

The pressure is increased by 2.3 per cent, i.e. the change in pressure

$$\delta p = \frac{2.3}{100} \cdot p = 0.023 \, p.$$

The volume is decreased by 0.84 per cent, i.e. the change in volume

$$\delta V = -\frac{0.84}{100} \cdot V = -0.008 \, 4 \, V.$$

Hence, the approximate change in C,

$$\delta C \simeq V^{1.4} \, (0.023 \, p) + 1.4 \, p \, V^{0.4} \, (-0.008 \, 4 \, V)$$

$$\simeq p \, V^{1.4} \, (0.023 - 1.4 \, (0.008 \, 4))$$

$$\simeq 0.011 \, 2 \, p \, V^{1.4}$$

i.e. $\delta C \simeq \dfrac{1.12}{100} \cdot C$

That is, the approximate change in C is a 1.12 per cent increase.

Problem 3. The side a of a triangle is calculated from:

$a^2 = b^2 + c^2 - 2 \, bc \cos A$. If b, c and A are measured as 2 mm, 4 mm and $60°$ respectively and the measurement errors which occur are $+ 0.1$ mm, $+ 0.15$ mm and $+ 2°$ respectively, determine the error in the calculated value of a

Since $a^2 = b^2 + c^2 - 2 \, bc \cos A$

$$a = (b^2 + c^2 - 2 \, bc \cos A)^{\frac{1}{2}}$$

From equation (7), $\delta a \simeq \dfrac{\partial a}{\partial b} \, \delta b + \dfrac{\partial a}{\partial c} \, \delta c + \dfrac{\partial a}{\partial A} \, \delta A$.

$$\frac{\partial a}{\partial b} = \frac{1}{2} \, (b^2 + c^2 - 2 \, bc \cos A)^{-\frac{1}{2}} \, (2b - 2c \cos A)$$

$$\frac{\partial a}{\partial c} = \frac{1}{2} \, (b^2 + c^2 - 2bc \cos A)^{-\frac{1}{2}} \, (2c - 2b \cos A)$$

and $\dfrac{\partial a}{\partial A} = \dfrac{1}{2} \, (b^2 + c^2 - 2bc \cos A)^{-\frac{1}{2}} \, (2bc \sin A)$

Since $b = 2$, $c = 4$ and $A = 60° = \dfrac{\pi}{3}$ rad.

$$b^2 + c^2 - 2bc \cos A = 4 + 16 - 2 \times 8 \times \frac{1}{2}$$

$$= 12$$

and $\dfrac{1}{2} \, (b^2 + c^2 - 2bc \cos A)^{-\frac{1}{2}} = \dfrac{1}{2\sqrt{12}} \simeq 0.144 \, 3$

$$\therefore \quad \frac{\partial a}{\partial b} \:\hat=\: 0.144\,3\,(4-4) = 0$$

$$\frac{\partial a}{\partial c} \:\hat=\: 0.144\,3\,(8-2) \:\hat=\: 0.866$$

$$\frac{\partial a}{\partial A} \:\hat=\: 0.144\,3\,(16\,\frac{\sqrt{3}}{2}\,) \:\hat=\: 1.999$$

Also $\delta b = 0.1$, $\delta c = 0.15$ and $\delta A = 2° = \dfrac{2 \times \pi}{180} = 0.035$ rad.

Hence, $\delta a \:\hat=\: 0 + 0.866 \times 0.15 + 1.999 \times 0.035$,
$$\hat=\: 0.199 \text{ mm}$$

i.e. **the approximate error in the calculated value of a is $+ 0.20$ mm.**

Further problems on small changes may be found in the following Section (4) (Problems 11 to 22).

4 Further problems

Total differential

1. Determine the total differential, dz, given $x = f(x, y)$ and
   $$z = 3xy^2 - \frac{\sqrt{x}}{y^3}. \quad \left[dz = (3y^2 - \frac{1}{2\sqrt{x}\,y^3}) \, dx + (6\,xy + \frac{3\sqrt{x}}{y^4}) \, dy \right]$$

2. Find the total differential, dz, if $z = f(r, \theta)$ and $z = e^r (\cos \theta + j \sin \theta)$.
   $$[dz = e^r (\cos \theta + j \sin \theta) \, dr + e^r (- \sin \theta + j \cos \theta) \, d\theta]$$

3. If $z = f(u, v, w)$ and $z = 2u^3 + 3v^2 - 4w$, find the total differential, dz.
   $$[dz = 6u^2 \, du + 6v dv - 4dw]$$

4. If $z = p\,e^q + \dfrac{1}{p} \ln q$, where both p and q are variables, show that
   $$dz = (e^q - \frac{1}{p^2} \ln q) \, dp + (p\,e^q + \frac{1}{pq}) \, dq.$$

5. If $xyz = k$, where k is a constant, prove that $dz = -z \left(\dfrac{dx}{x} + \dfrac{dy}{y} \right)$.

Rates of change

6. If $z = 2p^3 \tan \dfrac{q}{2}$, p is decreasing at 5 units per second and q is decreasing at 0.04 units per second, find the rate of change of z when $p = 1$ and $q = \pi/4$, correct to two decimal places.   [z is decreasing at 12.47 units per second]

7. Determine the rate of change of m, correct to four significant figures, given the following data:
   $$m = f(u, v\ w)$$

$$m = 3 \ln \frac{v}{3} + w^3 \sec u$$

v is decreasing at 0.4 cm s^{-1}
u is increasing at 0.2 cm s^{-1}
w is increasing at 0.7 cm s^{-1}
$u = 3$ cm, $v = 3.7$ cm and $w = 4.4$ cm.
[m is decreasing at 38.94 cm s^{-1}, correct to four significant figures]

8. If $z = f(x, y)$ and $z = e^{\frac{y}{2}} \ln (2x + 3y)$, find the rate of increase of z, correct to four significant figures, when $x = 2$ cm, $y = 3$ cm, x is increasing at 5 cm s^{-1} and y is increasing at 4 cm s^{-1}. [30.58 cm s^{-1}]

9. The radius of a right circular cone is increasing at a rate of 1 mm s^{-1} and its height is increasing at 2 mm s^{-1}. Determine, correct to three decimal places, the rate at which the volume is increasing in cm^3 s^{-1} when the radius is 1.2 cm and the height is 3.6 cm. [1.206 cm^3 s^{-1}]

10. The area of a triangle is given by: area $= \frac{1}{2} bc \sin A$, where A is the angle between sides b and c. If b is increasing at 0.4 units s^{-1}, c is decreasing at 0.25 units s^{-1} and A is increasing at 0.1 units s^{-1}, find the rate of change of area of the triangle, correct to four decimal places when b is 2 units, c is 3 units and A is 0.5 units.
[Increasing at 0.431 1 units s^{-1}]

Small changes

11. An error of 3.5 per cent too large is made when measuring the radius of a sphere. Determine the approximate error in calculating: (a) the volume and (b) the surface area when they are calculated using the correct radius measurement. (a) [10.5 per cent too large]
(b) [7 per cent too large]

12. The area of a triangle is given by $A = \frac{1}{2} a b \sin C$, where C is the angle between sides a and b of the triangle. Calculate the approximate change in area when: (a) both a and b are increased by 2 per cent and (b) when a is increased by 2 per cent and b is reduced by 2 per cent.
(a) [4 per cent increase]   (b) [no change.]

13. The moment of inertia of a body about an axis is given by $I = k b d^3$ where k is a constant and b and d are the dimensions of the body. If b and d are measured as 2 m and 0.8 m respectively and the measurement errors are 10 cm in b and -8 mm in d, determine the error in the calculated value of the moment of inertia using the measured values, in terms of k. [0.02 k m^4]

14. The radius of a cone is reduced from 10 cm to 9.55 cm and its height is increased from 20 cm to 20.3 cm. Determine the approximate percentage change in its volume. [7.5 per cent reduction]

15. The power developed by an engine is given by $I = b\,PLAN$, where b is a constant. Find the approximate change in power in terms of b (Pa m^3 min^{-1} units) when:

P is increased from 2×10^5 to 2.1×10^5 Pa
L is reduced from 8.3 to 8.1 cm
A is reduced from 2.7 cm^2 to 2.6 cm^2, and
N is increased from 300 to 302 rev min^{-1}.
[A reduction of $6b$ (Pa m^3 min^{-1}) units]

16. The modulus of rigidity G is given by $G = \dfrac{R^4 \theta}{L}$, where R is the radius,
θ the angle of twist and L the length. Find the approximate percentage
change in G when R is increased by 1.5 per cent and θ is reduced by 5 per
cent. [An increase of 1 per cent]

17. In triangle ABC, AB = 50 mm, AC = 60 mm and angle BAC = 45°. If
length can be measured correct to the nearest 0.1 mm, and angles correct
to the nearest 0.5°, determine the approximate value of maximum error
which can occur in the calculated value of BC based on the measured
results. [0.5 mm]

18. Side b of a triangle is calculated using the cosine rule. Determine the
approximate error in b using the measured values if the measured values
of a, c and B are 120 mm, 180 mm and 32° respectively, and the actual
values are 121 mm, 179 mm and 32.25° respectively. [−0.6 mm]

19. The power consumed in an electrical resistor is given by $P = \dfrac{E^2}{R}$ watts,
where E is the voltage drop across the resistor in volts and R the resistance
of the resistor in ohms. Determine the approximate change in power
when E increases by 4 per cent and R decreases by 0.05 per cent, if the
original values of E and R are 100 volts and 10 ohms respectively.
[An increase of 79.5 watts]

20. The resonant frequency of a series-connected electrical circuit is given by
$f = \dfrac{1}{2 \pi \sqrt{(LC)}}$ Hz, where L is the inductance in henrys and C the
capacitance in farads. Determine the approximate percentage change in
the resonant frequency when L is increased by 1.7 per cent and C is
reduced by 3.4 per cent. [An increase of 0.85 per cent.]

21. The rate of flow of gas in a pipe is given by $v = C d^{\frac{1}{2}} T^{-\frac{5}{6}}$, where C is a
constant, d is the diameter of the pipe and T is the thermodynamic
temperature of the gas. When determining the rate of flow experimentally,
d is measured and subsequently found to be in error by + 1.6 per cent,
and T has an error of −1.2 per cent. Determine the percentage error in
the rate of flow based on the measured values of d and T.
[1.8 per cent too large]

22. The volume (V) of a liquid of viscosity coefficient (η) delivered after
time, t, when passed through a tube of length l and diameter d by a
pressure p is given by:
$$V = \frac{p d^4 t}{128 \eta l}$$
If the error in V, p and l are 1 per cent, 2 per cent and 3 per cent respectively, determine the error in η. [−2 per cent]

Chapter 17

Maxima, minima and saddle points for functions of two variables (CC 21)

1 Functions of two independent variables

If a relation between two real variables, x and y, is such that when x is given, y is determined, then y is said to be a function of x and is usually denoted by $y = f(x)$. x is called the independent variable and y the dependent variable. However, it is not necessary that y should be a function of only one independent variable. If $y = f(u, v)$, then y is said to be a function of two independent variables u and v. The value of y is only determined when definite values are given to both u and v.

For example, let $y = f(u, v) = 2u^2 + 3v$. When $u = 1$ and $v = 2$ then $y = 2(1)^2 + 3(2) = 8$. This may be written as $f(1, 2) = 8$.

Similarly, if $(u, v) = (2, 5)$, then $f(2, 5) = 23$. Consider a function of two variables x and y defined by $z = f(x, y) = 3x^2 - 2y$.

Then if $(x, y) = (0, 0)$ we have $f(0, 0) = 0$
and if $(x, y) = (1, 2)$ we have $f(1, 2) = -1$.

Every pair of numbers, (x, y), may be represented by a point A in the (x, y) plane of a rectangular Cartesian co-ordinate system as shown in Fig. 1(a).

The corresponding value of $z = f(x, y)$ may be represented by the length of the line AA' drawn parallel to the z-axis. Thus if, for example, $z = 3x^2 - 2y$, as above, and A is the co-ordinate $(2, 3)$ then the length of AA' is $3(2)^2 - 2(3)$, i.e. 6. Figure 1(b) shows that when a large number of (x, y) co-ordinates are taken for a function $f(x, y)$ and then $f(x, y)$ calculated for each, a large number of lines such as AA' can be constructed, and, in the limit, when all points in the (x, y) plane are considered, a surface is seen to result (shown by

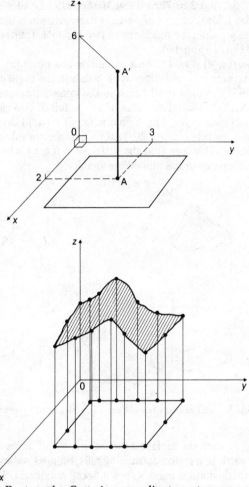

Fig 1 (a) Rectangular Cartesian co-ordinate system representing

$f(x, y) = 3x^2 - 2y$ at $f(2, 3)$.

(b) Representing $f(x, y)$ as a surface.

the shaded section in Fig. 1(b)). Thus the function $z = f(x, y)$ represents a surface and not a curve.

2 Maxima, minima and saddle points

When finding maximum and minimum values for functions of one variable the function is differentiated, equated to zero and the resulting equation solved. This gives the co-ordinates of the stationary points. To determine the nature of the turning points the second differential coefficient may be obtained (see Chapter 5, Section 2 of *Technician Mathematics Level 3* by J. O. Bird and A. J. C. May). The concept of partial differentiation is used when determining stationary points for functions of two variables (partial differentiation is discussed in Chs. 15 and 16).

A function $f(x, y)$ is said to be a maximum at a point (x, y) if the value of the function there is greater than at all points in the immediate vicinity and a minimum if less than at all points in the immediate vicinity. Figure 2(a) shows geometrically a maximum value of a function of two variables and it is seen that the surface $z = f(x, y)$ is higher at $(x, y) = (\alpha, \beta)$ than at any point in the immediate vicinity. Figure 2(b) shows a minimum value of a function of two variables and it is seen that the surface $z = f(x, y)$ is lower at $(x, y) = (\alpha, \beta)$ than at any point in the immediate vicinity.

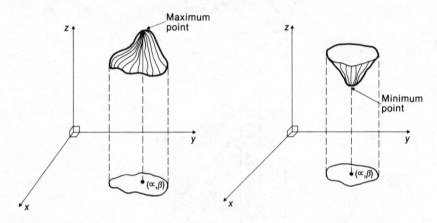

Fig. 2   (a) Maximum value.   (b) Minimum value.

If $z = f(x, y)$, then the surface z can be thought of as resembling the surface of the earth in a region containing hills, hollows, valleys, mountains and passes. A mathematical maximum is a 'local' maximum. The highest point of a small hillock in a valley bordered by high hills is a maximum although it lies below the surrounding hills. Similarly, a hollow high up in the hills gives a minimum. Whenever there is a lake, the lowest point of the bed of the lake is considered as a minimum point.

If $z = f(x, y)$ and a maximum occurs at (α, β), the curves lying in the two planes $x = \alpha$ and $y = \beta$ must also have a maximum at (α, β) as shown in Fig. 3. Consequently the tangents t_1 and t_2 to those curves at (α, β) must be parallel to Ox and Oy respectively. This requires:

$$\frac{\partial z}{\partial x} = 0 \text{ and } \frac{\partial z}{\partial y} = 0$$

at all maximum and minimum points. The solution of these two equations give the stationary (or critical) points of z.

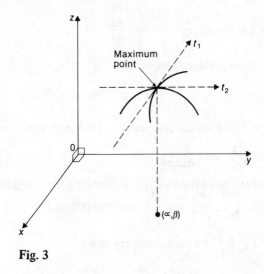

Fig. 3

With functions of two variables there are three types of stationary points possible, these being a maximum point, a minimum point and a saddle point. A saddle point S is shown in Fig. 4 and is such that the point S is a maximum for curve 1 and a minimum for curve 2. A saddle point may be thought of as 'a pass between two mountains'.

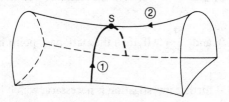

Fig. 4 A saddle point.

3 Procedure to find maxima, minima and saddle points for functions of two variables

The criteria for the existence of maxima, minima and saddle points may be deduced from Taylor's theorem for functions of two independent variables. The deductions are somewhat more sophisticated than in the simpler case of just one variable. The proof of Taylor's theorem for functions of two variables is beyond the scope of this text and we will use and accept the relevant results in the following procedure.

Given $z = f(x, y)$:

1. Find $\dfrac{\partial z}{\partial x}$ and $\dfrac{\partial z}{\partial y}$.

2. For stationary points, $\dfrac{\partial z}{\partial x} = 0$ and $\dfrac{\partial z}{\partial y} = 0$.

3. Solve the two simultaneous equations: $\dfrac{\partial z}{\partial x} = 0$

$$\text{and} \quad \dfrac{\partial z}{\partial y} = 0$$

for x and y. This gives the co-ordinates of the stationary points.

4. Find $\dfrac{\partial^2 z}{\partial x^2}$, $\dfrac{\partial^2 z}{\partial y^2}$ and $\dfrac{\partial^2 z}{\partial x \partial y}$.

5. For each of the co-ordinates of the stationary points substitute values of x and y into $\dfrac{\partial^2 z}{\partial x^2}$, $\dfrac{\partial^2 z}{\partial y^2}$ and $\dfrac{\partial^2 z}{\partial x \partial y}$ and evaluate each.

6. Evaluate $\left(\dfrac{\partial^2 y}{\partial x \partial y}\right)^2$ for each stationary point.

7. Substitute the values of $\dfrac{\partial^2 z}{\partial x^2}$, $\dfrac{\partial^2 z}{\partial y^2}$ and $\dfrac{\partial^2 z}{\partial x \partial y}$ into the equation:

$$\Delta = \left(\dfrac{\partial^2 z}{\partial x \partial y}\right)^2 - \left(\dfrac{\partial^2 z}{\partial x^2}\right)\left(\dfrac{\partial^2 z}{\partial y^2}\right) \quad \text{and evaluate } \Delta.$$

8. (i) If $\Delta > 0$, then the stationary point is **a saddle point**.

   (ii) If $\Delta < 0$, and $\dfrac{\partial^2 z}{\partial x^2} < 0$, then the stationary point is **a maximum point**.

   (iii) If $\Delta < 0$, and $\dfrac{\partial^2 z}{\partial x^2} > 0$, then the stationary point is **a minimum point**.

(Note that if $\Delta = 0$ further investigation is necessary, which is beyond the scope of this text.)

Worked problems on finding maxima, minima and saddle points for functions of two variables

Problem 1. Show that the function $z = x^2 + y^2$ has one stationary point only and determine its nature. Sketch the surface represented by z.

Following the above procedure:

1. $\dfrac{\partial z}{\partial x} = 2x$; $\dfrac{\partial z}{\partial y} = 2y$.

2. $2x = 0$ \hfill (1)

   $2y = 0$ \hfill (2)

3. From equations (1) and (2), $x = 0$ and $y = 0$. Hence only one stationary point exists at (0,0).

4. $\dfrac{\partial^2 z}{\partial x^2} = 2; \quad \dfrac{\partial^2 z}{\partial y^2} = 2; \quad \dfrac{\partial^2 z}{\partial x \partial y} = 0.$

5. $\dfrac{\partial^2 z}{\partial x^2} = \dfrac{\partial^2 z}{\partial y^2} = 2$ and $\dfrac{\partial^2 z}{\partial x \partial y} = 0.$

6. $\left(\dfrac{\partial^2 z}{\partial x \partial y}\right)^2 = 0.$

7. $\Delta = (0)^2 - (2)(2) = -4.$

8. Since $\Delta < 0$ and $\dfrac{\partial^2 z}{\partial x^2} > 0$, **the stationary point $(0, 0)$ is a minimum.**

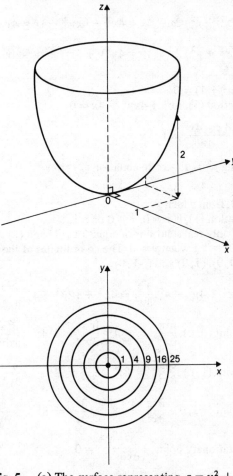

Fig. 5 (a) The surface representing $z = x^2 + y^2$.
(b) A contour map for $z = x^2 + y^2$.

The surface $z = x^2 + y^2$ is shown in three dimensions in Fig. 5(a). Looking down towards the xy plane from above it is possible to produce a **contour map**. A contour is a line on a map which gives places having the same vertical height above a datum line (usually the mean sea-level on a geographical map). A contour map for $z = x^2 + y^2$ is shown in Fig. 5(b). The values of z are shown on the map and these give an indication of the steepness of the rise or fall to the stationary point.

Problem 2. Determine the stationary points on the surface $z = (x^2 + y^2)^2 - 2(x^2 - y^2)$ and distinguish between them. Sketch the approximate contour map associated with z.

Following the procedure given on page 221:

1. $\dfrac{\partial z}{\partial x} = 2(x^2 + y^2) 2x - 4x = 4x^3 + 4xy^2 - 4x = 4x(x^2 + y^2 - 1)$

    $\dfrac{\partial z}{\partial y} = 2(x^2 + y^2) 2y + 4y = 4x^2 y + 4y^3 + 4y = 4y(x^2 + y^2 + 1)$

2. $4x(x^2 + y^2 - 1) = 0$ (1)

    $4y(x^2 + y^2 + 1) = 0$ (2)

3. From equation (1), $4x^3 + 4xy^2 - 4x = 0$

    i.e. $y^2 = \dfrac{4x - 4x^3}{4x} = 1 - x^2$

    Substituting $y^2 = 1 - x^2$ in equation (2) gives

    $4y(x^2 + 1 - x^2 + 1) = 0$
    i.e. $8y = 0$, from which $y = 0$.
    From equation (1), if $y = 0, x = 0$ or ± 1.
    Hence the only real solutions of equations (1) and (2) are $x = 0$ when $y = 0$ and $x = \pm 1$ when $y = 0$. **The co-ordinates of the stationary points are thus $(0, 0)$, $(1, 0)$ and $(-1, 0)$.**

4. $\dfrac{\partial^2 z}{\partial x^2} = 12x^2 + 4y^2 - 4; \quad \dfrac{\partial^2 z}{\partial y^2} = 4x^2 + 12y^2 + 4; \quad \dfrac{\partial^2 z}{\partial x \partial y} = 8xy.$

5. For the point $(0, 0)$, $\dfrac{\partial^2 z}{\partial x^2} = -4; \quad \dfrac{\partial^2 z}{\partial y^2} = +4; \quad \dfrac{\partial^2 z}{\partial x \partial y} = 0.$

    For the point $(1, 0)$, $\dfrac{\partial^2 z}{\partial x^2} = 8; \quad \dfrac{\partial^2 z}{\partial y^2} = 8; \quad \dfrac{\partial^2 z}{\partial x \partial y} = 0.$

    For the point $(-1, 0)$, $\dfrac{\partial^2 z}{\partial x^2} = 8; \quad \dfrac{\partial^2 z}{\partial y^2} = 8; \quad \dfrac{\partial^2 z}{\partial x \partial y} = 0.$

6. For each stationary point, $\left(\dfrac{\partial^2 z}{\partial x \partial y}\right)^2 = 0.$

7. $\Delta_{(0, 0)} = (0)^2 - (-4)(4) = +16$
    $\Delta_{(1, 0)} = (0)^2 - (8)(8) = -64$
    $\Delta_{(-1, 0)} = (0)^2 - (8)(8) = -64$

8. Since $\Delta_{(0, 0)} > 0$, the point $(0, 0)$ is a **saddle point**.

Since $\Delta_{(1,0)} < 0$ and $\left(\dfrac{\partial^2 z}{\partial x^2}\right)_{(1,0)} > 0$, the point $(1,0)$ is a maximum point.

Since $\Delta_{(-1,0)} < 0$ and $\left(\dfrac{\partial^2 z}{\partial x^2}\right)_{(-1,0)} > 0$, the point $(-1,0)$ is a minimum point.

Looking down towards the xy plane from above an approximate contour map may be constructed to represent the values of z. Such a map is shown in Fig. 6. To produce a contour map requires a large number of xy co-ordinates to be chosen and the value of z at each co-ordinate calculated. This can be a time-consuming exercise.

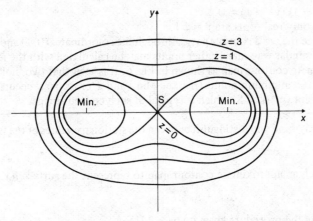

Fig. 6 Approximate contour map for $z = (x^2 + y^2)^2 - 2(x^2 - y^2)$.

Some points used to construct the contour $z = 0$ in Fig. 6 include:

(a) When $z = 0, 0 = (x^2 + y^2)^2 - 2(x^2 - y^2)$.
Also, in addition when $y = 0$, (i.e. on the x-axis), $0 = x^4 - 2x^2$
i.e. $x^2(x^2 - 2) = 0$
from which $x = 0$ or $x = \pm\sqrt{2}$
Hence the contour $z = 0$ crosses the x-axis at 0 and $\pm\sqrt{2}$, i.e. at co-ordinates $(0,0)$, $(1.414,0)$ and $(-1.414,0)$.

(b) When $z = 0$ and $x = 1$ then $0 = (1 + y^2)^2 - 2(1 - y^2)$
$$0 = 1 + 2y^2 + y^4 - 2 + 2y^2$$

$y^4 + 4y^2 - 1 = 0$
Let $y^2 = p$, then $p^2 + 4p - 1 = 0$
$$p = \frac{-4 \pm \sqrt{[4^2 - 4(-1)]}}{2} = \frac{-4 \pm \sqrt{20}}{2}$$

$p = 0.236$ or -4.236.

But $y = \sqrt{p} = \sqrt{0.236}$ (or $\sqrt{-4.236}$, which gives complex roots)
i.e. $y = \pm 0.486$.

Hence the $z = 0$ contour passes through the co-ordinates $(1, 0.486)$ and $(1, -0.486)$.

Similarly for the $z = 3$ contour:

(c) When $y = 0$ (i.e. on the x-axis), $3 = x^4 - 2x^2$
i.e. $x^4 - 2x^2 - 3 = 0$
$(x^2 - 3)(x^2 + 1) = 0$
The only real roots are $x = \pm\sqrt{3}$.
Hence the $z = 3$ contour passes through the co-ordinates $(1.732,0)$ and $(-1.732,0)$.

(d) When $x = 0$, (i.e. on the y-axis), $3 = y^4 + 2y^2$
i.e. $y^4 + 2y^2 - 3 = 0$
$(y^2 + 3)(y^2 - 1) = 0$
The only real roots are $y = \pm 1$.
Hence the $z = 3$ contour passes through the co-ordinates $(0, 1)$ and $(0, -1)$.

In a similar way many other points may be calculated with the resulting approximate contour map as shown in Fig. 6. It is seen that two 'hollows' occur at the minimum points (marked Min.) and a 'cross-over' occurs at the saddle point (marked S) which is typical of such contour maps.

Problem 3. Find and distinguish between the stationary values of the function

$$f(x, y) = x^3 - 6x^2 - 8y^2$$

and sketch an approximate contour map to represent the surface $f(x, y)$.

Let $z = f(x, y) = x^3 - 6x^2 - 8y^2$.

Following the procedure given on page 221:

1. $\dfrac{\partial z}{\partial x} = 3x^2 - 12x$; $\dfrac{\partial z}{\partial y} = -16y$.

2. $3x^2 - 12x = 0$ (1)
   $-16y = 0$ (2)

3. From equation (1), $3x(x - 4) = 0$, from which $x = 0$ or $x = 4$.
   From equation (2), $y = 0$.
   Hence the stationary points are at $(0, 0)$ and $(4, 0)$.

4. $\dfrac{\partial^2 z}{\partial x^2} = 6x - 12$; $\dfrac{\partial^2 z}{\partial y^2} = -16$; $\dfrac{\partial^2 z}{\partial x \partial y} = 0$.

5. For the point $(0, 0)$, $\dfrac{\partial^2 z}{\partial x^2} = -12$; $\dfrac{\partial^2 z}{\partial y^2} = -16$; $\dfrac{\partial^2 z}{\partial x \partial y} = 0$

   For the point, $(4, 0)$, $\dfrac{\partial^2 z}{\partial x^2} = 12$; $\dfrac{\partial^2 z}{\partial y^2} = -16$; $\dfrac{\partial^2 z}{\partial x \partial y} = 0$

6. For both stationary points $\left(\dfrac{\partial^2 z}{\partial x \partial y}\right)^2 = 0$.

7. $\Delta_{(0,0)} = (0)^2 - (-12)(-16) = -192$
   $\Delta_{(4,0)} = (0)^2 - (12)(-16) = +192$

8. Since $\Delta_{(0,0)} < 0$ and $\left(\dfrac{\partial^2 z}{\partial x^2}\right)_{(0,0)} < 0$, **the point (0,0) is a maximum point.**

Since $\Delta_{(4,0)} > 0$, **the point (4, 0) is a saddle point.**
The maximum value of $f(x, y)$ is 0 and the value of $f(x, y)$ at the saddle point $(4, 0)$ is $(4)^3 - 6 (4)^2 - 8 (0)^2 = -32$

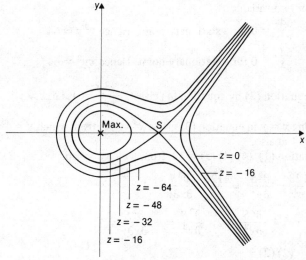

Fig. 7   Approximate contour map for $f(x, y) = x^3 - 6x^2 - 8y^2$.

An approximate contour map representing the surface $f(x, y)$ is shown in Fig. 7 where a 'hollow' effect is seen surrounding the maximum point (marked Max.) and a 'cross-over' occurs at the saddle point (marked S). The position of the contours may be calculated in a similar way to problem 2.

Problem 4. An open rectangular container is to have a volume of $32m^3$. Determine the dimensions and the total surface area so that the total surface area is a minimum.

Let the dimensions of the container be x, y and z, as shown in Fig. 8.

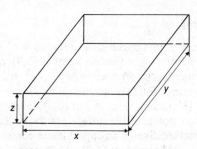

Fig. 8

Volume $V = xyz = 32$ (1)

Surface area $S = xy + 2yz + 2xz$ (2)

From equation (1), $z = \dfrac{32}{xy}$

Substituting for z in equation (2) gives: $S = xy + \dfrac{64}{x} + \dfrac{64}{y}$, which is a function of two variables.

$\dfrac{\partial S}{\partial x} = y - \dfrac{64}{x^2} = 0$ for a stationary point. Hence $x^2 y = 64$ (3)

$\dfrac{\partial S}{\partial y} = x - \dfrac{64}{y^2} = 0$ for a stationary point. Hence $xy^2 = 64$ (4)

Dividing equation (3) by equation (4) gives: $\dfrac{x^2 y}{xy^2} = 1$, i.e. $x = y$.

Substituting $x = y$ in equation (3) gives: $x^3 = 64$, from which $x = 4$ m.
Hence $y = 4$ m also.
From equation (1), $(4)(4)(z) = 32$, from which $z = 2$m.

$\dfrac{\partial^2 S}{\partial x^2} = \dfrac{128}{x^3}$; $\dfrac{\partial^2 S}{\partial y^2} = \dfrac{128}{y^3}$, $\dfrac{\partial^2 S}{\partial x \partial y} = 1.$

When $x = y = 4$, $\dfrac{\partial^2 S}{\partial x^2} = 2$; $\dfrac{\partial^2 S}{\partial y^2} = 2$; $\dfrac{\partial^2 S}{\partial x \partial y} = 1.$

$\Delta = (1)^2 - (2)(2) = -3.$

Since $\Delta < 0$ and $\dfrac{\partial^2 S}{\partial x^2} > 0$ then the surface area S is a minimum.

Hence the minimum dimensions of the container to have a volume of 32 m³ are 4 m by 4 m by 2 m.
From equation (2), **minimum surface area** $S = (4)(4) + 2(4)(2) + 2(4)(2)$ $= 48$m^2 .

Further problems on finding maxima, minima and saddle points for functions of two variables may be found in the following Section (4) (Problems 1 to 13).

4 Further problems

1. Find the stationary points of the surface $f(x, y) = x^3 - 6xy + y^3$. Determine their nature.   [Saddle point at (0,0); minimum at (2,2).]

2. Show that the function $f(x, y) = x^3 - 3x^2 - 4y^2 + 2$ has one saddle point and one maxima. Find the maximum value.
              [Maximum value = 2 at (0,0); saddle point at (2,0).]

3. Find the stationary points of the surface $z = 2x^3 - 3xy + \frac{3}{4}y^2$ and determine their nature. Sketch an approximate contour map to represent the surface z.   [Minimum at (1,2); saddle point at (0,0)]

4. Find the maximum value of the function $f(x, y) = x^3 - 3x^2 - 4y^2 + 2$. Sketch an approximate contour map of $f(x, y)$.

[Maximum value = 2 at (0,0) (Also, saddle point at (2,0)]

5. Find the minimum value of the function $z = x^3 + y^3 - 6xy$. Sketch an approximate contour map to represent z.

[Minimum value = -8 at (2,2)(Also, saddle point (0,0)]

6. An open rectangular container is to have a volume of 62.5 m^3. Find the least surface area of material required. [75 m^2]

7. Determine the location and nature of the stationary points of the following functions:

(a) $z = 24x^2 - 12xy + 30y^2$ (b) $z = 2x^3 - 6x^2 - 2y^2$.

[(a) [Minimum at (0,0)] (b) [Maximum at (0,0); saddle point at (2,0)]

8. Show that the function $z = (x^3 + y^3) - 2(x^2 + y^2) + 3xy$ has stationary values at the points (0,0) and $\left(\dfrac{1}{3} \cdot \dfrac{1}{3}\right)$ and investigate their nature.

$$\left[\text{Maximum at (0,0); saddle point at } \left(\frac{1}{3}, \frac{1}{3}\right) \right]$$

9. Determine the stationary points of the surface $f(x, y) = 2x^4 + 4x^2 y^2 - 4x^2 + 4y^2 - 2$ and distinguish between them. Find the value of $f(x, y)$ at the stationary points. [Minimum value = -4 at (1,0); minimum value = -4 at $(-1,0)$; saddle point of value -2 at (0,0)]

10. Show that the function $f(x, y) = x^3 + y^3 - 3x - 12y + 20$ has four stationary points and distinguish between them. [Minimum value = 2 at (1,2); maximum value = 38 at $(-1,-2)$; saddle points at $(-1,2)$ and $(1,-2)$]

11. Find the maximum and minimum values of $f(u, v) = 3u - u^3 - uv^2$

$$\left[\begin{array}{l} \text{Maximum value = 2 at (1,0)} \\ \text{Minimum value = } -2 \text{ at } (-1, 0) \end{array} \right]$$

12. A shelter is to be constructed in the form of a rectangular box-like space with canvas covering on the top, back and sides. Find the minimum surface area of canvas necessary if the volume of the shelter is to be 128 m^3.

[96 m^2]

13. Determine the minimum capacity C of a capacitor given that

$$C = \frac{ax}{x - a} + \frac{xy}{y - b} + \frac{yb}{b - y},$$

where a and b are fixed values and x and y vary independently such that

$a < x < y < b$. $\left[\dfrac{ab}{b - a}\right]$

Chapter 18

Integration using substitutions and partial fractions (CB 15)

1 Introduction

Revision of standard integrals.

1. $\int ax^n \, dx \quad\quad = \quad a\dfrac{x^{n+1}}{n+1} + c \text{ (except where } n = -1)$

2. $\int \cos ax \, dx \quad\quad = \quad \dfrac{1}{a} \sin ax + c$

3. $\int \sin ax \, dx \quad\quad = \quad -\dfrac{1}{a} \cos ax + c$

4. $\int \sec^2 ax \, dx \quad\quad = \quad \dfrac{1}{a} \tan ax + c$

5. $\int \operatorname{cosec}^2 ax \, dx \quad\quad = \quad -\dfrac{1}{a} \cot ax + c$

6. $\int \operatorname{cosec} ax \cot ax \, dx \quad = \quad -\dfrac{1}{a} \operatorname{cosec} ax + c$

7. $\int \sec ax \tan ax \, dx \quad\quad = \quad \dfrac{1}{a} \sec ax + c$

8. $\int e^{ax} \, dx \quad\quad = \quad \dfrac{1}{a} e^{ax} + c$

9. $\displaystyle\int \dfrac{1}{x} \, dx \quad\quad = \quad \ln ax + c \; (= \ln x + \ln a + c = \ln x + c')$

Each of the above standard integrals may be readily checked by differentiation. Functions which require integrating are not usually in the standard integral form shown above. In such cases it may be possible by one of four methods to change the function into a form which can be readily integrated.

The methods available are:

  (a) by using algebraic substututions,
  (b) by using trigonometrical and hyperbolic identities and substitutions,
  (c) by using partial fractions,
and   (d) integration by parts (see Chapter 19).

The first three methods are discussed in this chapter.

It should be realised that many mathematical functions cannot be integrated by analytical means at all and approximate methods have then to be used.

2 Integration using algebraic substitutions

An integral may often be reduced to a standard integral by making an algebraic substitution. The substitution usually made is to let u be equal to $f(x)$, such that $f(u)\,\mathrm{d}u$ is a standard integral.

A most important point in the use of substitution is that once it has been made the original variable must be removed completely, because a variable can only be integrated with respect to itself, i.e. we cannot integrate, for example, a function of t with respect to x.

A concept that $\dfrac{\mathrm{d}u}{\mathrm{d}x}$ is a single entity (indicating the change of u with respect to x) has been established in the work done on differentiation.

Frequently, in work on integration and on differential equations, $\dfrac{\mathrm{d}u}{\mathrm{d}x}$ is split. Provided that when this is done the original differential coefficient can be reformed by applying the rules of algebra, then it is in order to do so. For example, if $\dfrac{\mathrm{d}y}{\mathrm{d}x} = x$, then it is in order to write $\mathrm{d}y = x\,\mathrm{d}x$ since dividing both sides by $\mathrm{d}x$ re-forms the original differential coefficient. This principle is shown in the following worked problems.

Worked problems on integration by algebraic substitutions

Type 1

Problem 1. Find (a) $\int \sin(6x + 5)\,\mathrm{d}x$,   (b) $\int (3t - 2)^8\,\mathrm{d}t$

(c) $\displaystyle\int \frac{1}{(9x + 4)}\,\mathrm{d}x$

(a) $\int \sin(6x + 5)\,\mathrm{d}x$

   Let $u = 6x + 5$, then $\dfrac{\mathrm{d}u}{\mathrm{d}x} = 6$, i.e. $\mathrm{d}x = \dfrac{\mathrm{d}u}{6}$

   Hence $\int \sin(6x + 5)\,\mathrm{d}x = \int \sin u\,\dfrac{\mathrm{d}u}{6} = \dfrac{1}{6}\int \sin u\,\mathrm{d}u = \dfrac{1}{6}(-\cos u) + c$

Since the original integral is given in terms of x the result should be stated in terms of x.

Since $u = 6x + 5$, $\int \sin (6x + 5) \, dx = -\dfrac{1}{6} \cos (6x + 5) + c$

(b) $\int (3t - 2)^8 \, dt$

Let $u = 3t - 2$, then $\dfrac{du}{dt} = 3$, i.e. $dt = \dfrac{du}{3}$

Hence $\int (3t - 2)^8 \, dt = \int u^8 \, \dfrac{du}{3} = \dfrac{1}{3} \int u^8 \, du = \dfrac{1}{3} \left(\dfrac{u^9}{9} \right) + c = \dfrac{1}{27} u^9 + c$

Since $u = 3t - 2$, $\int (3t - 2)^8 \, dt = \dfrac{1}{27} (3t - 2)^9 + c$

(c) $\displaystyle\int \dfrac{1}{(9x + 4)} \, dx$

Let $u = 9x + 4$, then $\dfrac{du}{dx} = 9$, i.e. $dx = \dfrac{du}{9}$

Hence $\displaystyle\int \dfrac{1}{(9x + 4)} \, dx = \int \dfrac{1}{u} \, \dfrac{du}{9} = \dfrac{1}{9} \ln u + c = \dfrac{1}{9} \ln (9x + 4) + c$

It may be seen from problem 1 that:

If 'x' in a standard integral is replaced by $(ax + b)$, where a and b are constants, then $(ax + b)$ is written for x in the result and the result is multiplied by $\dfrac{1}{a}$.

In general, $\int (ax + b)^n \, dx = \dfrac{1}{a(n + 1)} (ax + b)^{n+1} + c$ [except when $n = -1$ (see (c) above)]

With practice integrals of this type may be determined by inspection. A check can always be made by differentiating the result.

Problem 2. Determine the following by inspection:

(a) $\int 2 \sin (4x - 1) \, dx$ (b) $\int 3 e^{7t-2} \, dt$ (c) $\displaystyle\int \dfrac{4}{6\theta + 7} \, d\theta$

(a) $\int 2 \sin (4x - 1) \, dx = \dfrac{2}{4} [- \cos (4x - 1)] + c = -\dfrac{1}{2} \cos (4x - 1) + c$

(b) $\int 3 e^{7t-2} \, dt = 3 (e^{7t-2})(\dfrac{1}{7}) + c = \dfrac{3}{7} e^{7t-2} + c$

(c) $\displaystyle\int \dfrac{4}{6\theta + 7} \, d\theta = 4 [\ln (6\theta + 7)] \, (\dfrac{1}{6}) + c = \dfrac{2}{3} \ln (6\theta + 7) + c$

Type 2.

Problem 3. Find (a) $\int 2x (3x^2 + 5)^6 \, dx$ (b) $\int \sin^3 \theta \cos \theta \, d\theta$

(a) $\int 2x (3x^2 + 5)^6 \, dx$

There appears to be two possible choices for the substitution, i.e. either to let $u = 2x$ or to let $u = (3x^2 + 5)^6$. Choosing the latter substitution, if $u = (3x^2 + 5)$, then $\dfrac{du}{dx} = 6x$, i.e. $dx = \dfrac{du}{6x}$

Hence $\int 2x\,(3x^2 + 5)^6\,dx = \int 2x\,(u)^6\,\dfrac{du}{6x} = \int \dfrac{2\,u^6}{6}\,du = \dfrac{1}{3}\int u^6\,du$

The original variable, x, has been removed completely and the integral is now only in terms of u.

Hence $\dfrac{1}{3}\int u^6\,du = \dfrac{1}{3}\left(\dfrac{u^7}{7}\right) + c = \dfrac{1}{21}\,(3x^2 + 5)^7 + c$

(b) $\int \sin^3 \theta \cos \theta\,d\theta$

Let $u = \sin \theta$, then $\dfrac{du}{d\theta} = \cos \theta$, i.e. $d\theta = \dfrac{du}{\cos \theta}$

Hence $\int \sin^3 \theta \cos \theta\,d\theta = \int u^3 \cos \theta\,\dfrac{du}{\cos \theta} = \int u^3\,du = \dfrac{u^4}{4} + c$

$\qquad\qquad = \dfrac{1}{4}\,\sin^4 \theta + c$

Whenever a power of a sine is multiplied by a cosine of power 1, or vice versa, the integral may be determined by inspection.

For example, $\int \cos^5 x \sin x\,dx = -\dfrac{1}{6}\cos^6 x + c$,

since $\dfrac{d}{dx}\left\{ -\dfrac{1}{6}\cos^6 x + c \right\} = \cos^5 x \sin x$.

It may be seen from problem 3 that:

Integrals of the form $k \int [f(x)]^n\,f'(x)\,dx$ (where k and n are constants) can be integrated by substituting u for $f(x)$.

Type 3
Problem 4. Find (a) $\displaystyle\int \dfrac{x}{1 + x^2}\,dx$ $\qquad$ (b) $\displaystyle\int \dfrac{4x}{\sqrt{(5x^2 - 1)}}\,dx$

(c) $\int \tan \theta\,d\theta$

(a) $\displaystyle\int \dfrac{x}{1 + x^2}\,dx$

Let $u = 1 + x^2$, then $\dfrac{du}{dx} = 2x$, i.e. $\qquad dx = \dfrac{du}{2x}$

Hence $\displaystyle\int \dfrac{x}{1 + x^2}\,dx = \int \dfrac{x}{u}\,\dfrac{du}{2x} = \dfrac{1}{2}\int \dfrac{1}{u}\,du \qquad = \dfrac{1}{2}\ln u + c$

$\qquad\qquad\qquad\qquad\qquad\qquad\qquad\qquad\qquad = \dfrac{1}{2}\ln (1 + x^2) + c$

(b) $\int \dfrac{4x}{\sqrt{(5x^2-1)}} \, dx$

Let $u = 5x^2 - 1$, then $\dfrac{du}{dx} = 10x$, i.e. $dx = \dfrac{du}{10x}$

Hence $\int \dfrac{4x}{\sqrt{(5x^2-1)}} \, dx = \int \dfrac{4x}{\sqrt{u}} \, \dfrac{du}{10x} = \dfrac{2}{5} \int u^{-1/2} \, du = \dfrac{2}{5} \dfrac{u^{1/2}}{1/2} + c$

$$= \dfrac{4}{5} \sqrt{u} + c$$

$$= \dfrac{4}{5} \sqrt{(5x^2-1)} + c$$

(c) $\int \tan \theta \, d\theta = \int \dfrac{\sin \theta}{\cos \theta} \, d\theta$

Let $u = \cos \theta$, then $\dfrac{du}{d\theta} = -\sin \theta$, i.e. $d\theta = -\dfrac{du}{\sin \theta}$

Hence $\int \dfrac{\sin \theta}{\cos \theta} \, d\theta = \int \dfrac{\sin \theta}{u} \left(\dfrac{-du}{\sin \theta} \right) = -\int \dfrac{1}{u} \, du = -\ln u + c$

$$= -\ln (\cos \theta) + c$$

$$= \ln (\cos \theta)^{-1} + c$$

i.e. $\int \tan \theta \, d\theta = \ln (\sec \theta) + c$

It may be seen from Problem 4 that:

Integrals of the form $k \int \dfrac{f'(x)}{[f(x)]^n} \, dx$ (where k and n are constants) can be intergrated by substituting u for $f(x)$.

If, in an integral, it is recognised that the numerator is the differential coefficient of the denominator (times a constant) then it may be integrated by inspection.

General problem on definite integrals

Problem 5. Evaluate the following: (a) $\int_1^3 3 \sec^2 (2\theta - 1) \, d\theta$
(b) $\int_0^3 3x \sqrt{(3x^2 + 9)} \, dx$, taking positive values of roots only.
(c) $\int_1^2 \dfrac{2e^t}{3 + e^t} \, dt$

(a) $\int_1^3 3 \sec^2 (2\theta - 1) \, d\theta = \left[\dfrac{3}{2} \tan (2\theta - 1) \right]_1^3 = \dfrac{3}{2} [\tan 5 - \tan 1]$

$$= \dfrac{3}{2} [\tan 286° \, 29' - \tan 57° \, 18'], \text{ since 'tan 5'}$$

means 'the tangent of 5 radians',

i.e. $\tan 286° \, 29'$.

$$= \frac{3}{2}\ [-3.380\ 5 - 1.557\ 4] \ = \ -7.406\ 9$$

(b) $\int_0^3 3x\ \sqrt{(3x^2 + 9)}\ dx$

Let $u = 3x^2 + 9$, then $\dfrac{du}{dx} = 6x$, i.e. $dx = \dfrac{du}{6x}$

Then $\int_0^3 3x\ \sqrt{(3x^2 + 9)}\ dx \ = \ \int 3x\ \sqrt{u}\ \dfrac{du}{6x}$

$$= \frac{1}{2}\ \int \sqrt{u}\ du = \frac{1}{3}\sqrt{u^3}$$

$$= \frac{1}{3}\ [\sqrt{(3x^2 + 9)^3}\ ]_0^3$$

$$= \frac{1}{3}\ [\sqrt{(36)^3} - \sqrt{(9)^3}]$$

$$= \frac{1}{3}\ [216 - 27]\text{, taking positive values of roots}$$
only.

$$= \textbf{63}$$

(c) $\displaystyle\int_1^2 \dfrac{2e^t}{3 + e^t}\ dt = \ [2\ln(3 + e^t)]_1^2$ (by inspection, or by using the substitution $u = 3 + e^t$)

$$= \ 2\ [\ln(3 + e^2) - \ln(3 + e^1)] \ = \ 2\ln\left(\frac{3 + e^2}{3 + e^1}\right)$$

$$= \ \textbf{1.194 2}$$

Further problems on integration by algebraic substitutions may be found in Section 6, (Problems 1 to 38) page 253.

3 Integration using trigonometrical and hyperbolic identities and substitutions

(A) Integration of $\cos^2 x$, $\sin^2 x$, $\tan^2 x$ and $\cot^2 x$

It may be shown that the compound angle addition formula $\cos(A + B)$ is given by:

$$\cos(A + B) \ = \ \cos A \cos B \ - \ \sin A \sin B.$$

If $A = B$, then $\cos 2A = \cos^2 A - \sin^2 A$ \hfill (1)

Since $\cos^2 A + \sin^2 A = 1$ then $\sin^2 A = 1 - \cos^2 A$

  Hence $\cos 2A \ = \ \cos^2 A - (1 - \cos^2 A)$

  i.e. $\cos 2A \ = \ 2\cos^2 A - 1$ \hfill (2)

Also $\cos^2 A = 1 - \sin^2 A$

  Hence $\cos 2A \ = \ (1 - \sin^2 A) - \sin^2 A$

  i.e. $\cos 2A \ = \ 1 - 2\sin^2 A$ \hfill (3)

From equation (2), $\cos^2 A \ = \ \frac{1}{2}(1 + \cos 2A)$ \hfill (4)

From equation (3), $\sin^2 A \ = \ \frac{1}{2}\ (1 - \cos 2A)$ \hfill (5)

Thus $\int\cos^2 x\, dx = \int \frac{1}{2} (1 + \cos 2x)\, dx = \frac{1}{2} \left(x + \frac{\sin 2x}{2} \right) + c$

Similarly, $\int\sin^2 x\, dx = \int \frac{1}{2} (1 - \cos 2x)\, dx = \frac{1}{2} \left(x - \frac{\sin 2x}{2} \right) + c$

From $1 + \tan^2 x = \sec^2 x$, $\tan^2 x = \sec^2 x - 1$

  Thus $\int\tan^2 x\, dx = \int(\sec^2 x - 1)\, dx = \tan x - x + c$

From $\cot^2 x + 1 = \operatorname{cosec}^2 x$, $\cot^2 x = \operatorname{cosec}^2 x - 1$

  Thus $\int \cot^2 x\, dx = \int (\operatorname{cosec}^2 x - 1)\, dx = -\cot x - x + c$

Problem 1. Find (a) $\int \cos^2 5t\, dt$ (b) $\int \tan^2 3\theta\, d\theta$

(a) $\int \cos^2 5t\, dt = \int \frac{1}{2} (1 + \cos 10t)\, dt = \frac{1}{2} \left(t + \frac{\sin 10t}{10} \right) + c$

(b) $\int \tan^2 3\theta\, d\theta = \int(\sec^2 3\theta - 1)\, d\theta = \frac{1}{3} \tan 3\theta - \theta + c$

(B) Powers of sines and cosines of the form $\int \cos^m x \sin^n x\, dx$

*(i) To evaluate $\int \cos^m x \sin^n x\, dx$ when either m or n is **odd** (but not both), use is made of the trigonometric identity $\cos^2 x + \sin^2 x = 1$ as shown in Problems 1 and 2.*

Problem 1. Find $\int \cos^5 x\, dx$

$\int \cos^5 x\, dx = \int \cos x\, (\cos^2 x)^2\, dx$

$\quad = \int \cos x\, (1 - \sin^2 x)^2\, dx = \int \cos x\, (1 - 2 \sin^2 x + \sin^4 x)\, dx$

$\quad = \int \cos x - 2 \sin^2 x \cos x + \sin^4 x \cos x\, dx$

Hence the integral has been reduced to one which may be determined by inspection (see Section 2, type 2).

Hence $\int \cos^5 x\, dx = \sin x - \frac{2}{3} \sin^3 x + \frac{1}{5} \sin^5 x + c$

Problem 2. Find $\int \sin^3 x \cos^2 x\, dx$

$\int \sin^3 x \cos^2 x\, dx = \int \sin x\, (\sin^2 x)\, (\cos^2 x)\, dx$

$\quad = \int \sin x\, (1 - \cos^2 x)\, (\cos^2 x)\, dx$

$\quad = \int(\sin x \cos^2 x - \sin x \cos^4 x)\, dx$

$\quad = -\frac{1}{3} \cos^3 x + \frac{1}{5} \cos^5 x + c$

---

*(ii) To evaluate $\int \cos^m x \sin^n x\, dx$ when m and n are both **even** use is made of equations (4) and (5) of Section 3(A), i.e.*

$\cos^2 x = \frac{1}{2}(1 + \cos 2x)$ *and* $\sin^2 x = \frac{1}{2}(1 - \cos 2x)$.

(It follows from these equations that $\cos^2 2x = \frac{1}{2}(1 + \cos 4x)$, $\sin^2 3x = \frac{1}{2}(1 - \cos 6x)$, and so on.)

Problem 3. Find $\int \cos^4 x \, dx$

$$\int \cos^4 x \, dx = \int (\cos^2 x)^2 \, dx = \int [\tfrac{1}{2}(1 + \cos 2x)]^2 \, dx$$
$$= \tfrac{1}{4} \int (1 + 2 \cos 2x + \cos^2 2x) \, dx$$
$$= \tfrac{1}{4} \int 1 + 2 \cos 2x + \tfrac{1}{2}(1 + \cos 4x) \, dx$$
$$\text{(since } \cos 4x = 2 \cos^2 2x - 1)$$
$$= \tfrac{1}{4} \int [\tfrac{3}{2} + 2 \cos 2x + \tfrac{1}{2} \cos 4x] \, dx$$

Hence $\qquad \int \cos^4 x \, dx = \tfrac{1}{4} [\tfrac{3}{2} x + \sin 2x + \tfrac{1}{8} \sin 4x] + c$

Problem 4. Find $\int \sin^4 x \cos^2 x \, dx$

$$\int \sin^4 x \cos^2 x \, dx = \int (\sin^2 x)^2 \cos^2 x \, dx = \int \left(\frac{1 - \cos 2x}{2} \right)^2 \left(\frac{1 + \cos 2x}{2} \right) dx$$
$$= \tfrac{1}{8} \int (1 - 2 \cos 2x + \cos^2 2x)(1 + \cos 2x) \, dx$$
$$= \tfrac{1}{8} \int (1 - 2 \cos 2x + \cos^2 2x + \cos 2x - 2 \cos^2 2x$$
$$+ \cos^3 2x) \, dx$$
$$= \tfrac{1}{8} \int (1 - \cos 2x - \cos^2 2x + \cos^3 2x) \, dx$$
$$= \tfrac{1}{8} \int 1 - \cos 2x - \left(\frac{1 + \cos 4x}{2} \right) + \cos 2x (1 - \sin^2 2x) \, dx$$
$$= \tfrac{1}{8} \int [\tfrac{1}{2} - \cos 2x - \tfrac{1}{2} \cos 4x + \cos 2x - \cos 2x \sin^2 2x] \, dx$$
$$= \tfrac{1}{8} \int [\tfrac{1}{2} - \tfrac{1}{2} \cos 4x - \cos 2x \sin^2 2x] \, dx$$
$$= \tfrac{1}{8} \left[\frac{x}{2} - \frac{\sin 4x}{8} - \frac{\sin^3 2x}{6} \right] + c$$

Hence $\int \sin^4 x \cos^2 x \, dx = \tfrac{1}{16} (x - \tfrac{1}{4} \sin 4x - \tfrac{1}{3} \sin^3 2x) + c$

(C) Products of sines and cosines

It is shown in *Technician Mathematics level 3* by J.O. Bird and A.J.C. May, Chapter 1, Section 11, that:

$$\sin A \cos B = \tfrac{1}{2} [\sin (A + B) + \sin (A - B)] \qquad (1)$$
$$\cos A \sin B = \tfrac{1}{2} [\sin (A + B) - \sin (A - B)] \qquad (2)$$
$$\cos A \cos B = \tfrac{1}{2} [\cos (A + B) + \cos (A - B)] \qquad (3)$$
$$\sin A \sin B = -\tfrac{1}{2} [\cos (A + B) - \cos (A - B)] \qquad (4)$$

These formulae are used when integrating products of sines and cosines.

Problem 1. Find $\int \sin 4\theta \cos 3\theta \, d\theta$.

$\int \sin 4\theta \cos 3\theta \, d\theta = \int \frac{1}{2} (\sin 7\theta + \sin \theta) \, d\theta$ from equation (1).

$$= \frac{1}{2} \left(-\frac{\cos 7\theta}{7} - \cos \theta \right) + c$$

$$= -\frac{1}{14} (\cos 7\theta + 7 \cos \theta) + c$$

Problem 2. Find $\int \cos 6x \sin 2x \, dx$

$\int \cos 6x \sin 2x \, dx = \int \frac{1}{2} (\sin 8x - \sin 4x) \, dx$,   from equation (2)

$$= \frac{1}{2} \left(\frac{-\cos 8x}{8} + \frac{\cos 4x}{4} \right) + c$$

$$= \frac{1}{16} (2 \cos 4x - \cos 8x) + c$$

Problem 3. Find $\int \cos 3t \cos t \, dt$

$\int \cos 3t \cos t \, dt = \int \frac{1}{2} (\cos 4t + \cos 2t) \, dt$, from equation (3)

$$= \frac{1}{2} \left(\frac{\sin 4t}{4} + \frac{\sin 2t}{2} \right) + c$$

$$= \frac{1}{8} (\sin 4t + 2 \sin 2t) + c$$

Problem 4. Find $\int \sin 6x \sin 3x \, dx$

$\int \sin 6x \sin 3x \, dx = \int -\frac{1}{2} (\cos 9x - \cos 3x) \, dx$, from equation (4)

$$= -\frac{1}{2} \left(\frac{\sin 9x}{9} - \frac{\sin 3x}{3} \right) + c$$

$$= -\frac{1}{18} (\sin 9x - 3 \sin 3x) + c$$

(D) Integrals containing $\sqrt{(a^2 - x^2)}$ – the 'sine θ' substitution
When an integral contains a term $\sqrt{(a^2 - x^2)}$, the substitution $x = a \sin \theta$
is used. The reasons for this are made obvious by the following worked
problems.

Problem 1. Find $\displaystyle \int \frac{1}{\sqrt{(a^2 - x^2)}} \, dx$

Let $x = a \sin \theta$, then $\dfrac{dx}{d\theta} = a \cos \theta$, i.e. $dx = a \cos \theta \, d\theta$

Hence $\int \dfrac{1}{\sqrt{(a^2 - x^2)}} \, dx = \int \dfrac{1}{\sqrt{(a^2 - a^2 \sin^2 \theta)}} \, a \cos \theta \, d\theta$

$$= \int \dfrac{a \cos \theta \, d\theta}{\sqrt{[(a^2 (1 - \sin^2 \theta)]}}$$

$$= \int \dfrac{a \cos \theta}{a \cos \theta} \, d\theta, \text{ since } 1 - \sin^2 \theta = \cos^2 \theta$$

$$= \int d\theta = \theta + c$$

Since $x = a \sin \theta$ then $\sin \theta = \dfrac{x}{a}$ and $\theta = \arcsin \dfrac{x}{a}$

Hence $\int \dfrac{1}{\sqrt{(a^2 - x^2)}} \, dx = \arcsin \dfrac{x}{a} + c$

Problem 2. Evaluate $\displaystyle\int_0^2 \dfrac{1}{\sqrt{(16 - x^2)}} \, dx$

Using the result of problem 1, $\displaystyle\int_0^2 \dfrac{1}{\sqrt{(16 - x^2)}} \, dx = \left[\arcsin \dfrac{x}{4} \right]_0^2$

$$= (\arcsin \tfrac{1}{2} - \arcsin 0)$$

$$= \dfrac{\pi}{6} \text{ or } 0.523\,6$$

Problem 3. Find $\int \sqrt{(a^2 - x^2)} \, dx$

Let $x = a \sin \theta$, then $\dfrac{dx}{d\theta} = a \cos \theta$, i.e. $dx = a \cos \theta \, d\theta$

Thus $\int \sqrt{(a^2 - x^2)} \, dx = \int \sqrt{(a^2 - a^2 \sin^2 \theta)} \, (a \cos \theta \, d\theta)$

$$= \int \sqrt{[a^2 (1 - \sin^2 \theta)]} \, (a \cos \theta \, d\theta)$$

$$= \int (a \cos \theta) (a \cos \theta) \, d\theta$$

$$= \int a^2 \cos^2 \theta \, d\theta$$

$$= a^2 \int \tfrac{1}{2} (1 + \cos 2\theta) \, d\theta$$

$$= \dfrac{a^2}{2} \left(\theta + \dfrac{\sin 2\theta}{2} \right) + c$$

In the compound angle addition formula $\sin (A + B) = \sin A \cos B + \cos A \sin B$. If $A = B$ then $\sin 2A = 2 \sin A \cos A$.

Hence $\int \sqrt{(a^2 - x^2)} \, dx = \dfrac{a^2}{2} \left(\theta + \dfrac{2 \sin \theta \cos \theta}{2} \right) + c$

$$= \dfrac{a^2}{2} (\theta + \sin \theta \cos \theta) + c$$

Since $x = a \sin \theta$ then $\sin \theta = \dfrac{x}{a}$ and $\theta = \arcsin \dfrac{x}{a}$

Also $\cos^2 \theta + \sin^2 \theta = 1$ from which $\cos \theta = \sqrt{(1 - \sin^2 \theta)}$

i.e. $\cos \theta = \sqrt{\left[1 - \left(\dfrac{x}{a} \right)^2 \right]} = \sqrt{\left(\dfrac{a^2 - x^2}{a^2} \right)} = \dfrac{\sqrt{(a^2 - x^2)}}{a}$

Hence $\int \sqrt{(a^2 - x^2)}\, dx = \dfrac{a^2}{2} \left[\arcsin \dfrac{x}{a} + \left(\dfrac{x}{a} \right) \dfrac{\sqrt{(a^2 - x^2)}}{a} \right] + c$

$$= \dfrac{a^2}{2} \arcsin \dfrac{x}{a} + \dfrac{x}{2} \sqrt{(a^2 - x^2)} + c$$

Problem 4. Evaluate $\int_0^3 \sqrt{(9 - x^2)}\, dx$

Using the result of Problem 3:

$\int_0^3 \sqrt{(9 - x^2)}\, dx = \left[\dfrac{9}{2} \arcsin \dfrac{x}{3} + \dfrac{x}{2} \sqrt{(9 - x^2)} \right]_0^3$

$$= \left[\dfrac{9}{2} \arcsin 1 + \dfrac{3}{2} \sqrt{(9 - 9)} \right] - \left[\dfrac{9}{2} \arcsin 0 + 0 \right]$$

$$= \dfrac{9}{2} \arcsin 1$$

'arcsin 1' means 'the angle whose sine is equal to 1', i.e. $\dfrac{\pi}{2}$ radians.

Hence $\int_0^3 \sqrt{(9 - x^2)}\, dx = \dfrac{9}{2} \times \dfrac{\pi}{2} = \dfrac{9\pi}{4}$ **or 7.068 6.**

Problem 5. Find $\displaystyle\int \dfrac{1}{\sqrt{(5 + 4x - x^2)}}\, dx$

$\displaystyle\int \dfrac{1}{\sqrt{(5 + 4x - x^2)}}\, dx = \int \dfrac{1}{\sqrt{[-(x^2 - 4x - 5)]}}\, dx$

$$= \int \dfrac{1}{\sqrt{[-\{(x - 2)^2 - 9\}]}}\, dx$$

$$= \int \dfrac{1}{\sqrt{[(3)^2 - (x - 2)^2]}}\, dx = \arcsin \left(\dfrac{x - 2}{3} \right) + c$$

(E) $\displaystyle\int \dfrac{1}{a^2 + x^2}\, dx$; the '$\tan \theta$' substitution

When an integral is of the form $\dfrac{1}{a^2 + x^2}$ the substitution $x = a \tan \theta$ is used. The reason for this is made obvious by the following worked problem.

Problem 1. Find $\int \dfrac{1}{a^2 + x^2}\ \mathrm{d}x$

Let $x = a\tan\theta$, then $\dfrac{\mathrm{d}x}{\mathrm{d}\theta} = a\sec^2\theta$, i.e. $\mathrm{d}x = a\sec^2\theta\ \mathrm{d}\theta$

$$\int \dfrac{1}{a^2 + x^2}\ \mathrm{d}x = \int \dfrac{1}{a^2 + a^2\tan^2\theta}\ a\sec^2\theta\ \mathrm{d}\theta = \int \dfrac{a\sec^2\theta\ \mathrm{d}\theta}{a^2(1 + \tan^2\theta)}$$

$$= \int \dfrac{a\sec^2\theta}{a^2\sec^2\theta}\ \mathrm{d}\theta,\ \text{since } 1 + \tan^2\theta = \sec^2\theta$$

$$= \int \dfrac{1}{a}\ \mathrm{d}\theta = \dfrac{1}{a}\ (\theta) + c$$

Hence $\int \dfrac{1}{a^2 + x^2}\ \mathrm{d}x = \dfrac{1}{a}\ \arctan \dfrac{x}{a} + c$, since $x = a\tan\theta$.

Problem 2. Evaluate $\displaystyle\int_0^1 \dfrac{1}{9 + 4x^2}\ \mathrm{d}x$, correct to four significant figures.

$$\int_0^1 \dfrac{1}{9 + 4x^2}\ \mathrm{d}x = \int_0^1 \dfrac{1}{4(\frac{9}{4} + x^2)}\ \mathrm{d}x = \dfrac{1}{4}\int_0^1 \dfrac{1}{(\frac{3}{2})^2 + x^2}\ \mathrm{d}x$$

$$= \dfrac{1}{4}\left[\dfrac{1}{(\frac{3}{2})}\ \arctan \dfrac{x}{(\frac{3}{2})}\right]_0^1$$

from the result of Problem 1

$$= \dfrac{1}{6}\left[\arctan \dfrac{2x}{3}\right]_0^1$$

$$= \dfrac{1}{6}\left[\arctan \dfrac{2}{3} - \arctan 0\right]$$

$$= \dfrac{1}{6}\ (0.588\ 0) = \mathbf{0.098\ 0}$$

(F) Integrals containing $\sqrt{(x^2 + a^2)}$ – the 'sinh θ' substitution

When an integral contains a term $\sqrt{(x^2 + a^2)}$ the substitution $x = a\sinh\theta$ is used. The reason for this is made obvious by the following worked problems.

Problem 1. Find (a) $\displaystyle\int \dfrac{1}{\sqrt{(x^2 + a^2)}}\ \mathrm{d}x$, and (b) evaluate $\displaystyle\int_0^3 \dfrac{1}{\sqrt{(x^2 + 9)}}\ \mathrm{d}x$

(a) Let $x = a\sinh\theta$ then $\dfrac{\mathrm{d}x}{\mathrm{d}\theta} = a\cosh\theta$, i.e. $\mathrm{d}x = a\cosh\theta\ \mathrm{d}\theta$

$$\int \dfrac{1}{\sqrt{(x^2 + a^2)}}\ \mathrm{d}x = \int \dfrac{1}{\sqrt{(a^2\sinh^2\theta + a^2)}}\ a\cosh\theta\ \mathrm{d}\theta$$

$$= \int \frac{a \cosh \theta \, d\theta}{\sqrt{[a^2 (\sinh^2 \theta + 1)]}}$$

$$= \int \frac{a \cosh \theta}{a \cosh \theta} \, d\theta, \text{ since } \sinh^2 \theta + 1 = \cosh^2 \theta$$

$$= \int d\theta = \theta = \mathbf{arsinh} \, \frac{x}{a} + c \text{ since } x = a \sinh \theta$$

From Chapter 14, $\text{arsinh} \, \dfrac{x}{a} = \ln \left\{ \dfrac{x + \sqrt{(x^2 + a^2)}}{a} \right\}$ which gives an alternative solution to $\displaystyle\int \frac{1}{\sqrt{(x^2 + a^2)}} \, dx$.

(b) $\displaystyle\int_0^3 \frac{1}{\sqrt{(x^2 + 9)}} \, dx = \left[\text{arsinh} \, \frac{x}{3} \right]_0^3$ or $\left[\ln \left\{ \dfrac{x + \sqrt{(x^2 + 9)}}{3} \right\} \right]_0^3$

from part (a)

$$\left[\text{arsinh} \, \frac{x}{3} \right]_0^3 = (\text{arsinh } 1 - \text{arsinh } 0)$$

To evaluate arsinh 1: If $x = \text{arsinh } 1$

$$\begin{aligned} \text{then } \sinh x &= 1 \\ \frac{e^x - e^{-x}}{2} &= 1 \\ e^x - e^{-x} &= 2 \end{aligned}$$

Multiplying each term by e^x gives: $(e^x)^2 - 2e^x - 1 = 0$

$$e^x = \frac{-(-2) \pm \sqrt{[(-2)^2 - 4(1)(-1)]}}{2(1)}$$

$$= \frac{2 \pm \sqrt{8}}{2}$$

Taking natural logarithms of both sides gives:

$$x = \ln \left(\frac{2 + \sqrt{8}}{2} \right), \text{ since the logarithms of a negative number cannot be determined in real terms}$$

$$= 0.881 \, 4$$

Hence $\text{arsinh } 1 = 0.881 \, 4$

Hence $\displaystyle\int_0^3 \frac{1}{\sqrt{(x^2 + 9)}} \, dx = \mathbf{0.881 \, 4}$

Using the logarithmic form:

$$\int_0^3 \frac{1}{\sqrt{(x^2 + 9)}} \, dx = \left[\ln \left\{ \frac{x + \sqrt{(x^2 + 9)}}{3} \right\} \right]_0^3$$

$$= \ln \left(\frac{3 + \sqrt{18}}{3} \right) = \mathbf{0.881 \, 4}, \text{ as above.}$$

The logarithmic form is thus seen to be easier for evaluating definite integrals.

Problem 2. Find $\int \sqrt{(x^2 + a^2)}\, dx$

Let $x = a \sinh \theta$, then $dx = a \cosh \theta\, d\theta$

$$\int \sqrt{(x^2 + a^2)}\, dx = \int \sqrt{(a^2 \sinh^2 \theta + a^2)}\, a \cosh \theta\, d\theta$$

$$= \int \sqrt{[a^2 (\sinh^2 \theta + 1)]}\, a \cosh \theta\, d\theta$$

$$= \int (a \cosh \theta)(a \cosh \theta)\, d\theta, \text{ since } \sinh^2 \theta + 1 = \cosh^2 \theta$$

$$= \int a^2 \cosh^2 \theta\, d\theta = a^2 \int \left(\frac{1 + \cosh 2\theta}{2}\right) d\theta$$

since $\cosh 2\theta = 2 \cosh^2 \theta - 1$ (see Chapter 4)

$$= \frac{a^2}{2}\left[\theta + \frac{\sinh 2\theta}{2}\right] + c$$

Hence $\int \sqrt{(x^2 + a^2)}\, dx = \frac{a^2}{2}\,[\theta + \sinh \theta \cosh \theta] + c$, since

$$\sinh 2\theta = 2 \sinh \theta \cosh \theta$$

Since $x = a \sinh \theta$ then $\sinh \theta = \frac{x}{a}$ and $\theta = \operatorname{arsinh} \frac{x}{a}$.

Since $\cosh^2 \theta - \sinh^2 \theta = 1$, then $\cosh \theta = \sqrt{(1 + \sinh^2 \theta)}$

$$= \sqrt{\left[1 + \left(\frac{x}{a}\right)^2\right]} = \frac{1}{a}\sqrt{(a^2 + x^2)}$$

Hence $\int \sqrt{(x^2 + a^2)}\, dx = \frac{a^2}{2}\left[\operatorname{arsinh} \frac{x}{a} + \left(\frac{x}{a}\right)\left(\frac{1}{a}\sqrt{(a^2 + x^2)}\right)\right] + c$

$$= \frac{a^2}{2} \operatorname{arsinh} \frac{x}{a} + \frac{x}{2}\sqrt{(x^2 + a^2)} + c$$

Problem 3. Evaluate $\displaystyle\int_{-1}^{1} \frac{1}{x^2 \sqrt{(1 + x^2)}}\, dx$

Since the integral contains a term in $\sqrt{(a^2 + x^2)}$, let $x = \sinh \theta$, then $dx = \cosh \theta\, d\theta$.

Hence $\displaystyle\int \frac{1}{x^2 \sqrt{(1 + x^2)}}\, dx = \int \frac{1}{(\sinh^2 \theta) \sqrt{(1 + \sinh^2 \theta)}} \cosh \theta\, d\theta$

$$= \int \operatorname{cosech}^2 \theta\, d\theta = -\coth \theta + c$$

Since $\cosh^2 \theta - \sinh^2 \theta = 1$ then $\cosh \theta = \sqrt{(1 + \sinh^2 \theta)} = \sqrt{(1 + x^2)}$

Hence $-\coth \theta = \dfrac{-\cosh \theta}{\sinh \theta} = \dfrac{-\sqrt{(1 + x^2)}}{x}$

Therefore $\displaystyle\int_{-1}^{1} \frac{1}{x^2 \sqrt{(1 + x^2)}}\, dx = -\left[\frac{\sqrt{(1 + x^2)}}{x}\right]_{-1}^{1} = -\left[\frac{\sqrt{2}}{1} - \frac{\sqrt{2}}{-1}\right]$

$$= -2\sqrt{2} = -2.828\,4$$

(G) Intergrals containing $\sqrt{(x^2 - a^2)}$; the 'cosh θ' substitution

When an integral contains a term $\sqrt{(x^2 - a^2)}$ the substitution $x = a \cosh \theta$ is used. The reason for this is made obvious by the following worked problems.

Problem 1. Find $\displaystyle\int \frac{1}{\sqrt{(x^2 - a^2)}} \, dx$

Let $x = a \cosh \theta$, then $dx = a \sinh \theta \, d\theta$

$$\int \frac{1}{\sqrt{(x^2 - a^2)}} \, dx = \int \frac{1}{\sqrt{(a^2 \cosh^2 \theta - a^2)}} \, a \sinh \theta \, d\theta$$

$$= \int \frac{a \sinh \theta \, d\theta}{\sqrt{[a^2 (\cosh^2 \theta - 1)]}} = \int \frac{a \sinh \theta \, d\theta}{a \sinh \theta}$$

since $\cosh^2 \theta - 1 = \sinh^2 \theta$

$$= \int d\theta = \theta + c$$

Hence $\displaystyle\int \frac{1}{\sqrt{(x^2 - a^2)}} \, dx = \text{arcosh} \frac{x}{a} + c$, since $x = a \cosh \theta$

From Chapter 14, $\text{arcosh} \dfrac{x}{a} = \ln \left\{ \dfrac{x + \sqrt{(x^2 - a^2)}}{a} \right\}$ which gives an alternative solution to $\displaystyle\int \frac{1}{\sqrt{(x^2 - a^2)}} \, dx$.

Problem 2. Find $\int \sqrt{(x^2 - a^2)} \, dx$

Let $x = a \cosh \theta$, then $dx = a \sinh \theta \, d\theta$.

$$\int \sqrt{(x^2 - a^2)} \, dx = \int \sqrt{(a^2 \cosh^2 \theta - a^2)} \, a \sinh \theta \, d\theta$$

$$= \int (a \sinh \theta)(a \sinh \theta) \, d\theta = \int a^2 \sinh^2 \theta \, d\theta$$

$$= a^2 \int \left(\frac{\cosh 2\theta - 1}{2} \right) d\theta, \text{ since } \cosh 2\theta = 1 + 2 \sinh^2 \theta$$

$$= \frac{a^2}{2} \left[\frac{\sinh 2\theta}{2} - \theta \right] + c$$

$$= \frac{a^2}{2} \left[\sinh \theta \cosh \theta - \theta \right] + c, \text{ since } \sinh 2\theta = 2 \sinh \theta \cosh \theta$$

Since $x = a \cosh \theta$ then $\cosh \theta = \dfrac{x}{a}$ and $\theta = \text{arcosh} \dfrac{x}{a}$.

Since $\cosh^2 \theta - \sinh^2 \theta = 1$ then $\sinh \theta = \sqrt{(\cosh^2 \theta - 1)} = \sqrt{\left[\left(\frac{x}{a} \right)^2 - 1 \right]}$

$$= \frac{1}{a} \sqrt{(x^2 - a^2)}$$

Hence $\int \sqrt{(x^2 - a^2)}\,dx = \dfrac{a^2}{2}\left[\dfrac{1}{a}\sqrt{(x^2 - a^2)}\left(\dfrac{x}{a}\right) - \text{arcosh}\,\dfrac{x}{a}\right] + c$

$$= \dfrac{x}{2}\sqrt{(x^2 - a^2)} - \dfrac{a^2}{2}\,\text{arcosh}\,\dfrac{x}{a} + c$$

Problem 3. Find $\displaystyle\int \dfrac{4x - 1}{\sqrt{(x^2 + 4x - 12)}}\,dx$

With this type, two fractions are created, one of the form $k\displaystyle\int \dfrac{f'(x)}{f(x)}\,dx$, and the other to keep the ratio the same.

Thus $\displaystyle\int \dfrac{4x - 1}{\sqrt{(x^2 + 4x - 12)}}\,dx$

$$= \int \left\{ \dfrac{2(2x + 4)}{\sqrt{(x^2 + 4x - 12)}} - \dfrac{7}{\sqrt{(x^2 + 4x - 12)}} \right\} dx$$

$$= 4\sqrt{(x^2 + 4x - 12)} - 7\int \dfrac{1}{\sqrt{[(x + 2)^2 - (4)^2]}}\,dx$$

$$= 4\sqrt{(x^2 + 4x - 12)} - 7\,\text{arcosh}\left(\dfrac{x + 2}{4}\right) + c$$

Further problems on integration using trigonometric and hyperbolic identities and substitutions may be found in Section 6 (Problems 39 to 58) page 255.

Summary

$f(x)$	$\int f(x)\,dx$	Method
$\cos^2 x$	$\dfrac{1}{2}\left(x + \dfrac{\sin 2x}{2}\right) + c$	Use $\cos 2x = 2\cos^2 x - 1$
$\sin^2 x$	$\dfrac{1}{2}\left(x - \dfrac{\sin 2x}{2}\right) + c$	Use $\cos 2x = 1 - 2\sin^2 x$
$\tan^2 x$	$\tan x - x + c$	Use $1 + \tan^2 x = \sec^2 x$
$\cot^2 x$	$-\cot x - x + c$	Use $\cot^2 x + 1 = \text{cosec}^2 x$
$\cos^m x \sin^n x$	If m or n is odd, use $\cos^2 x + \sin^2 x = 1$ If both m and n are even, use either $\cos 2x = 2\cos^2 x - 1$ or $\cos 2x = 1 - 2\sin^2 x$	
$\sin A \cos B$		Use $\dfrac{1}{2}\left[\sin (A + B) + \sin (A - B)\right]$
$\cos A \sin B$		Use $\dfrac{1}{2}\left[\sin (A + B) + \sin (A - B)\right]$

$f(x)$	$\int f(x)\,dx$	Method
$\cos A \cos B$	Use $\frac{1}{2}\left[\cos(A+B)+\cos(A-B)\right]$	
$\sin A \sin B$	Use $-\frac{1}{2}\left[\cos(A+B)-\cos(A-B)\right]$	
$\dfrac{1}{\sqrt{(a^2-x^2)}}$	$\arcsin\dfrac{x}{a}+c$	Use $x=a\sin\theta$ substitution
$\sqrt{(a^2-x^2)}$	$\dfrac{a^2}{2}\arcsin\dfrac{x}{a}+\dfrac{x}{2}\sqrt{(a^2-x^2)}+c$	Use $x=a\sin\theta$ substitution
$\dfrac{1}{a^2+x^2}$	$\dfrac{1}{a}\arctan\dfrac{x}{a}+c$	Use $x=\tan\theta$ substitution
$\dfrac{1}{\sqrt{(x^2+a^2)}}$	$\operatorname{arsinh}\dfrac{x}{a}+c$ or $\ln\left\{\dfrac{x+\sqrt{(x^2+a^2)}}{a}\right\}+c$	Use $x=a\sinh\theta$ substitution
$\sqrt{(x^2+a^2)}$	$\dfrac{a^2}{2}\operatorname{arsinh}\dfrac{x}{a}+\dfrac{x}{2}\sqrt{(x^2+a^2)}+c$	Use $x=a\sinh\theta$ substitution
$\dfrac{1}{\sqrt{(x^2-a^2)}}$	$\operatorname{arcosh}\dfrac{x}{a}+c$ or $\ln\left\{\dfrac{x+\sqrt{(x^2-a^2)}}{a}\right\}+c$	Use $x=a\cosh\theta$ substitution
$\sqrt{(x^2-a^2)}$	$\dfrac{x}{2}\sqrt{(x^2-a^2)}-\dfrac{a^2}{2}\operatorname{arcosh}\dfrac{x}{a}+c$	Use $x=a\cosh\theta$ substitution

4. Change of limits of integration by a substitution

When evaluating definite integrals involving substitutions it is often easier to change the limits of the integral as shown in the following worked problems.

Problem 1. Evaluate $\displaystyle\int_0^2 \frac{t}{\sqrt{(2t^2+1)}}\,dt$, taking positive values of square roots only.

Let $u=2t^2+1$, then $\dfrac{du}{dt}=4t$, i.e. $dt=\dfrac{du}{4t}$

Hence $\displaystyle\int_0^2 \frac{t}{\sqrt{(2t^2+1)}}\,dt = \int_{t=0}^{t=2}\frac{t}{\sqrt{u}}\frac{du}{4t} = \frac{1}{4}\int_{t=0}^{t=2} u^{-1/2}\,du$

However if $u = 2t^2 + 1$ then: when $t = 2$, $u = 9$ and when $t = 0$, $u = 1$.

Hence $\dfrac{1}{4} \displaystyle\int_{t=0}^{t=2} u^{-1/2} \, du = \dfrac{1}{4} \displaystyle\int_{u=1}^{u=9} u^{-1/2} \, du$, i.e. the limits have been changed

$$= \frac{1}{4} \left[\frac{u^{-1/2}}{\frac{1}{2}} \right]_1^9 = \tfrac{1}{2} (\sqrt{9} - \sqrt{1}) = 1$$

taking positive values of square roots only.

When the limits are changed it makes it unnecessary to change back to the original variable (in this case, t) after integration.

Problem 2. Evaluate $\displaystyle\int_0^5 \frac{1}{\sqrt{(25 - x^2)}} \, dx$

Let $x = 5 \sin \theta$, then $dx = 5 \cos \theta \, d\theta$

When $x = 5$, $\sin \theta = 1$ and $\theta = \dfrac{\pi}{2}$

When $x = 0$, $\sin \theta = 0$ and $\theta = 0$.

Hence $\displaystyle\int_{x=0}^{x=5} \frac{1}{\sqrt{(25 - x^2)}} \, dx = \int_{\theta=0}^{\theta=\pi/2} \frac{5 \cos \theta \, d\theta}{\sqrt{[25 (1 - \sin^2 \theta)]}}$

$$= [\theta]_0^{\pi/2} = \frac{\pi}{2} \text{ or } \mathbf{1.570\ 8}$$

Problem 3. Evaluate $\displaystyle\int_0^\infty \frac{1}{(x^2 + 4)} \, dx$

Let $x = 2 \tan \theta$, then $dx = 2 \sec^2 \theta \, d\theta$

When $x = + \infty$, $\theta = \dfrac{\pi}{2}$ and when $x = 0$, $\theta = 0$.

Hence $\displaystyle\int_{x=0}^{x=\infty} \frac{1}{(x^2 + 4)} \, dx = \int_{\theta=0}^{\theta=\pi/2} \frac{2 \sec^2 \theta \, d\theta}{4(\tan^2 \theta + 1)} = \int_0^{\pi/2} \frac{1}{2} \, d\theta$

$$= \left[\frac{\theta}{2} \right]_0^{\pi/2} = \frac{1}{2} \left(\frac{\pi}{2} - 0 \right) = \frac{\pi}{4} \text{ or } \mathbf{0.785\ 4}$$

Problem 4. Evaluate $\displaystyle\int_0^1 \sqrt{(4 - x^2)} \, dx$

Let $x = 2 \sin \theta$, then $dx = 2 \cos \theta \, d\theta$

When $x = 1$, $\sin \theta = \dfrac{1}{2}$ and $\theta = \dfrac{\pi}{6}$

When $x = 0$, $\sin \theta = 0$ and $\theta = 0$

Hence $\displaystyle\int_{x=0}^{x=1} \sqrt{(4 - x^2)} \, dx = \int_{\theta=0}^{\theta=\pi/6} \sqrt{[4 (1 - \sin^2 \theta)]} \, 2 \cos \theta \, d\theta$

$$= \int_0^{\pi/6} 4 \cos^2 \theta \, d\theta$$

$$= \int_0^{\pi/6} 4 \left(\frac{1 + \cos 2\theta}{2} \right) \, d\theta$$

$$= 2 \left[\theta + \frac{\sin 2\theta}{2} \right]_0^{\pi/6}$$

$$= 2 \left[\left(\frac{\pi}{6} + \frac{\sin \pi/3}{2} \right) - (0) \right]$$

$$= \frac{\pi}{3} + \frac{\sqrt{3}}{2} \quad \text{or} \quad 1.913 \, 2$$

Problem 5. Evaluate $\int_1^5 \frac{dx}{2x^2 + 4x + 26}$

$$\int_1^5 \frac{dx}{2x^2 + 4x + 26} = \frac{1}{2} \int_1^5 \frac{dx}{x^2 + 2x + 13} = \frac{1}{2} \int_1^5 \frac{dx}{(x+1)^2 + 12}$$

$$= \frac{1}{2} \int_1^5 \frac{dx}{(x+1)^2 + (2\sqrt{3})^2}$$

Let $x + 1 = 2\sqrt{3} \tan \theta$, then $dx = 2\sqrt{3} \sec^2 \theta \, d\theta$

When $x = 5, \theta = \frac{\pi}{3}$ and when $x = 1, \theta = \frac{\pi}{6}$

Hence $\frac{1}{2} \int_1^5 \frac{dx}{(x+1)^2 + (2\sqrt{3})^2} = \frac{1}{2} \int_{\pi/6}^{\pi/3} \frac{2\sqrt{3} \sec^2 \theta \, d\theta}{(2\sqrt{3} \tan \theta)^2 + (2\sqrt{3})^2}$

$$= \frac{1}{2} \int_{\pi/6}^{\pi/3} \frac{2\sqrt{3} \sec^2 \theta \, d\theta}{(2\sqrt{3})^2 (\tan^2 \theta + 1)}$$

$$= \frac{1}{2} \int_{\pi/6}^{\pi/3} \frac{1}{2\sqrt{3}} \, d\theta = \frac{1}{4\sqrt{3}} \left[\theta \right]_{\pi/6}^{\pi/3}$$

$$= \frac{\pi}{24\sqrt{3}} \quad \text{or} \quad 0.075 \, 57$$

Further problems on evaluating definite integrals may be found in Section 6 (Problems 59 to 69) page 258.

5 Integration using partial fractions

The process of expressing a fraction in terms of simple fractions — called partial fractions — was explained in Chapter 2. Certain functions can only be integrated when they have been resolved into partial fractions.

Worked problems on integration using partial fractions

(a) Type 1. Denominator containing linear factors

Problem 1. Find $\int \dfrac{x-8}{x^2-x-2}\ dx$

It was shown on page 15 that $\dfrac{x-8}{x^2-x-2} \equiv \dfrac{3}{x+1} - \dfrac{2}{x-2}$

Hence $\int \dfrac{x-8}{x^2-x-2}\ dx = \int \left(\dfrac{3}{x+1} - \dfrac{2}{x-2} \right) dx$

$\qquad\qquad = 3 \ln (x+1) - 2 \ln (x-2) + c$

$\qquad\qquad$ or $\ln \left\{ \dfrac{(x+1)^3}{(x-2)^2} \right\} + c$

Problem 2. Determine $\int \dfrac{6x^2+7x-25}{(x-1)(x+2)(x-3)}\ dx$

It was shown on page 16 that $\dfrac{6x^2+7x-25}{(x-1)(x+2)(x-3)}$

$\qquad\qquad\qquad \equiv \dfrac{2}{(x-1)} - \dfrac{1}{(x+2)} + \dfrac{5}{(x-3)}$

Hence $\int \dfrac{6x^2+7x-25}{(x-1)(x+2)(x-3)}\ dx = \int \left(\dfrac{2}{(x-1)} - \dfrac{1}{(x+2)} + \dfrac{5}{(x-3)} \right) dx$

$\qquad\qquad = 2 \ln (x-1) - \ln (x+2) + 5 \ln (x-3) + c$

$\qquad\qquad$ or $\ln \left\{ \dfrac{(x-1)^2\ (x-3)^5}{(x+2)} \right\} + c$

Problem 3. Evaluate $\int_3^4 \dfrac{x^3-x^2-5x}{x^2-3x+2}\ dx$

It was shown on page 17 that $\dfrac{x^3-x^2-5x}{x^2-3x+2} \equiv x+2+\dfrac{5}{(x-1)} - \dfrac{6}{(x-2)}$

Hence $\int_3^4 \dfrac{x^3-x^2-5x}{x^2-3x+2}\ dx = \int_3^4 \left(x+2+\dfrac{5}{(x-1)} - \dfrac{6}{(x-2)} \right) dx$

$\qquad\qquad = \left[\dfrac{x^2}{2} + 2x + 5 \ln (x-1) - 6 \ln (x-2) \right]_3^4$

$\qquad\qquad = (8 + 8 + 5 \ln 3 - 6 \ln 2) - (\tfrac{9}{2} + 6 + 5 \ln 2 - 6 \ln 1)$

$\qquad\qquad = 3.368\ 4$

(b) Type 2. Denominator containing repeated linear factors

Problem 4. Find $\int \dfrac{x+5}{(x+3)^2}\ dx$

It was shown on page 17 that $\dfrac{x+5}{(x+3)^2} \equiv \dfrac{1}{(x+3)} + \dfrac{2}{(x+3)^2}$

Hence $\int \dfrac{x+5}{(x+3)^2}\ dx = \int \left(\dfrac{1}{(x+3)} + \dfrac{2}{(x+3)^2} \right) dx$

$$= \ln(x+3) - \dfrac{2}{(x+3)} + c$$

Problem 5. Find $\int \dfrac{5x^2 - 19x + 3}{(x-2)^2\ (x+1)}\ dx$

It was shown on page 18 that $\dfrac{5x^2 - 19x + 3}{(x-2)^2\ (x+1)}$

$$\equiv \dfrac{2}{(x-2)} - \dfrac{5}{(x-2)^2} + \dfrac{3}{(x+1)}$$

Hence $\int \dfrac{5x^2 - 19x + 3}{(x-2)^2\ (x+1)}\ dx = \int \left(\dfrac{2}{(x-2)} - \dfrac{5}{(x-2)^2} + \dfrac{3}{(x+1)} \right)\ dx$

$$= 2 \ln(x-2) + \dfrac{5}{(x-2)} + 3 \ln(x+1) + c$$

$$= \ln(x-2)^2\ (x+1)^3 + \dfrac{5}{(x-2)} + c$$

Problem 6. Evaluate $\int_5^6 \dfrac{2x^2 - 13x + 13}{(x-4)^3}\ dx$

It was shown on page 19 that $\dfrac{2x^2 - 13x + 13}{(x-4)^3}$

$$\equiv \dfrac{2}{(x-4)} + \dfrac{3}{(x-4)^2} - \dfrac{7}{(x-4)^3}$$

Hence $\int_5^6 \dfrac{2x^2 - 13x + 13}{(x-4)^3} = \int_5^6 \left(\dfrac{2}{(x-4)} + \dfrac{3}{(x-4)^2} - \dfrac{7}{(x-4)^3} \right)\ dx$

$$= \left[2 \ln(x-4) - \dfrac{3}{(x-4)} + \dfrac{7}{2(x-4)^2} \right]_5^6$$

$$= (2 \ln 2 - \tfrac{3}{2} + \tfrac{7}{8}) - (2 \ln 1 - \tfrac{3}{1} + \tfrac{7}{2})$$

$$= 0.261\ 3$$

(c) Type 3. Denominator containing a quadratic factor

Problem 7. Find $\int \dfrac{8x^2 - 3x + 19}{(x^2 + 3)(x - 1)} \, dx$

It was shown on page 20 that $\dfrac{8x^2 - 3x + 19}{(x^2 + 3)(x - 1)} \equiv \dfrac{2x - 1}{(x^2 + 3)} + \dfrac{6}{(x - 1)}$

Hence $\int \dfrac{8x^2 - 3x + 19}{(x^2 + 3)(x - 1)} \, dx = \int \left(\dfrac{2x - 1}{(x^2 + 3)} + \dfrac{6}{(x - 1)} \right) \, dx$

$= \int \left(\dfrac{2x}{(x^2 + 3)} - \dfrac{1}{(x^2 + 3)} + \dfrac{6}{(x - 1)} \right) \, dx$

$= \ln (x^2 + 3) - \dfrac{1}{\sqrt{3}} \arctan \dfrac{x}{\sqrt{3}} + 6 \ln (x - 1) + c$

$= \ln (x^2 + 3)(x - 1)^6 - \dfrac{1}{\sqrt{3}} \arctan \dfrac{x}{\sqrt{3}} + c$

Problem 8. Find $\int \dfrac{2 + x + 6x^2 - 2x^3}{x^2 (x^2 + 1)} \, dx$

It was shown on page 20 that $\dfrac{2 + x + 6x^2 - 2x^3}{x^2 (x^2 + 1)} \equiv \dfrac{1}{x} + \dfrac{2}{x^2} + \dfrac{4 - 3x}{x^2 + 1}$

Hence $\int \dfrac{2 + x + 6x^2 - 2x^3}{x^2 (x^2 + 1)} \, dx = \int \left(\dfrac{1}{x} + \dfrac{2}{x^2} + \dfrac{4 - 3x}{x^2 + 1} \right) \, dx$

$= \int \left(\dfrac{1}{x} + \dfrac{2}{x^2} + \dfrac{4}{x^2 + 1} - \dfrac{3x}{x^2 + 1} \right) \, dx$

$= \ln x - \dfrac{2}{x} + 4 \arctan x - \dfrac{3}{2} \ln (x^2 + 1) + c$

$= \ln \left\{ \dfrac{x}{(x^2 + 1)^{3/2}} \right\} - \dfrac{2}{x} + 4 \arctan x + c$

Problem 9. Find $\int \dfrac{5(x^2 + x + 3)}{x(x^2 + 2x + 5)} \, dx$

Let $\dfrac{5(x^2 + x + 3)}{x(x^2 + 2x + 5)} \equiv \dfrac{A}{x} + \dfrac{Bx + C}{x^2 + 2x + 5}$

$\equiv \dfrac{A(x^2 + 2x + 5) + (Bx + C)(x)}{x(x^2 + 2x + 5)}$

Hence $5x^2 + 5x + 15 \equiv A(x^2 + 2x + 5) + (Bx + C)(x)$ by equating numerators.

Let $x = 0$, then $A = 3$
Equating the coefficients of x^2 gives $5 = A + B$, i.e. $B = 2$
Equating the coefficients of x gives: $5 = 2A + C$, i.e. $C = -1$.

Hence $\int \dfrac{5(x^2+x+3)}{x(x^2+2x+5)} \, dx = \int \left(\dfrac{3}{x} + \dfrac{2x-1}{x^2+2x+5} \right) \, dx$

Now $\int \dfrac{2x-1}{x^2+2x+5} \, dx = \int \dfrac{2x+2}{(x^2+2x+5)} \, dx - \int \dfrac{3}{(x^2+2x+5)} \, dx$

i.e. the numerator of the first integral on the right-hand side has deliberately been made equal to the differential coefficient of the denominator so that the integral will integrate as $\ln(x^2+2x+5)$.

$$\int \dfrac{3}{(x^2+2x+5)} \, dx = \int \dfrac{3}{(x+1)^2+4} \, dx = \int \dfrac{3}{(x+1)^2+(2)^2} \, dx$$

$$= \dfrac{3}{2} \arctan \dfrac{(x+1)}{2}$$

Hence $\int \dfrac{5(x^2+x+3)}{x(x^2+2x+5)} \, dx = 3 \ln x + \ln(x^2+2x+5)$

$$- \dfrac{3}{2} \arctan \dfrac{(x+1)}{2} + c$$

$$= \ln x^3 (x^2+2x+5) - \dfrac{3}{2} \arctan \dfrac{(x+1)}{2} + c$$

(d) Integrals of the form $\displaystyle\int \dfrac{1}{x^2-a^2} \, dx$ *and* $\displaystyle\int \dfrac{1}{a^2-x^2} \, dx$

Problem 10. Determine (a) $\displaystyle\int \dfrac{1}{x^2-a^2} \, dx$, (b) $\displaystyle\int_1^2 \dfrac{1}{(9-x^2)} \, dx$.

(a) Let $\dfrac{1}{x^2-a^2} \equiv \dfrac{A}{(x-a)} + \dfrac{B}{(x+a)} \equiv \dfrac{A(x+a)+B(x-a)}{(x-a)(x+a)}$

Hence $1 = A(x+a) + B(x-a)$ by equating numerators

Let $x=a$, then ; $A = \dfrac{1}{2a}$. Let $x=-a$, then $B = -\dfrac{1}{2a}$

Hence $\int \dfrac{1}{x^2-a^2} \, dx = \dfrac{1}{2a} \int \left(\dfrac{1}{(x-a)} - \dfrac{1}{(x+a)} \right) dx$

$$= \dfrac{1}{2a} \, [\ln(x-a) - \ln(x+a)] + c$$

$$= \dfrac{1}{2a} \ln \left(\dfrac{x-a}{x+a} \right) + c$$

Similarly, it may be shown that $\displaystyle\int \dfrac{1}{a^2-x^2} \, dx = \dfrac{1}{2a} \ln \left(\dfrac{a+x}{a-x} \right) + c$

or $\dfrac{1}{a} \operatorname{artanh} \dfrac{x}{a} + c$ (by using the substitution $x = a \tanh \theta$)

(b) $\int_1^2 \dfrac{1}{9-x^2}\, dx = \int_1^2 \dfrac{1}{3^2-x^2}\, dx = \dfrac{1}{2(3)}\left[\ln\left(\dfrac{3+x}{3-x}\right)\right]_1^2$ from (a)

$$= \dfrac{1}{6}\ln\dfrac{5}{2} \text{ or } 0.152\,7$$

Further problems on integration using partial fractions may be found in the following Section (6) (Problems 70 to 100) page 259.

6. Further problems

Integrating using algebraic substitution
In Problems 1 to 24 integrate with respect to the variable.

1. $\sin(5x-3)$. $[-\frac{1}{5}\cos(5x-3)+c]$
2. $4\cos(3t-1)$. $[\frac{4}{3}\sin(3t-1)+c]$
3. $5\sec^2(7\theta-2)$. $[\frac{5}{7}\tan(7\theta-2)+c]$
4. $(3x+2)^9$. $[\frac{1}{30}(3x+2)^{10}+c]$
5. $14(7t+4)^6$. $[\frac{2}{7}(7t+4)^7+c]$

6. $\dfrac{1}{3d+1}$. $[\frac{1}{3}\ln(3d+1)+c]$

7. $\dfrac{-2}{8x+3}$. $[-\frac{1}{4}\ln(8x+3)+c]$

8. $3e^{4x-5}$. $[\frac{3}{4}e^{4x-5}+c]$

9. $6e^{3-2t}$. $[-3e^{3-2t}+c]$

10. $4x(2x^2-4)^6$. $[\frac{1}{7}(2x^2-4)^7+c]$

11. $18t(1-3t^2)^5$. $[-\frac{1}{2}(1-3t^2)^6+c]$

12. $\frac{2}{3}\sin^3\theta\cos\theta$. $[\frac{1}{6}\sin^4\theta+c]$

13. $3\cos^4 t\sin t$. $[-\frac{3}{5}\cos^5 t+c]$

14. $\sec^2 2x\tan 2x$. $[\frac{1}{4}\tan^2 2x+c]$

15. $4x\sqrt{(4x^2+1)}$. $[\frac{1}{3}\sqrt{(4x^2+1)^3}+c]$

16. $(9x^2-4x)\sqrt{(3x^3-2x^2+1)}$. $[\frac{2}{3}\sqrt{(3x^3-2x^2+1)^3}+c]$

17. $\dfrac{2\ln t}{t}$. $[(\ln t)^2+c]$

18. $\dfrac{4x+3}{(2x^2+3x-1)^4}$. $\left[-\dfrac{1}{3(2x^2+3x-1)^3}+c\right]$

19. $\dfrac{t}{\sqrt{(t^2-3)}}$. $[\sqrt{(t^2-3)}+c]$

20. $\dfrac{x^2-1}{\sqrt{(x^2-3x+4)}}$. $[\frac{2}{3}\sqrt{(x^3-3x+4)}+c]$

21. $\dfrac{4e^t}{\sqrt{(2+e^t)}}$. $\quad[\,8\sqrt{(2+e^t)}+c\,]$

22. $5\tan 2\alpha$. $\quad[\frac{5}{2}\ln(\sec 2\alpha)+c]$

23. $(5\theta+2)\sec^2(5\theta^2+4\theta)$. $\quad[\frac{1}{2}\tan(5\theta^2+4\theta)+c]$

24. $2t\,e^{3t^2-2}$. $\quad[\frac{1}{3}e^{3t^2-2}+c]$

In Problems 25 to 35 evaluate the definite integrals.

25. $\int_0^1 (2x-1)^6\,dx$. $\quad[\frac{1}{7}]$

26. $\int_0^2 t\sqrt{(2t^2+1)}\,dt$. $\quad[4\frac{1}{3}]$

27. $\displaystyle\int_0^{\pi/2}\sin\left(2\theta+\dfrac{\pi}{3}\right)d\theta$. $\quad[\frac{1}{2}]$

28. $\int_1^3 2\cos(2t-1)\,dt$. $\quad[-1.800\,4]$

29. $\int_0^2 (2x^2-1)\sqrt{(2x^3-3x)}\,dx$. $\quad[7.027\,3]$

30. $\displaystyle\int_1^2 \dfrac{5\ln x}{x}\,dx$. $\quad[1.201\,1]$

31. $\displaystyle\int_1^3 \dfrac{(x-2)}{(x^2-4x+1)^4}\,dx$. $\quad[0]$

32. $\int_0^1 2t\,e^{4t^2-3}\,dt$. $\quad[0.667\,1]$

33. $\int_0^{\pi/3} 2\sin^5\theta\cos\theta\,d\theta$. $\quad[0.140\,6]$

34. $\int_0^1 \theta\sec^2(5\theta^2)\,d\theta$. $\quad[-0.338\,1]$

35. $\displaystyle\int_1^2\left[\dfrac{e^{(2x-1)}-e^{(-2x+1)}}{2}\right]dx$. $\quad[4.262\,3]$

36. The entropy change for an ideal gas is given by the equation

$$\Delta S = \int_{T_1}^{T_2} C_v\,\frac{dT}{T} - R\int_{V_1}^{V_2}\frac{dV}{V}$$

where T is the thermodynamic temperature, V is the volume and $R = 8.314$.

  Determine the entropy change when a gas expands from 2 litres to 3 litres for a temperature rise from 200 K to 400 K given that

$$C_V = 45 + 6\times 10^{-3}\,T + 8\times 10^{-6}\,T^2. \qquad [29.5]$$

37. The electrostatic potential on all parts of a conducting circular disc of radius a is given by the equation

$$V = 2\pi\sigma \int_0^9 \frac{R\,dR}{\sqrt{(R^2 + r^2)}}$$

Solve the equation by determining the integral.

$$[V = 2\pi\sigma\,[\sqrt{(9^2 + r^2)} - r]]$$

38. In the study of a rigid rotor the following integration occurs:

$$Z_r = \int_0^\infty (2J + 1)\,e^{-J(J+1)h^2/8\pi^2 Ikt}\,dJ$$

Determine Z_r for constant temperature T assuming h, I, and k are constants.

$$\left[\frac{8\pi^2 IkT}{h^2}\right]$$

Integration using trigonometric and hyperbolic identities and substitutions
In Problems 39 to 58 integrate with respect to the variable.

39. (a) $\cos^2 2x$ (b) $\sin^2 3x$ (c) $\tan^2 4x$ (d) $\cot^2 5x$

  (a) $\left[\frac{1}{2}\left(x + \frac{\sin 4x}{4}\right) + c\right]$ (b) $\left[\frac{1}{2}\left(x - \frac{\sin 6x}{6}\right) + c\right]$

  (c) $\left[\frac{1}{4}\tan 4x - x + c\right]$ (d) $\left[-\frac{1}{5}\cot 5x - x + c\right]$

Powers of sines and cosines
40. (a) $\cos^3\theta$ (b) $\sin^3 2\theta$.

  (a) $\left[\sin\theta - \frac{\sin^3\theta}{3} + c\right]$ (b) $\left[\frac{1}{6}\cos^3 2\theta - \frac{1}{2}\cos 2\theta + c\right]$

41. (a) $\cos^3 t \sin^2 t$ (b) $\sin^4 x \cos^3 x$.

  (a) $[\frac{1}{3}\sin^3 t - \frac{1}{5}\sin^5 t + c]$ (b) $[\frac{1}{5}\sin^5 x - \frac{1}{7}\sin^7 x + c]$

42. (a) $\sin^4 2x$ (b) $\cos^2 x \sin^2 x$.

  (a) $\left[\frac{3x}{8} - \frac{1}{8}\sin 4x + \frac{1}{64}\sin 8x + c\right]$ (b) $\left[\frac{x}{8} - \frac{1}{32}\sin 4x + c\right]$

43. (a) $3\cos^4 3t \sin^2 3t$ (b) $\frac{1}{2}\sin^4 2\theta \cos^2 2\theta$.

  (a) $[\frac{3}{16}(t - \frac{1}{12}\sin 12t + \frac{1}{9}\sin^3 6t) + c]$

  (b) $[\frac{1}{32}(\theta - \frac{1}{8}\sin 8\theta - \frac{1}{6}\sin^3 4\theta) + c]$

Products of sines and cosines
44. (a) $\sin 3t \cos t$ (b) $2\cos 4t \cos 2t$.

  (a) $[-\frac{1}{8}(\cos 4t + 2\cos 2t) + c]$ (b) $[\frac{1}{6}\sin 6t + \frac{1}{2}\sin 2t + c]$

45. (a) $3 \cos 6x \cos 2x$ (b) $4 \sin 5x \sin 2x$.

    (a) $\left[\frac{3}{16} \left(\sin 8x + 2 \sin 4x \right) + c \right]$ (b) $\left[2 \left(\frac{1}{3} \sin 3x - \frac{1}{7} \sin 7x \right) + c \right]$

46. (a) $\frac{1}{2} \sin 9t \sin 3t$ (b) $2 \sin 4\theta \cos 3\theta$.

    (a) $\left[\frac{1}{48} \left(2 \sin 6t - \sin 12t \right) + c \right]$ (b) $\left[-\frac{1}{7} \cos 7\theta - \cos \theta + c \right]$

47. (a) $9 \cos 5t \cos 4t$ (b) $\frac{3}{2} \cos 2x \sin x$

    (a) $\left[\frac{9}{2} \left(\frac{1}{9} \sin 9t + \sin t \right) + c \right]$ (b) $\left[\frac{3}{4} \left(\cos x - \frac{1}{3} \cos 3x \right) + c \right]$

'Sine θ' substitution

48. (a) $\dfrac{2}{\sqrt{(4 - x^2)}}$ (b) $\dfrac{1}{\sqrt{(9 - 4x^2)}}$.

    (a) $\left[2 \arcsin \frac{x}{2} + c \right]$ (b) $\left[\frac{1}{2} \arcsin \frac{2x}{3} + c \right]$

49. (a) $\sqrt{(16 - x^2)}$ (b) $\sqrt{(16 - 9x^2)}$.

    (a) $\left[8 \arcsin \frac{x}{4} + \frac{x}{2} \sqrt{(16 - x^2)} + c \right]$

    (b) $\left[\frac{8}{3} \arcsin \frac{3x}{4} + \frac{x}{2} \sqrt{(16 - 9x^2)} + c \right]$

50. (a) $\dfrac{1}{\sqrt{(4 + 2x - x^2)}}$ (b) $\sqrt{(4 + 2x - x^2)}$

    (a) $\left[\arcsin \frac{(x - 1)}{\sqrt{5}} + c \right]$

    (b) $\left[\frac{5}{2} \arcsin \frac{(x - 1)}{\sqrt{5}} + \left(\frac{x - 1}{2} \right) \sqrt{(4 + 2x - x^2)} + c \right]$

'Tan θ' substitution

51. (a) $\dfrac{2}{1 + x^2}$ (b) $\dfrac{3}{16 + x^2}$.

    (a) $[2 \arctan x + c]$ (b) $\left[\frac{3}{4} \arctan \frac{x}{4} + c \right]$

52. (a) $\dfrac{1}{9 + 16x^2}$ (b) $\dfrac{1}{2x^2 + 4x + 18}$.

    (a) $\left[\frac{1}{12} \arctan \frac{4x}{3} + c \right]$ (b) $\left[\frac{1}{4\sqrt{2}} \arctan \frac{x + 1}{2\sqrt{2}} + c \right]$

'Sinh θ' substitution

53. (a) $\dfrac{1}{\sqrt{(x^2 + 4)}}$ (b) $\dfrac{1}{\sqrt{(9 + 2x^2)}}$.

(a) $\left[\text{arsinh } \dfrac{x}{2} + c \text{ or } \ln \left\{ \dfrac{x + \sqrt{(x^2 + 4)}}{2} \right\} + c \right]$

(b) $\left[\dfrac{1}{\sqrt{2}} \text{ arsinh } \dfrac{\sqrt{2}x}{3} + c \text{ or } \dfrac{1}{\sqrt{2}} \ln \left\{ \dfrac{x + \sqrt{(\frac{9}{2} + x^2)}}{3/\sqrt{2}} \right\} + c \right]$

54. (a) $\sqrt{(x^2 + 16)}$ (b) $\sqrt{(4x^2 + 25)}$.

(a) $\left[8 \text{ arsinh } \dfrac{x}{4} + \dfrac{x}{2} \sqrt{(x^2 + 16)} + c \right]$

(b) $\left[\dfrac{25}{4} \text{ arsinh } \dfrac{2x}{5} + x \sqrt{(x^2 + \dfrac{25}{4})} + c \right]$

55. (a) $\sqrt{(x^2 + 2x + 5)}$ (b) $\dfrac{2}{3x^2 \sqrt{(4 + x^2)}}$ (by letting $x = 2 \sinh \theta$).

(a) $\left[2 \text{ arsinh } \dfrac{x+1}{2} + \left(\dfrac{x+1}{2} \right) \sqrt{(x^2 + 2x + 5)} + c \right]$

(b) $\left[\dfrac{-\sqrt{(4 + x^2)}}{6x} + c \right]$

'Cosh θ' substitution

56. (a) $\dfrac{1}{\sqrt{(x^2 - 9)}}$ (b) $\dfrac{1}{\sqrt{(4x^2 - 9)}}$.

(a) $\left[\text{arcosh } \dfrac{x}{3} + c \text{ or } \ln \left\{ \dfrac{x + \sqrt{(x^2 - 9)}}{3} \right\} + c \right]$

(b) $\left[\dfrac{1}{2} \text{ arcosh } \dfrac{2x}{3} + c \text{ or } \dfrac{1}{2} \left| \ln \left\{ \dfrac{x + \sqrt{(x^2 - \frac{9}{4})}}{\frac{3}{2}} \right\} + c \right. \right]$

57. (a) $\sqrt{(x^2 - 4)}$ (b) $\sqrt{(9x^2 - 16)}$.

(a) $\left[\dfrac{x}{2} \sqrt{(x^2 - 4)} - 2 \text{ arcosh } \dfrac{x}{2} + c \right]$

(b) $\left[3 \left\{ \dfrac{x}{2} \sqrt{(x^2 - \dfrac{16}{9})} - \dfrac{8}{9} \text{ arcosh } \dfrac{3x}{4} \right\} + c \right]$

58. (a) $\dfrac{1}{\sqrt{(x^2 + 4x - 5)}}$ (b) $\sqrt{(x^2 + 6x + 5)}$.

(a) $\left[\text{arcosh } \dfrac{x+2}{3} + c \text{ or } \ln \left\{ \dfrac{(x+2) + \sqrt{(x^2 + 4x - 5)}}{3} \right\} + c \right]$

(b) $\left[\left(\dfrac{x+3}{2} \right) \sqrt{(x^2 + 6x + 5)} - 2 \text{ arcosh} \left(\dfrac{x+3}{2} \right) + c \right]$

Definite integrals

In Problems 59 to 69 evaluate the definite integrals.

59. (a) $\int_0^{\pi/2} \sin^2 \theta \, d\theta$ (b) $\int_0^1 2 \cos^2 \theta \, d\theta$

    (a) $[\frac{\pi}{4}$ or $0.785\,4\,]$ (b) $[1.454\,6\,]$

60. (a) $\int_0^{\pi/4} \sin^3 t \cos t \, dt$ (b) $\int_0^{\pi/3} \cos^3 2t \, dt.$
    (a) $[\,0.062\,5\,]$ (b) $[\,0.324\,8]$

61. (a) $\int_{\pi/4}^{\pi/2} 4 \sin^2 \theta \cos^2 \theta \, d\theta$ (b) $\int_0^{\pi/2} 3 \sin^2 2t \cos^4 2t \, dt.$

    (a) $[\frac{\pi}{8}$ or $0.392\,7\,]$ (b) $[\frac{3\pi}{32}$ or $0.294\,5\,]$

62. (a) $\int_0^{\pi/4} \sin 5\theta \cos 3\theta \, d\theta$ (b) $\int_0^{\pi/3} 4 \cos 6t \sin 3t \, dt$
    (a) $[\frac{1}{4}\,]$ (b) $[-\frac{8}{9}\,]$

63. (a) $\int_0^1 2 \cos 7x \cos 2x \, dx$ (b) $\int_0^{\pi/2} \sin 8\alpha \sin 5\alpha \, d\alpha.$
    (a) $[-0.146\,0]$ (b) $[-0.205\,1]$

64. (a) $\int_0^3 \frac{1}{\sqrt{(9-x^2)}} \, dx$ (b) $\int_0^3 \sqrt{(9-x^2)} \, dx.$

    (a) $[\frac{\pi}{2}$ or $1.570\,8]$ (b) $[\frac{9\pi}{4}$ or $7.068\,6]$

65. (a) $\int_0^{1/2} \frac{2}{\sqrt{(1-x^2)}} \, dx$ (b) $\int_0^{3/4} \sqrt{(9-16x^2)} \, dx.$

    (a) $[\frac{\pi}{3}$ or $1.047\,2]$ (b) $[\frac{9\pi}{16}$ or $1.767\,1]$

66. (a) $\int_0^1 \frac{1}{1+x^2} \, dx$ (b) $\int_0^2 \frac{3}{4+x^2} \, dx.$

    (a) $[\frac{\pi}{4}$ or $0.785\,4]$ (b) $[\frac{3\pi}{8}$ or $1.178\,1]$

67. (a) $\int_{-1}^2 \frac{1}{2x^2 + 4x + 20} \, dx$ (b) $\int_{-3}^0 \frac{5x-2}{x^2 + 6x + 13} \, dx.$

    (a) $[\frac{\pi}{24}$ or $0.130\,9]$ (b) $[-5.407\,1]$

68. (a) $\int_0^2 \frac{1}{\sqrt{(x^2+4)}} \, dx$ (b) $\int_0^1 \sqrt{(x^2+9)} \, dx.$
    (a) $[0.881\,4]$ (b) $[3.054\,7]$

69. (a) $\displaystyle\int_1^2 \frac{1}{\sqrt{(x^2-1)}}\ dx$ (b) $\int_2^3 \sqrt{(x^2-4)}\ dx.$

(a) [1.317 0]   (b) [1.429 3]

Integration using partial fractions

In Problems 70 to 89 integrate after resolving into partial fractions.

70. $\dfrac{8}{x^2-4}$. $\left[\ 2\ln(x-2)-2\ln(x+2)+c\ \text{ or }\ 2\ln\left(\dfrac{x-2}{x+2}\right)+c\ \right]$

71. $\dfrac{3x+5}{x^2+2x-3}$.

$$\left[\ 2\ln(x-1)+\ln(x+3)+c\ \text{ or }\ \ln\{(x-1)^2\ (x+3)\}\ +\ c\right]$$

72. $\dfrac{y-13}{y^2-y-6}$.

$$\left[\ 3\ln(y+2)-2\ln(y-3)+c\ \text{ or }\ \ln\left\{\dfrac{(y+2)^3}{(y-3)^2}\right\}\ +c\ \right]$$

73. $\dfrac{17x^2-21x-6}{x\,(x+1)\,(x-3)}$.

$$\left[2\ln x+8\ln(x+1)+7\ln(x-3)+c\ \text{ or }\ \ln\{x^2\,(x+1)^8\,(x-3)^7\}+c\right]$$

74. $\dfrac{6x^2+7x-49}{(x-4)\,(x+1)\,(2x-3)}$.

$$[3\ln(x-4)-2\ln(x+1)+4\ln(2x-3)+c\ \text{ or }$$

$$\ln\left\{\dfrac{(x-4)^3(2x-3)^4}{(x+1)^2}\right\}\ +c\ ]$$

75. $\dfrac{x^2+2}{(x+4)\,(x-2)}$.

$$\left[x-3\ln(x+4)+\ln(x-2)+c\ \text{ or }\ x+\ln\left\{\dfrac{(x-2)}{(x+4)^3}\right\}+c\ \right]$$

76. $\dfrac{2x^2+4x+19}{2\,(x-3)\,(x+4)}$.

$$\left[x+\tfrac{7}{2}\ln(x-3)-\tfrac{5}{2}\ln(x+4)+c\ \text{ or }\ x+\ln\left\{\dfrac{(x-3)^{7/2}}{(x+4)^{5/2}}\right\}\ +c\ \right]$$

77. $\dfrac{2x^3+7x^2-2x-27}{(x-1)\,(x+4)}$.

$$\left[x^2+x-4\ln(x-1)+7\ln(x+4)+c\ \text{ or }\ x^2+x+\ln\left\{\dfrac{(x+4)^7}{(x-1)^4}\right\}\ +c\ \right]$$

78. $\dfrac{2t-1}{(t+1)^2}$. $\left[\, 2 \ln{(t+1)} + \dfrac{3}{(t+1)} + c \,\right]$

79. $\dfrac{8x^2+12x-3}{(x+2)^3}$. $\left[\, 8 \ln{(x+2)} + \dfrac{20}{(x+2)} - \dfrac{5}{2(x+2)^2} + c \,\right]$

80. $\dfrac{6x+1}{(2x+1)^2}$. $\left[\, \tfrac{3}{2} \ln{(2x+1)} + \dfrac{1}{(2x+1)} + c \,\right]$

81. $\dfrac{1}{x^2\,(x+2)}$.

$$\left[\, -\dfrac{1}{2x} - \dfrac{1}{4} \ln{x} + \dfrac{1}{4} \ln{(x+2)} + c \;\; \text{or} \;\; \dfrac{1}{4} \ln{\left(\dfrac{x+2}{x}\right)} - \dfrac{1}{2x} + c \,\right]$$

82. $\dfrac{9x^2 - 73x + 150}{(x-7)\,(x-3)^2}$.

$$\left[\, 5 \ln{(x-7)} + 4 \ln{(x-3)} + \dfrac{3}{(x-3)} + c \;\; \text{or} \,\right.$$

$$\left. \ln{\{(x-7)^5\,(x-3)^4\}} + \dfrac{3}{x-3} + c \,\right]$$

83. $\dfrac{-(9x^2 + 4x + 4)}{x^2(x^2 - 4)}$.

$$\left[\, \ln{x} - \dfrac{1}{x} + 2 \ln{(x+2)} - 3 \ln{(x-2)} + c \;\; \text{or} \;\; \ln{\left\{\dfrac{x(x+2)^2}{(x-2)^3}\right\}} - \dfrac{1}{x} + c \,\right]$$

84. $\dfrac{-(a^2 + 5a + 13)}{(a^2 + 5)\,(a-2)}$.

$$\left[\, \ln{(a^2+5)} - \dfrac{1}{\sqrt{5}} \arctan{\dfrac{a}{\sqrt{5}}} - 3 \ln{(a-2)} + c \;\; \text{or} \,\right.$$

$$\left. \ln{\left\{\dfrac{(a^2+5)}{(a-2)^3}\right\}} - \dfrac{1}{\sqrt{5}} \arctan{\dfrac{a}{\sqrt{5}}} + c \,\right]$$

85. $\dfrac{3-x}{(x^2+3)\,(x+3)}$.

$$\left[\, \dfrac{1}{2\sqrt{3}} \arctan{\dfrac{x}{\sqrt{3}}} - \dfrac{1}{4} \ln{(x^2+3)} + \dfrac{1}{2} \ln{(x+3)} + c \;\; \text{or} \,\right.$$

$$\left. \ln{\left\{\dfrac{(x+3)^{1/2}}{(x^2+3)^{1/4}}\right\}} + \dfrac{1}{2\sqrt{3}} \arctan{\dfrac{x}{\sqrt{3}}} + c \,\right]$$

86. $\dfrac{12 - 2x - 5x^2}{(x^2 + x + 1)\,(3-x)}$.

$$\left[\, \ln{(x^2+x+1)} + \dfrac{8}{\sqrt{3}} \arctan{\dfrac{2\left(x+\tfrac{1}{2}\right)}{\sqrt{3}}} + 3 \ln{(3-x)} + c \,\right]$$

87. $\dfrac{x^3 + 7x^2 + 8x + 10}{x(x^2 + 2x + 5)}$.

$[\, x + 2 \ln x + \tfrac{3}{2} \ln (x^2 + 2x + 5) - 2 \arctan \left(\dfrac{x + 1}{2} \right) + c \,]$

88. $\dfrac{5x^3 - 3x^2 + 41x - 64}{(x^2 + 6)(x - 1)^2}$.

$\left[\dfrac{2}{\sqrt{6}} \arctan \dfrac{x}{\sqrt{6}} - \dfrac{3}{2} \ln (x^2 + 6) + 8 \ln (x - 1) + \dfrac{3}{(x - 1)} + c \right]$

89. $\dfrac{6x^3 + 5x^2 + 4x + 3}{(x^2 + x + 1)(x^2 - 1)}$.

$[\ln (x^2 + x + 1) - \dfrac{4}{\sqrt{3}} \arctan \dfrac{2 (x + \frac{1}{2})}{\sqrt{3}} + 3 \ln (x - 1) + \ln (x + 1) + c]$

In Problems 90 to 99 evaluate the definite integrals correct to four decimal places.

90. $\displaystyle\int_4^6 \dfrac{x - 7}{x^2 - 2x - 3} \, dx.$ $[-0.425\ 7]$

91. $\displaystyle\int_2^3 \dfrac{x^3 - 2x^2 - 3x - 2}{(x + 2)(x - 1)} \, dx.$ $[-0.993\ 7]$

92. $\displaystyle\int_3^4 \dfrac{4x^2 + 15x - 1}{(x + 1)(x - 2)(x + 3)} \, dx.$ $[2.371\ 6]$

93. $\displaystyle\int_3^5 \dfrac{x^2 + 1}{x^2 + x - 6} \, dx.$ $[2.523\ 2]$

94. $\displaystyle\int_6^8 \dfrac{1}{(x^2 - 25)} \, dx.$ $[0.093\ 2]$

95. $\displaystyle\int_2^3 \dfrac{1}{(16 - x^2)} \, dx.$ $[0.105\ 9]$

96. $\displaystyle\int_1^2 \dfrac{2 + x + 6x^2 - 2x^3}{x^2 (x^2 + 1)} \, dx.$ $[1.605\ 7]$

97. $\displaystyle\int_2^3 \dfrac{2x^2 - x - 2}{(x - 1)^3} \, dx.$ $[2.511\ 3]$

98. $\displaystyle\int_3^4 \dfrac{2x^3 + 4x^2 - 8x - 4}{x^2 (x^2 - 4)} \, dx.$ $[1.041\ 8]$

99. $\displaystyle\int_0^1 \dfrac{-(4x^2 + 9x + 8)}{(x + 1)^2 (x + 2)} \, dx.$ $[-2.546\ 5]$

100. The velocity constant k of a given chemical reaction is given by

$$kt = \int \frac{\mathrm{d}x}{(3 - 0.4x)(2 - 0.6x)}$$

where $x = 0$ when $t = 0$.

Determine kt.

$$\left[\ln \left\{ \frac{2(3 - 0.4x)}{3(2 - 0.6x)} \right\} \right]$$

Chapter 19

Integration by parts (CB 16)

1 Introduction

When differentiating the product uv, where u and v are both functions of x, then:

$$\frac{d}{dx}(uv) = v\frac{du}{dx} + u\frac{dv}{dx}$$

This is known as the **product rule for differentiation**.
Rearranging this formula gives:

$$u\frac{dv}{dx} = \frac{d}{dx}(uv) - v\frac{du}{dx}$$

Integrating both sides with respect to x gives:

$$\int u\frac{dv}{dx}dx = \int \frac{d}{dx}(uv)\,dx - \int v\frac{du}{dx}\,dx$$

Since integration is the reversal of the differentiation process this becomes:

$$\int u\,dv = uv - \int v\,du$$

This formula enables products of certain simple functions to be integrated in cases where it is possible to evaluate $\int v\,du$. This is known as the **integration by parts formula** and is a useful method of integration enabling such integrals as $\int x\,e^x\,dx$, $\int x\cos x\,dx$, $\int t^2\sin t\,dt$, $\int x^3\ln x\,dx$, $\int \ln x\,dx$, $\int e^{ax}\sin bx\,dx$, etc., to be determined.

2 Application of the integration by parts formula

Problem 1. Find $\int x\, e^x\, dx$.

In the integration by parts formula we must let one function of our product be equal to u and the other be equal to dv.

Let $u = x$ and $dv = e^x\, dx$.

Then $\dfrac{du}{dx} = 1$, i.e. $du = dx$ and $v = \int e^x\, dx = e^x$.

There are now expressions for u, du, dv and v which are substituted into the formula:

$$\int u \quad dv \;=\; u \cdot v \;-\; \int v \quad du$$
$$\int x \quad e^x\, dx = x \cdot e^x - \int e^x\, dx$$
$$= x \cdot e^x - e^x + c$$

Hence, $\int x\, e^x\, dx \;=\; e^x\,(x-1) + c$.

The following four points should be noted:

(i) The above result may be checked by differentiation.

Thus $\dfrac{d}{dx}\,[e^x\,(x-1) + c] = e^x\,(1) + (x-1)\,e^x + 0$
$$= e^x + x\,e^x - e^x = x\,e^x$$

(ii) Given that $dv = e^x\, dx$ then $v = \int e^x\, dx$, which is strictly equal to $e^x + a$ constant. However, the constant is omitted at this stage. If a constant, say k, were included, then:

$\int x\, e^x\, dx \;=\; x(e^x + k) - \int(e^x + k)\, dx$
$$= x\,e^x + x\,k - (e^x + kx + c)$$
$$= x\,e^x - e^x + c, \text{ as before.}$$

Thus the constant k is an unnecessary addition. A constant is added only after the final integration (i.e. c in the above problem).

(iii) If instead of choosing to let $u = x$ and $dv = e^x\, dx$ we let $u = e^x$ and $dv = x\, dx$ then $du = e^x\, dx$ and $v = \int x\, dx = \dfrac{x^2}{2}$.

Hence $\int x\, e^x\, dx \;=\; (e^x)\left(\dfrac{x^2}{2}\right) - \int\left(\dfrac{x^2}{2}\right) e^x\, dx$.

The integral on the far right-hand side is seen to be more complicated than the original, thus the original choice of letting $u = e^x$ and $dv = x\, dx$ was wrong. The choice must be such that the 'u part' becomes a constant after differentiation and the 'dv' part can be integrated easily. (It will be seen later that for the 'u part' to become a constant often requires more than one differentiation (see Problem 5).)

(iv) If a product to be integrated contains an 'x' term then this term is chosen as the '*u* part' and the other function as the '*dv* part' except where a logarithmic term is involved (see Problems 6 and 7).

Problem 2. Determine $\int x \sin x \, dx$.

Let $u = x$ and $dv = \sin x \, dx$
Then $du = dx$ and $v = \int \sin x \, dx = -\cos x$
Substituting into $\int u \, dv = uv - \int v \, du$ gives:

$$\int x \sin dx = (x)(-\cos x) - \int (-\cos x) \, dx$$

$$= -x \cos x + \int \cos x \, dx$$

$$= -x \cos x + \sin x + c.$$

This result can be checked by differentiating.

$$\text{Thus } \frac{d}{dx}(-x \cos x + \sin x + c) = (-x)(-\sin x) + (\cos x)(-1) + \cos x$$

$$= x \sin x - \cos x + \cos x = x \sin x$$

Problem 3. Evaluate $\int_0^1 2x \, e^{3x} \, dx$

Let $u = 2x$ and $dv = e^{3x} \, dx$
Then $\dfrac{du}{dx} = 2$, i.e. $du = 2 \, dx$ and $v = \int e^{3x} \, dx = \dfrac{e^{3x}}{3}$

Substituting into $\int u \, dv = uv - \int v \, du$ gives:

$$\int 2x \, e^{3x} \, dx = (2x)\left(\frac{e^{3x}}{3}\right) - \int \left(\frac{e^{3x}}{3}\right)(2 \, dx)$$

$$= \frac{2}{3} x \, e^{3x} - \frac{2}{3} \int e^{3x} \, dx$$

$$= \frac{2}{3} x \, e^{3x} - \frac{2}{3}\left(\frac{e^{3x}}{3}\right) + c$$

$$= \frac{2}{3} x \, e^{3x} - \frac{2}{9} e^{3x} + c$$

$$\text{Hence } \int_0^1 2x \, e^{3x} \, dx = \left[\frac{2}{3} x \, e^{3x} - \frac{2}{9} e^{3x}\right]_0^1 =$$

$$= \left(\frac{2}{3} e^3 - \frac{2}{9} e^3\right) - \left(0 - \frac{2}{9} e^0\right)$$

$$= \frac{4}{9} e^3 + \frac{2}{9} = 8.926\ 9 + 0.222\ 2$$

$$= \textbf{9.149 correct to three decimal places.}$$

Problem 4. Evaluate $\displaystyle\int_0^{\frac{\pi}{2}} 3t \cos 2t \, dt$.

Let $u = 3t$ and $dv = \cos 2t \, dt$

Then $du = 3dt$ and $v = \int\cos 2t \, dt = \dfrac{1}{2} \sin 2t$

Substituting into $\int u \, dv = uv - \int v \, du$ gives:

$$\int 3t \cos 2t \, dt = (3t)\left(\frac{\sin 2t}{2}\right) - \int \left(\frac{\sin 2t}{2}\right)(3 \, dt)$$

$$= \frac{3t}{2} \sin 2t - \frac{3}{2}\int \sin 2t \, dt$$

$$= \frac{3t}{2} \sin 2t - \frac{3}{2}\left(\frac{-\cos 2t}{2}\right) + c$$

$$= \frac{3t}{2} \sin 2t + \frac{3}{4} \cos 2t + c$$

Hence $\displaystyle\int_0^{\frac{\pi}{2}} 3t \cos 2t \, dt = \left[\frac{3t}{2} \sin 2t + \frac{3}{4} \cos 2t\right]_0^{\frac{\pi}{2}} =$

$$= \left[\frac{3}{2}\left(\frac{\pi}{2}\right) \sin \pi + \frac{3}{4} \cos \pi\right] - \left[0 + \frac{3}{4} \cos 0\right]$$

$$= \left(0 + \frac{3}{4}(-1)\right) - \left(0 + \frac{3}{4}\right) = -\frac{3}{4} - \frac{3}{4} = -1\tfrac{1}{2}.$$

Problem 5. Determine $\int x^2 \cos x \, dx$.

Let $u = x^2$ and $dv = \cos x \, dx$
Then $du = 2x \, dx$ and $v = \int\cos x \, dx = \sin x$
Substituting into $\int u \, dv = uv - \int v \, du$ gives:

$$\int x^2 \cos x \, dx = (x^2)(\sin x) - \int(\sin x)(2x \, dx)$$

$$= x^2 \sin x - 2[\int x \sin x \, dx]$$

The integral on the right-hand side in the bracket is not a standard and cannot be determined 'on sight'. Since it is a product of two simple functions we may use integration by parts again.

Now $\int x \sin x \, dx = -x \cos x + \sin x$ (from Problem 2)

Thus $\int x^2 \cos x \, dx = x^2 \sin x - 2[-x \cos x + \sin x] + c$

$$= x^2 \sin x + 2x \cos x - 2 \sin x + c$$

$$= (x^2 - 2) \sin x + 2x \cos x + c$$

Generally, if the term in x is of power n then the integration by parts formula is applied n times, provided one of the functions is not $\ln x$, as in Problem 6 following.

Problem 6. $\int x^2 \ln x\, dx$.

Whenever a product consists of a term in x and a logarithmic function, as in this problem, it is always the logarithmic function that is chosen as the 'u part'. The reason for this is that $\frac{d}{dx}(\ln x)$ is $\frac{1}{x}$, but $\int \ln x\, dx$ is not normally remembered as a standard integral.

Thus if $u = \ln x$ and $dv = x^2\, dx$

then $du = \frac{1}{x}\, dx$ and $v = \int x^2\, dx = \frac{x^3}{3}$

Substituting into $\int u\, dv = uv - \int v\, du$ gives:

$$\int x^2 \ln x\, dx = (\ln x)\left(\frac{x^3}{3}\right) - \int \left(\frac{x^3}{3}\right)\left(\frac{1}{x}\, dx\right)$$

$$= \frac{x^3}{3}\ln x - \frac{1}{3}\int x^2\, dx$$

$$= \frac{x^3}{3}\ln x - \frac{1}{3}\left(\frac{x^3}{3}\right) + c$$

$$= \frac{x^3}{3}\ln x - \frac{x^3}{9} + c$$

$$= \frac{x^3}{9}\ (3\ln x - 1) + c$$

Problem 7. Determine $\int \ln x\, dx$.

In each of the previous problems the components of the product have been obvious. However, $\int \ln x\, dx$ is a special case for initially it appears not to be a product. However, $\int \ln x\, dx$ is the same as $\int 1 \times \ln x\, dx$.

Let $u = \ln x$ and $dv = 1\, dx$

Then $du = \frac{1}{x}\, dx$ and $v = \int 1\, dx = x$.

Hence $\int \ln x\, dx = (\ln x)(x) - \int (x)\left(\frac{1}{x}\, dx\right)$

$$= x \ln x - \int dx$$

$$= x \ln x - x + c$$

$$= x(\ln x - 1) + c$$

Problem 8. Find $\int e^{ax} \sin bx \, dx$.

With an integral of a product of an exponential function and a sine or cosine function it does not matter which function is made equal to 'u'. Thus let $u = e^{ax}$ and $dv = \sin bx \, dx$

then $du = a\, e^{ax}\, dx$ and $v = \int \sin bx \, dx = \dfrac{-\cos bx}{b}$

Thus
$$\int e^{ax} \sin bx \, dx = (e^{ax}) \left(\frac{-\cos bx}{b}\right) - \int \left(\frac{-\cos bx}{b}\right)(a\, e^{ax}\, dx)$$

$$= -\frac{1}{b}\, e^{ax} \cos bx + \frac{a}{b}\left[\int e^{ax} \cos bx \, dx\right] \qquad (1)$$

It would seem that we are no nearer a solution of the initial integral. However, the integration by parts formula may be applied to the integral in the bracket.

Let $u = e^{ax}$ and $dv = \cos bx \, dx$

Then $du = a\, e^{ax}\, dx$ and $v = \int \cos bx \, dx = \dfrac{\sin bx}{b}$

Thus
$$\int e^{ax} \cos bx \, dx = (e^{ax})\left(\frac{\sin bx}{b}\right) - \int \left(\frac{\sin bx}{b}\right)(a\, e^{ax}\, dx)$$

$$= \frac{1}{b}\, e^{ax} \sin bx - \frac{a}{b}\int e^{ax} \sin bx \, dx$$

Substituting this result into equation (1) gives:

$$\int e^{ax} \sin bx \, dx = -\frac{1}{b}\, e^{ax} \cos bx + \frac{a}{b}\left[\frac{1}{b}\, e^{ax} \sin bx - \frac{a}{b}\int e^{ax} \sin bx \, dx\right]$$

$$= \frac{-1}{b}\, e^{ax} \cos bx + \frac{a}{b^2}\, e^{ax} \sin bx - \frac{a^2}{b^2}\left[\int e^{ax} \sin bx \, dx\right]$$

The integral in the bracket on the right-hand side is the same as the integral on the left-hand side thus they may be combined on the left-hand side of the equation. Thus:

$$\int e^{ax} \sin bx \, dx + \frac{a^2}{b^2}\int e^{ax} \sin bx \, dx = \frac{-1}{b}\, e^{ax} \cos bx + \frac{a}{b^2}\, e^{ax} \sin bx$$

i.e. $\left(1 + \dfrac{a^2}{b^2}\right)\displaystyle\int e^{ax} \sin bx \, dx = \dfrac{e^{ax}}{b^2}(a \sin bx - b \cos bx)$

$\left(\dfrac{b^2 + a^2}{b^2}\right)\displaystyle\int e^{ax} \sin bx \, dx = \dfrac{e^{ax}}{b^2}(a \sin bx - b \cos bx)$

Hence $\displaystyle\int e^{ax} \sin bx \, dx = \left(\dfrac{b^2}{a^2 + b^2}\right)\dfrac{e^{ax}}{b^2}(a \sin bx - b \cos bx)$

$$= \left(\frac{e^{ax}}{a^2 + b^2}\right)(a \sin bx - b \cos bx) + c$$

A product of an exponential function and a sine or cosine function thus involves integration by parts twice. If, in the above problem, the exponential function is made equal to dv instead of u the same result is obtained. It is left as a student exercise to prove this.

By a similar method to above it may be shown that:

$$\int e^{ax} \cos bx \ dx = \left(\frac{e^{ax}}{a^2 + b^2}\right)(b \sin bx + a \cos bx) + c$$

Further problems on integration by parts may be found in the following Section (3) (Problems 1 to 27).

3 Further problems

Find the following integrals using integration by parts

1. $\int x \ e^{3x} \ dx$ $\quad \left[\dfrac{e^{3x}}{9}(3x - 1) + c\right]$

2. $\int 3x \ e^{2x} \ dx.$ $\quad \left[\dfrac{3}{4} \ e^{2x}(2x - 1) + c\right]$

3. $\displaystyle\int \dfrac{5x}{e^{4x}} \ dx.$ $\quad \left[\dfrac{-5}{16} \ e^{-4x}(4x + 1) + c\right]$

4. $\int x \cos x \ dx.$ $\quad [x \sin x + \cos x + c]$

5. $\int x \ln x \ dx.$ $\quad \left[\dfrac{x^2}{4} \ (2 \ln x - 1) + c\right]$

6. $\int 3x \sin 2x \ dx.$ $\quad \left[\dfrac{-3x}{2} \ \cos 2x + \dfrac{3}{4} \sin 2x + c\right]$

7. $\int \ln 4x \ dx.$ $\quad [x (\ln 4x - 1) + c]$

8. $\displaystyle\int \dfrac{\ln y \ dy}{y^2}.$ $\quad \left[-\dfrac{1}{y} \ (\ln y + 1) + c\right]$

9. $\int 2x \cos 5x \ dx.$ $\quad \left[\dfrac{2}{25} \ (5x \sin 5x + \cos 5x) + c\right]$

10. $\int 2x^2 \ e^x \ dx.$ $\quad [2e^x (x^2 - 2x + 2) + c]$

11. $\int x^2 \sin 2x \ dx.$ $\quad \left[\left(\dfrac{1}{4} - \dfrac{x^2}{2}\right)\cos 2x + \dfrac{x}{2} \ \sin 2x + c\right]$

12. $\int e^{2x} \cos x \ dx.$ $\quad \left[\dfrac{e^{2x}}{5} \ (\sin x + 2 \cos x) + c\right]$

13. $\int 4 \ \theta^2 \cos 3\theta \ d\theta.$ $\quad \left[4\left(\dfrac{\theta^2}{3} - \dfrac{2}{27}\right)\sin 3\theta + \dfrac{8}{9} \ \theta \cos 3\theta + c\right]$

14. $\int 3e^x \sin 2x \, dx.$ $\left[\dfrac{3}{5} \, e^x \, (\sin 2x - 2 \cos 2x) + c\right]$

15. $\int 4x \sec^2 x \, dx.$ $\quad [4[x \tan x - \ln (\sec x)] + c]$

Evaluate the following integrals correct to three decimal places.

16. $\displaystyle\int_0^2 x \, e^x \, dx.$ $\quad [8.389]$

17. $\displaystyle\int_0^{\frac{\pi}{2}} x \cos x \, dx.$ $\quad [0.571]$

18. $\displaystyle\int_0^1 t \, e^{2t} \, dt.$ $\quad [2.097]$

19. $\displaystyle\int_0^{\frac{\pi}{4}} \phi \sin 2\phi \, d\phi.$ $\quad [0.250]$

20. $\displaystyle\int_1^2 \ln x^2 \, dx.$ $\quad [0.773]$

21. $\displaystyle\int_0^{\frac{\pi}{2}} 3x^2 \cos x \, dx.$ $\quad [1.402]$

22. $\displaystyle\int_0^1 x^2 \, e^x \, dx.$ $\quad [0.718]$

23. $\displaystyle\int_1^4 \sqrt{e^t} \, \ln t \, dt.$ $\quad [4.282]$

24. $\displaystyle\int_0^1 2e^t \cos 2t \, dt.$ $\quad [1.125]$

25. $\displaystyle\int_0^{\frac{\pi}{2}} e^\theta \sin \theta \, d\theta.$ $\quad [2.905]$

26. In the study of damped oscillations integrations of the following type are important:

$$C = \int_0^1 e^{-0.4\theta} \cos 1.2\theta \, d\theta$$

and $S = \int_0^1 e^{-0.4\theta} \sin 1.2\theta \, d\theta$

Determine C and S. $[C = 0.66, S = 0.41]$

27. If a string is plucked at a point $x = \dfrac{l}{3}$ with an amplitude a and released, the equation of motion is

$$K = \frac{2}{l}\left\{ \int_0^{\frac{l}{3}} \frac{3a}{l} x \sin \frac{n\pi}{l} x \, dx + \int_{\frac{l}{3}}^l \frac{3a}{2l}(l-x) \sin \frac{n\pi}{l} x \, dx \right\}$$

where n is a constant.

Show that $K = \dfrac{9a}{\pi^2 n^2} \sin \dfrac{n\pi}{3}$.

Chapter 20

First order differential equations by separation of variables (CD 17)

1 Introduction

Definition. A differential equation is one that contains differential coefficients. Examples include:

(i) $\dfrac{\mathrm{d}y}{\mathrm{d}x} = 5x$

(ii) $\dfrac{\mathrm{d}^2 y}{\mathrm{d}x^2} + 3\dfrac{\mathrm{d}y}{\mathrm{d}x} + 4y = 0$

(iii) $\left(\dfrac{\mathrm{d}^2 s}{\mathrm{d}t^2}\right)^3 + 2\left(\dfrac{\mathrm{d}s}{\mathrm{d}t}\right)^4 = 4.$

Order. Differential equations are classified according to the highest derivative which occurs in them. Hence example (i) above is a **first order differential equation** and examples (ii) and (iii) are both **second order differential equations**.

Degree. The degree of a differential equation is that of the highest power of the highest differential which the equation contains after any necessary simplification. Thus example (i) above is of first order, first degree, example (ii) is of second order, first degree and example (iii) is of second order, third degree.

Solutions of a differential equation

Starting with a differential equation it is possible, by integration and by being given enough data to determine the constants, to obtain the original function. The process of determining the original relationship between the two variables is called 'solving the differential equation'. A solution to a differential equation which contains one or more arbitrary constants of integration is called the **general solution** of the differential equation. When finding the general solution of a first order differential equation one arbitrary constant results; when finding the general solution of a second order differential equation two arbitrary constants result, and so on. When additional information is given so that constants may be calculated the **particular solution** of the differential equation is obtained. The additional information is called the **boundary conditions.**

In this text only differential equations of the first and second order and first degree are discussed and the method of solution of five types of first order and two types of second order are considered. The first order types discussed are:

(a) differential equations of the form $\dfrac{dy}{dx} = f(x)$,

(b) differential equations of the form $\dfrac{dy}{dx} = f(y)$,

(c) 'variable separable' types of differential equations,

(d) differential equations which are homogeneous in x and y, and

(e) differential equations of the form $\dfrac{dy}{dx} + Py = Q$, where P and Q are

functions of x.

Types (a), (b) and (c) are considered in this chapter, type (d) in Chapter 21 and type (e) in Chapter 22. Two second order types of differential equation are discussed in Chapters 23 and 24.

In Chapter 9, *Technician Mathematics level 3* by J.O. Bird and A.J.C. May, an introduction to differential equations is given, together with the method of solution of equations of the form $\dfrac{dy}{dx} = f(x)$ and $\dfrac{dQ}{dt} = kQ$. The former type is revised in the following section of this chapter.

2 Solution of differential equations of the form $\dfrac{dy}{dx} = f(x)$

An equation of the form $\dfrac{dy}{dx} = f(x)$ may be solved immediately by integration. The solution is $y = \int f(x)\, dx$.

Problem 1. Find the general solutions of the following differential equations.

(a) $\dfrac{dy}{dx} = 5x^2 + \cos 3x$ (b) $x\dfrac{dy}{dx} = 3 - 2x^2$.

274

(a) If $\dfrac{dy}{dx} = 5x^2 + \cos 3x$, then $y = \int(5x^2 + \cos 3x)\,dx$

i.e. $y = \dfrac{5x^3}{3} + \dfrac{1}{3}\sin 3x + c$

(b) If $x\dfrac{dy}{dx} = 3 - 2x^2$, then $\dfrac{dy}{dx} = \dfrac{3 - 2x^2}{x} = \dfrac{3}{x} - 2x$

Hence $y = \int\left(\dfrac{3}{x} - 2x\right)dx$

i.e. $y = 3\ln x - x^2 + c$

Problem 2. Find the particular solutions of the following differential equations satisfying the given boundary conditions:

(a) $3\dfrac{dy}{dx} + x = 6$ and $y = 5\tfrac{1}{2}$ when $x = 3$.

(b) $2\dfrac{dr}{d\theta} + \sin 2\theta = 0$ and $r = 2$ when $\theta = \dfrac{\pi}{2}$.

(a) If $3\dfrac{dy}{dx} + x = 6$, then $\dfrac{dy}{dx} = \dfrac{6 - x}{3} = 2 - \dfrac{x}{3}$

Hence $y = \int\left(2 - \dfrac{x}{3}\right)dx$

i.e. $y = 2x - \dfrac{x^2}{6} + c$

(This is the general solution.)

Substituting the boundary conditions, $y = 5\tfrac{1}{2}$ when $x = 3$ to evaluate c gives:

$5\tfrac{1}{2} = 6 - 1\tfrac{1}{2} + c$

i.e. $c = 1$.

Hence the particular solution is $y = 2x - \dfrac{x^2}{6} + 1$.

(b) If $2\dfrac{dr}{d\theta} + \sin 2\theta = 0$, then $\dfrac{dr}{d\theta} = \dfrac{-\sin 2\theta}{2}$

Hence $r = \int\dfrac{-\sin 2\theta}{2}\,d\theta$

i.e. $r = \dfrac{1}{4}\cos 2\theta + c$

Substituting the boundary conditions, $r = 2$ when $\theta = \dfrac{\pi}{2}$ to evaluate c

gives:

$$2 = \frac{1}{4} \cos \pi + c$$

i.e. $c = 2 - \left(-\frac{1}{4}\right) = 2\frac{1}{4}$

Hence the particular solution is $r = \frac{1}{4} \cos 2\theta + 2\frac{1}{4} = \frac{1}{4} (\cos 2\theta + 9)$.

Further problems on solving differentials equations of the form $\frac{dy}{dx} = f(x)$
may be found in Section 5 (Problems 1 to 14), page 280.

3 Solution of differential equations of the form $\frac{dy}{dx} = f(y)$

An equation of the form $\frac{dy}{dx} = f(y)$ may be rearranged to give:

$$\frac{dx}{dy} = \frac{1}{f(y)}$$

i.e. $dx = \frac{dy}{f(y)}$

Integrating both sides gives: $\int dx = \int \frac{dy}{f(y)}$

Hence the solution may be obtained by direct integration.

Problem 1. Solve the differential equations:

(a) $\frac{dy}{dx} = 2 + y$. (b) $3\frac{dy}{dx} = \sec 2y$.

(a) Rearranging $\frac{dy}{dx} = 2 + y$ gives $dx = \frac{dy}{(2 + y)}$

Integrating both sides gives: $\int dx = \int \frac{dy}{(2 + y)}$

i.e. $x = \ln (2 + y) + c$ \hfill (1)

The general solution of differential equations can sometimes be rearranged. In this case, for example, if $C = \ln D$, where D is a constant, then:

$$x = \ln (2 + y) + \ln D$$
i.e. $x = \ln D (2 + y)$, from the law of logarithms,
from which $e^x = D(2 + y)$ \hfill (2)

Equations (1) and (2) are both acceptable general solutions of the differential equation $\dfrac{dy}{dx} = 2 + y$.

(b) Rearranging $3 \dfrac{dy}{dx} = \sec 2y$ gives $dx = \dfrac{3}{\sec 2y} dy = 3 \cos 2y \, dy$

Integrating both sides gives: $\int dx = \int 3 \cos 2y \, dy$

Hence the general solution is $x = \dfrac{3}{2} \sin 2y + c.$

Problem 2. The rate at which a body cools is given by the equation $\dfrac{d\theta}{dt} = -k\theta$, where θ is the temperature of the body above its surroundings and k is a constant. Solve the equation for θ given that at $t = 0, \theta = \theta_0$

$\dfrac{d\theta}{dt} = -k\theta$ is of the form $\dfrac{dy}{dx} = f(y)$.

Rearranging gives $dt = \dfrac{-1}{k\theta} d\theta$

Integrating both sides gives: $\int dt = \dfrac{-1}{k} \int \dfrac{d\theta}{\theta}$

i.e. $t = \dfrac{-1}{k} \ln \theta + c$ (1)

Substituting the boundary conditions $t = 0$ and $\theta = \theta_0$ to find c gives:

$0 = \dfrac{-1}{k} \ln \theta_0 + c$

i.e. $c = \dfrac{1}{k} \ln \theta_0$

Substituting $c = \dfrac{1}{k} \ln \theta_0$ in equation (1) gives:

$t = -\dfrac{1}{k} \ln \theta + \dfrac{1}{k} \ln \theta_0$

$t = \dfrac{1}{k} (\ln \theta_0 - \ln \theta) = \dfrac{1}{k} \ln \left(\dfrac{\theta_0}{\theta} \right)$

$kt = \ln \left(\dfrac{\theta_0}{\theta} \right)$

$e^{kt} = \dfrac{\theta_0}{\theta}$

$e^{-kt} = \dfrac{\theta}{\theta_0}$

Hence $\theta = \theta_0 \, e^{-kt}$

Further problems on solving differential equations of the form $\dfrac{dy}{dx} = f(y)$
may be found in Section 5 (Problems 15 to 29), page 281.

4 Solution of 'variable separable' type of differential equations

An equation of the form $\dfrac{dy}{dx} = f(x) \cdot g(y)$, where $f(x)$ is a function of x only
and $g(y)$ is a function of y only, may be rearranged thus:

$$\frac{dy}{g(y)} = f(x) \, dx$$

Integrating both sides gives $\displaystyle\int \frac{dy}{g(y)} = \int f(x) \, dx$, i.e. the left-hand side is
the integral of a function of y with respect to y and the right-hand side is the
integral of a function of x with respect to x.

When two variables can be rearranged into two separate groups as shown
above, each consisting of only one variable, the variables are said to be
separable.

The equations of the type $\dfrac{dy}{dx} = f(x)$ and $\dfrac{dy}{dx} = f(y)$ discussed in Sections
2 and 3 are, in fact, merely special simple cases of 'separating the variables'.

Problem 1. Solve the differential equations:

(a) $\dfrac{dy}{dx} = \dfrac{3x^2 - 2}{2y - 1}$ (b) $2xy \dfrac{dy}{dx} = 1 + y^2$

(a) $\dfrac{dy}{dx} = \dfrac{3x^2 - 2}{2y - 1}$.

Separating the variables gives $(2y - 1) \, dy = (3x^2 - 2) \, dx$.
Integrating both sides gives $\int (2y - 1) \, dy = \int (3x^2 - 2) \, dx$.
Hence the general solution is $y^2 - y = x^3 - 2x + C$.

Note that when integrating both sides of an equation there is no need
to put an arbitrary constant on both sides of the result. In this case, if
this was done, then:

$$y^2 - y + A = x^3 - 2x + B$$
$$\text{and} \quad y^2 - y \quad = x^3 - 2x + C, \text{ where } C = B - A.$$

(b) $2xy \dfrac{dy}{dx} = 1 + y^2$.

Separating the variables gives: $\dfrac{2y}{1 + y^2} \, dy = \dfrac{1}{x} \, dx$.

Integrating both sides gives: $\displaystyle\int \frac{2y}{(1+y^2)}\ dy = \int \frac{1}{x}\ dx$

Hence the general solution is $\ln(1+y^2) = \ln x + C$ (1)

$$\text{or } \ln(1+y^2) - \ln x = C$$

from which $\ln\left(\dfrac{1+y^2}{x}\right) = C$

and $\dfrac{1+y^2}{x} = e^C$ (2)

Also, if in equation (1), $C = \ln A$, we have $\ln(1+y^2) = \ln x + \ln A$

$$\ln(1+y^2) = \ln(Ax)$$

i.e. $1+y^2 = Ax$ (3)

Equations (1), (2) and (3) are all valid general solutions to the differential equation $2xy\ \dfrac{dy}{dx} = 1+y^2$, none of them being any more correct than the others. Thus, by manipulation, it is possible to obtain several general solutions to a differential equation.

Problem 2. Find the particular solution of $\dfrac{dy}{dx} = 3e^{2x-3y}$ given that $y = 0$ when $x = 0$.

$\dfrac{dy}{dx} = 3e^{2x-3y} = 3(e^{2x})(e^{-3y})$ by the laws of indices.

Separating the variables gives $\dfrac{dy}{e^{-3y}} = 3e^{2x}\ dx$

i.e. $e^{3y}\ dy = 3e^{2x}\ dx$

Integrating both sides gives $\int e^{3y}\ dy = 3\int e^{2x}\ dx$

$$\frac{1}{3}\ e^{3y} = \frac{3}{2}\ e^{2x} + C.$$

(This is the general solution.)

When $y = 0, x = 0$, thus $\dfrac{1}{3}\ e^0 = \dfrac{3}{2}\ e^0 + C$

$$C = \frac{1}{3} - \frac{3}{2} = -\frac{7}{6}$$

Hence the particular solution is $\dfrac{1}{3}\ e^{3y} = \dfrac{3}{2}\ e^{2x} - \dfrac{7}{6}$

or $2e^{3y} = 9e^{2x} - 7.$

Problem 3. An electrical circuit contains inductance L and resistance R connected to a constant voltage source E. The current i is given by the differential equation $E - L \dfrac{di}{dt} = Ri$, where L and R are constants. Find the current in terms of time t given that when $t = 0, i = 0$.

$E - L \dfrac{di}{dt} = Ri.$ Rearranging gives $\dfrac{di}{dt} = \dfrac{E - Ri}{L}$

and $\dfrac{di}{E - Ri} = \dfrac{dt}{L}$

Integrating both sides gives: $\displaystyle\int \dfrac{di}{E - Ri} = \int \dfrac{dt}{L}$

$$-\dfrac{1}{R} \ln (E - Ri) = \dfrac{t}{L} + C.$$

(This is the general solution.)

$t = 0$ when $i = 0$ hence $-\dfrac{1}{R} \ln E = C.$

Thus, $-\dfrac{1}{R} \ln (E - Ri) = \dfrac{t}{L} - \dfrac{1}{R} \ln E.$

This particular solution must now be transposed to find i.

$$-\dfrac{1}{R} \ln (E - Ri) + \dfrac{1}{R} \ln E = \dfrac{t}{L}$$

$$\dfrac{1}{R} (\ln E - \ln (E - Ri)) = \dfrac{t}{L}$$

$$\ln \left(\dfrac{E}{E - Ri} \right) = \dfrac{Rt}{L}$$

$$\dfrac{E}{E - Ri} = e^{\frac{Rt}{L}}$$

$$\dfrac{E - Ri}{E} = e^{-\frac{Rt}{L}}$$

$$Ri = E - E e^{-\frac{Rt}{L}}$$

Hence current $i = \dfrac{E}{R}\left(1 - e^{-\frac{Rt}{L}}\right)$

This expression for current represents the natural law of growth of current in an inductive circuit.

280

Further problems on solving 'variables separable' types of differential equations may be found in the following Section (5) (Problems 30 to 46), page 283.

5 Further problems

Differential equations of the form $\dfrac{dy}{dx} = f(x)$

In problems 1 to 9 solve the differential equations.

1. $\dfrac{dy}{dx} = 2x^4$. $\quad \left[y = \dfrac{2}{5}x^5 + C \right]$

2. $\dfrac{dy}{dx} = 5x + \sin x$. $\quad \left[y = \dfrac{5}{2}x^2 - \cos x + C \right]$

3. $x\dfrac{dy}{dx} = 4 - x^2$. $\quad \left[y = 4 \ln x - \dfrac{x^2}{2} + C \right]$

4. $\dfrac{dy}{dx} - 2x^3 = e^{3x}$. $\quad [6y = 2e^{3x} + 3x^4 + C]$

5. $x^2 \dfrac{dy}{dx} = 2 + x$. $\quad \left[y = \ln x - \dfrac{2}{x} + C \right]$

6. $\dfrac{dy}{dx} + x = 2$ and $y = 3$ when $x = 2$. $\quad \left[y = 2x - \dfrac{x^2}{2} + 1 \right]$

7. $2\dfrac{dr}{d\theta} + \cos \theta = 0$ and $r = \dfrac{5}{2}$ when $\theta = \dfrac{\pi}{2}$. $\quad \left[r = 3 - \dfrac{1}{2}\sin \theta \right]$

8. $x \left(x - \dfrac{dy}{dx} \right) = 3$ and $y = 1$ when $x = 1$. $\quad [2y = x^2 - 6 \ln x + 1]$

9. $\dfrac{1}{2e^t} + 4 = t - 3\dfrac{d\theta}{dt}$ and $\theta = \dfrac{1}{6}$ when $t = 0$

$$\left[\theta = \dfrac{1}{3} \left(\dfrac{t^2}{2} + \dfrac{e^{-t}}{2} - 4t \right) \right]$$

10. The acceleration of a body a is equal to its rate of change of velocity, $\dfrac{dv}{dt}$. Determine an equation for v in terms of t given that the velocity is u when $t = 0$. $\quad [v = u + at]$

11. The velocity of a body, v is equal to its rate of change of distance, $\dfrac{dx}{dt}$. Determine an equation for x in terms of t, given $v = u + at$, where u and a are constants and $x = 0$ when $t = 0$. $\quad [x = ut + \frac{1}{2}at^2]$

12. The gradient of a curve is given by $\dfrac{dy}{dx} = 4x - \dfrac{x^3}{6}$. Determine the equation of the curve if it passes through the point $\left(2, 3\dfrac{1}{3}\right)$.

$$\left[y = 2x^2 - \frac{x^4}{24} - 4\right]$$

13. The bending moment of a beam, M, and shear force F are related by the equation $\dfrac{dM}{dx} = F$, where x is the distance from one end of the beam. Determine M in terms of x when $F = -w\,(l-x)$ where w and l are constants and $M = \dfrac{1}{2}\,w\,l^2$ when $x = 0$. $\left[M = \dfrac{1}{2}\,w(l-x)^2\right]$

14. The angular velocity ω of a flywheel of moment of inertia I is given by $I\dfrac{d\omega}{dt} + N = 0$, where N is a constant. Determine ω in terms of t given that $\omega = \omega_0$ when $t = 0$. $\left[\omega = \omega_0 - \dfrac{Nt}{I}\right]$

Differential equations of the form $\dfrac{dy}{dx} = f(y)$

In Problems 15 to 22 solve the differential equations.

15. $\dfrac{dy}{dx} = 3 + 2y.$ $\left[\dfrac{1}{2}\,\ln(3 + 2y) = x + c\right]$

16. $5\dfrac{dy}{dx} = \cot 2y.$ $\left[\dfrac{5}{2}\,\ln(\sec 2y) = x + c\right]$

17. $y\dfrac{dy}{dx} = 3 - y^2.$ $\left[-\dfrac{1}{2}\,\ln(3 - y^2) = x + c\right]$

18. $2\dfrac{dy}{dx} + 3y = 4.$ $\left[-\dfrac{2}{3}\,\ln(4 - 3y) = x + c\right]$

19. $\dfrac{dy}{dx} = 2\tan y.$ Find y. $[y = \arcsin(e^{2x+c})]$

20. $y\dfrac{dy}{dx} = 1 - y$, and $y = 0$ when $x = 1$. $[x + y + \ln(1 - y) = 1]$

21. $(y^2 + 1)\dfrac{dy}{dx} = 2y$ and $y = 1$ when $x = \dfrac{1}{4}$. $[y^2 + 2\ln y = 4x]$

22. $\sqrt{y}\,\dfrac{dy}{dx} - 1 = 0$ and $y = 4$ when $x = \dfrac{1}{3}$. $\left[\dfrac{2}{3}\,\sqrt{y^3} = x + 5\right]$

23. An equation of motion may be represented by the equation $\dfrac{dv}{dt} + kv^2 = 0$

where v is the velocity of a body travelling in a restraining medium. Show that $v = \dfrac{v_0}{1 + kt\, v_0}$ given that $v = v_0$ when $t = 0$.

24. The current in an electric circuit is given by $L\,\dfrac{di}{dt} + Ri = 0$ where L and R are constants. Solve for i given that $i = I$ when $t = 0$.

$$\left[i = I\, e^{-\frac{Rt}{L}} \right]$$

25. The difference in tension, T newtons, between two sides of a belt when in contact with a pulley over an angle of θ radians and when it is on the point of slipping, is given by $\dfrac{dT}{d\theta} = \mu\, T$, where μ is the coefficient of friction between the material of the belt and that of the pulley at the point of slipping. When $\theta = 0$ radians, the tension is 150 N and $\mu = 0.29$ as slipping starts. Find the tension at the point of slipping when $\theta = \dfrac{2\pi}{3}$ radians. Also determine the angle of lap correct to the nearest degree to give a tension of 300 N just before slipping starts. [275.3 N, $137°$]

26. The charge Q coulomb at time t seconds for a capacitor of capacitance C farads when discharging through a resistance of R ohms is given by $R\,\dfrac{dQ}{dt} + \dfrac{Q}{C} = 0$. Solve for Q given that $Q = Q_0$ when $t = 0$. A circuit contains a resistance of 400 kilohms and a capacitance of 7.3 microfarads, and after 225 milliseconds the charge falls to 7.0 coulombs. Find the initial charge and the charge after 2 seconds, correct to three significant figures.

$$[Q = Q_0\, e^{-\frac{t}{CR}};\ 7.56 \text{ C}; 3.81 \text{ C}]$$

27. In a chemical reaction in which x is the amount transformed in time t the velocity of the reaction is given by $\dfrac{dx}{dt} = k(a - x)$ where k is a constant and a is the concentration at time $t = 0$ when $x = 0$. Find x in terms of t. $[x = a(1 - e^{-kt})]$

28. The rate of decay of a radioactive substance is given by $\dfrac{dN}{dt} = -\lambda N$, where λ is the decay constant and λN is the number of radioactive atoms disintegrating per second. Determine the half-life of radium in years (i.e. the time for N to become one-half of its original value) taking the decay constant for radium as 1.36×10^{-11} atoms per second and assuming a '365 day' year. [1 616 years]

29. The variation of resistance R ohms, of a copper conductor with temperature, $\theta°$C, is given by $\dfrac{dR}{d\theta} = \alpha R$, where α is the temperature coefficient

of resistance of copper. If $R = R_0$ at $\theta = 0°C$, solve the equation for R. Taking α as 39×10^{-4} per $°C$, find the resistance of a copper conductor at $20°C$, correct to four significant figures, when its resistance at $80°C$ is 57.4 ohms. $\quad [R = R_0 e^{\alpha\theta}; 45.42 \text{ ohms}]$

'Variables separable' types of differential equations

In Problems 30 to 36 solve the differential equations.

30. $\dfrac{dy}{dx} = (2y)(x^2).$ $\qquad \left[\begin{array}{l} \dfrac{1}{2}\ \ln y = \dfrac{x^3}{3} + c \\[2mm] \text{or } \dfrac{1}{2}\ \ln 2y = \dfrac{x^3}{3} + k \end{array}\right]$

31. $\dfrac{dy}{dx} = y \cos x.$ $\quad [\ln y = \sin x + c]$

32. $(x + 2)\dfrac{dy}{dx} = (1 - y).$ $\quad [\ln (x + 2)(1 - y) = c]$

33. $\dfrac{dy}{dx} = \dfrac{2x^2 - 1}{3y + 2}$ and $x = 0$ when $y = 0$. $\quad \left[\dfrac{3y^2}{2} + 2y = \dfrac{2x^3}{3} - x\right]$

34. $\dfrac{dy}{dx} = e^{x-2y}$ and $x = 0$ when $y = 0$. $\quad [e^{2y} = 2e^x - 1]$

35. $\dfrac{1}{2}\dfrac{dy}{dx} = e^{x+3y}$ and $x = 0$ when $y = 0$. $\quad [7 - e^{-3y} = 6e^x]$

36. $y(1 + x) + x(1 - y)\dfrac{dy}{dx} = 0$ and $x = 1$ when $y = 1$. $\quad [\ln (xy) = y - x]$

37. Show that the solution of the equation $xy\ \dfrac{dy}{dx} = 1 + 2y^2$ may be of the form $y = \sqrt{\left(\dfrac{x^4 k - 1}{2}\right)}$, where k is a constant.

38. Show that the solution of the differential equation $\dfrac{y^2 + 2}{x^2 + 2} = \dfrac{y}{x}\dfrac{dy}{dx}$ is of the form $\sqrt{\left(\dfrac{x^2 + 2}{y^2 + 2}\right)}$ = constant.

39. Prove that $y = x$ is a solution of the equation $x\sqrt{(y^2 - 1)} - y\sqrt{(x^2 - 1)}\dfrac{dy}{dx} = 0$ when $x = 1$ and $y = 1$.

40. Solve $xy = (1 + x^2)\dfrac{dy}{dx}$ for y. $\quad \left[y = \dfrac{1}{k}\sqrt{(1 + x^2)}\right]$

41. Find the curve which satisfies the equation $2xy\ \dfrac{dy}{dx} = x^2 + 1$ and which passes through the point $(1, 2)$. $[2y^2 = x^2 + 2 \ln x + 7]$

42. Solve the equation $y \cos^2 x \dfrac{dy}{dx} = \tan x + 2$ given that $y = 2$ when

$x = \dfrac{\pi}{4}$. $[y^2 = \tan^2 x + 4 \tan x - 1]$

43. A capacitor C is charged by applying a steady voltage E through a resistance R. The p.d. between the plates, V, is given by the differential equation $CR \dfrac{dV}{dt} + V = E$. Solve the equation for V given that $V = 0$ when $t = 0$ and evaluate V when $E = 20$ volts, $C = 25$ microfarads, $R = 300$ kilohms and $t = 2$ seconds.

$$[V = E(1 - e^{-\frac{t}{CR}}); 4.681 \text{ volts}]$$

44. For an adiabatic expansion of a gas $C_p \dfrac{dv}{V} + C_v \dfrac{dp}{P} = 0$, where C_p and

C_v are constants. Show that $pv^n = $ constant, where $n = \dfrac{C_p}{C_v}$.

45. The streamlines of a cylinder of radius a in a stream of liquid of ambient velocity v are given by the equation: $\dfrac{dr}{d\theta} = \dfrac{r(a^2 - r^2)}{(a^2 + r^2)} \cot \theta$.

r is the distance of the centre of the cyclinder from the applied force, the line joining them being at an angle θ to the direction of the liquid flow. Solve the equation. (Hint: let $a^2 + r^2 = a^2 - r^2 + 2r^2$.)

$$\left[\left(\frac{a^2}{r} - r \right) \sin \theta = C \right]$$

46. The equilibrium constant (K) of a chemical reaction varies with temperature (T) according to the equation: $\dfrac{d (\ln K)}{dT} = \dfrac{\Delta H}{RT^2}$. If $\Delta H = 10^4$, $R = 8.3$ and $K = 4$ when $T = 600$, solve the equation completely.

$$\left[\ln K = \frac{-\Delta H}{RT} + 3.39 \right]$$

Chapter 21

Homogeneous first order differential equations (CD 18)

1 Solution of differential equations of the form $P \dfrac{dy}{dx} = Q$

Certain first order differential equations are not separable but can be made separable by a simple change of variables.

An equation of the form $P\dfrac{dy}{dx} = Q$, where P and Q are functions of both x and y of the same degree throughout, is said to be **homogeneous**. An example of a homogeneous function is $f(x, y) = x^3 + 2x^2 y + y^3$, since each of the three terms are of the degree 3. However, $f(x, y) = 2x^2 + xy^2 + y^2$ is not homogeneous since the second term is of degree 3 and the other two are of degree 2. Similarly, $f(x, y) = \dfrac{x + 2y}{3x - y}$ is homogeneous in x and y since each of the four terms are of degree 1.

Procedure to solve a differential equation of the form $P \dfrac{dy}{dx} = Q$

(i) Rearrange $P \dfrac{dy}{dx} = Q$ into the form $\dfrac{dy}{dx} = \dfrac{Q}{P}$.

(ii) Make the substitution $y = vx$, where v is a function of x. If $y = vx$ then, by the product rule, $\dfrac{dy}{dx} = v(1) + x\dfrac{dv}{dx}$. There is no particular reason for choosing the letter v, and in any case, it does not appear in the final solution.

(iii) Substitute for both y and $\dfrac{dy}{dx}$ in the equation $\dfrac{dy}{dx} = \dfrac{Q}{P}$. It will be found that the x terms will cancel leaving an equation in which the variables are separable.

(iv) Separate the variables and solve using the method shown in Chapter 20.

(v) The solution will be in terms of v and x. Thus, substitute $v = \dfrac{y}{x}$ to solve in terms of the original variables x any y.

Problem 1. Solve $xy \dfrac{dy}{dx} = x^2 + y^2$ given that $x = 1$ when $y = 2$.

Using the above procedure:

(i) Since $xy \dfrac{dy}{dx} = x^2 + y^2$ then $\dfrac{dy}{dx} = \dfrac{x^2 + y^2}{xy}$ which is homogeneous in x and y, since each of the three terms on the right-hand side are of the same degree (i.e. degree 2).

(ii) Let $y = vx$, then $\dfrac{dy}{dx} = v + x \dfrac{dv}{dx}$.

(iii) Substituting for y and $\dfrac{dy}{dx}$ in $\dfrac{dy}{dx} = \dfrac{x^2 + y^2}{xy}$ gives:

$$v + x \frac{dv}{dx} = \frac{x^2 + (vx)^2}{x(vx)} = \frac{x^2 + v^2 x^2}{x^2 v} = \frac{1 + v^2}{v}$$

Hence $\quad v + x \dfrac{dv}{dx} = \dfrac{1 + v^2}{v}$

(iv) Separating the variables gives: $x \dfrac{dv}{dx} = \dfrac{1 + v^2}{v} - v = \dfrac{1 + v^2 - v^2}{v} = \dfrac{1}{v}$

$$\text{i.e.} \qquad v \, dv = \frac{dx}{x}$$

Integrating both sides gives: $\int v \, dv = \int \dfrac{dx}{x}$

$$\therefore \quad \frac{v^2}{2} = \ln x + c$$

(v) Replacing v by $\dfrac{y}{x}$ gives: $\dfrac{y^2}{2x^2} = \ln x + c$

When $x = 1, y = 2$ thus $\quad \dfrac{4}{2} = \ln 1 + c$

$$\text{i.e.} \quad c = 2$$

Thus the particular solution of $xy \dfrac{dy}{dx} = x^2 + y^2$ is $\dfrac{y^2}{2x^2} = \ln x + 2$.

Problem 2. Solve the equation $(4y + 3x) \dfrac{dy}{dx} = (3x - y)$.

(i) If $(4y + 3x) \dfrac{dy}{dx} = (3x - y)$ then $\dfrac{dy}{dx} = \dfrac{(3x - y)}{(4y + 3x)}$, which is homogeneous in x and y since each of the four terms on the right-hand side
is of the same degree (i.e. degree 1).

(ii) Let $y = vx$ then $\dfrac{dy}{dx} = v + x \dfrac{dv}{dx}$.

(iii) Substituting for y and $\dfrac{dy}{dx}$ gives: $v + x \dfrac{dv}{dx} = \dfrac{3x - vx}{4vx + 3x} = \dfrac{3 - v}{4v + 3}$

(iv) Separating the variables gives: $x \dfrac{dv}{dx} = \dfrac{3 - v}{4v + 3} - v$

$$= \dfrac{(3 - v) - v(4v + 3)}{4v + 3} = \dfrac{3 - 4v - 4v^2}{4v + 3}$$

Hence $\dfrac{4v + 3}{(3 + 2v)(1 - 2v)} \, dv = \dfrac{dx}{x}$

Integrating both sides gives: $\displaystyle\int \dfrac{4v + 3}{(3 + 2v)(1 - 2v)} \, dv = \int \dfrac{dx}{x}$

Resolving $\dfrac{4v + 3}{(3 + 2v)(1 - 2v)}$ into partial fractions gives:

$\dfrac{5}{4(1 - 2v)} - \dfrac{3}{4(3 + 2v)}$

Hence $\displaystyle\int \left\{ \dfrac{5}{4(1 - 2v)} - \dfrac{3}{4(3 + 2v)} \right\} \, dv = \int \dfrac{dx}{x}$

$$-\dfrac{5}{8} \ln (1 - 2v) - \dfrac{3}{8} \ln (3 + 2v) = \ln x + c$$

(v) Replacing v by $\dfrac{y}{x}$ gives: $-\dfrac{5}{8} \ln \left(1 - \dfrac{2y}{x}\right) - \dfrac{3}{8} \ln \left(3 + \dfrac{2y}{x}\right) = \ln x + c$

or $\ln x + \dfrac{5}{8} \ln \left(\dfrac{x - 2y}{x}\right) + \dfrac{3}{8} \ln \left(\dfrac{3x + 2y}{x}\right) + c = 0$

Multiplying throughout by 8 gives:

$8 \ln x + 5 \ln \left(\dfrac{x - 2y}{x}\right) + 3 \ln \left(\dfrac{3x + 2y}{x}\right) + 8c = 0.$

By the laws of logarithms: $\ln x^8 + \ln \left(\dfrac{x - 2y}{x}\right)^5 + \ln \left(\dfrac{3x + 2y}{x}\right)^3 = K$

(where $K = -8c$)

Hence $\quad \ln \left\{ \dfrac{(x^8)(x-2y)^5\,(3x+2y)^3}{(x^5)(x^3)} \right\} = K$

i.e. $\quad \ln\,\{(x-2y)^5\,(3x+2y)^3\,\} = K$ **is the general solution.**

Problem 3. Solve $(y-x)\dfrac{dy}{dx} = \dfrac{y^2}{x} - y + \dfrac{x^2}{y}$ given that $x=1$ when $y=3$.

(i) Multiplying throughout by xy gives: $\quad xy\,(y-x)\,\dfrac{dy}{dx} = y^3 - xy^2 + x^3$

   from which $\quad \dfrac{dy}{dx} = \dfrac{y^3 - xy^2 + x^3}{xy(y-x)} = \dfrac{y^3 - xy^2 + x^3}{xy^2 - x^2 y}$,

   which is homogeneous in x and y.

(ii) Let $y = vx$, then $\quad \dfrac{dy}{dx} = v + x\dfrac{dv}{dx}$.

(iii) Substituting for y and $\dfrac{dy}{dx}$ gives:

$$v + x\dfrac{dv}{dx} = \dfrac{(vx)^3 - x(vx)^2 + x^3}{x(vx)^2 - x^2(vx)} = \dfrac{v^3 - v^2 + 1}{v^2 - v}$$

(iv) Separating the variables gives:

$$x\dfrac{dv}{dx} = \dfrac{v^3 - v^2 + 1}{v^2 - v} - v = \dfrac{(v^3 - v^2 + 1) - v(v^2 - v)}{v^2 - v}$$

$$= \dfrac{v^3 - v^2 + 1 - v^3 + v^2}{v^2 - v} = \dfrac{1}{v^2 - v}$$

Hence $(v^2 - v)\,dv = \dfrac{dx}{x}$

Integrating both sides gives: $\int(v^2 - v)\,dv = \displaystyle\int \dfrac{dx}{x}$

$$\dfrac{v^3}{3} - \dfrac{v^2}{2} = \ln x + c$$

(v) Replacing v by $\dfrac{y}{x}$ gives: $\quad \dfrac{y^3}{3x^3} - \dfrac{y^2}{2x^2} = \ln x + c.$

   (This is the general solution.)

   When $x=1, y=3$, thus: $\quad \dfrac{27}{3(1)} - \dfrac{9}{2(1)} = \ln 1 + c$

   i.e. $c = \dfrac{9}{2}$

Hence the particular solution is $\dfrac{y^3}{3x^3} - \dfrac{y^2}{2x^2} = \ln x + \dfrac{9}{2}$

i.e. $2y^3 - 3xy^2 = 6x^3 \ln x + 27 x^3$

Further problems on homogeneous differential equations may be found in the following Section (2) (Problems 1 to 10).

2 Further problems

In Problems 1 to 5 find the general solution of the differential equations.

1. $x\dfrac{dy}{dx} + \dfrac{y^2}{x} = y.$ $\left[\dfrac{x}{y} = \ln x + c \text{ or } x = y \ln k\, x \right]$

2. $(x + y)\dfrac{dy}{dx} = \dfrac{y}{x}\,(y - x).$

$\left[-\dfrac{1}{2} \ln \dfrac{y}{x} - \ln x = \dfrac{y}{2x} + c \text{ or } \dfrac{y}{2x} + \ln \sqrt{(xy)} = K \right]$

3. $y^2 = x^2\,\dfrac{dy}{dx}.$ $[y = x(1 + ky)]$

4. $xy\dfrac{dy}{dx} = \dfrac{x^2 + y^2}{2}.$ $[x^2 - y^2 = kx]$

5. $(x + y) = (x - y)\dfrac{dy}{dx}.$ $\left[\arctan \dfrac{y}{x} = \ln \sqrt{(x^2 + y^2)} + c \right]$

6. If $x + y + x\,\dfrac{dy}{dx} = 0$, show that $x(x + 2y) = k$, where k is a constant.

In Problems 7 to 9 find the particular solution of the differential equations.

7. $\dfrac{dy}{dx} = \dfrac{xy}{x^2 + y^2}$ given that $x = 1$ when $y = 1$. $\left[y = e^{\frac{x^2 - y^2}{2y^2}} \right]$

8. $(x + y)\dfrac{dy}{dx} = x + \dfrac{y^2}{x}$ given that $x = 4$ when $y = 2$.

$\left[2x \ln\left\{ \dfrac{(x - y)^2}{x} \right\} = x - 2y \right]$

9. $\dfrac{xy^2}{(x^3 + y^3)}\dfrac{dy}{dx} = 1$ given that $x = 1$ when $y = 3$. $[y^3 = 3x^3\,(\ln x + 9)]$

10. Given that $\dfrac{4xy}{(x^2 - y^2)}\dfrac{dy}{dx} = 1$, and $y = 0$ when $x = 1$, show that $(\sqrt{x})(x^2 - 5y^2) = 1$.

Chapter 22

Linear first order differential equations (CD 19)

1 Solution of differential equations of the form $\dfrac{dy}{dx} + Py = Q$

An equation of the form $\dfrac{dy}{dx} + Py = Q$, where P and Q are functions of x only, is called a **linear differential equation** since y and its derivative (i.e. $\dfrac{dy}{dx}$) are of the first degree.

The solution of $\dfrac{dy}{dx} + Py = Q$ is obtained by multiplying throughout by what is called an **integrating factor**.

Multiplying $\dfrac{dy}{dx} + Py = Q$ throughout by some function of x alone, say R, gives:

$$R \frac{dy}{dx} + RPy = RQ \tag{1}$$

Now the differential coefficient of a product Ry is given by the product rule and is:

$$\frac{d}{dx} (Ry) = R \frac{dy}{dx} + y \frac{dR}{dx}$$

$R \dfrac{dy}{dx} + y \dfrac{dR}{dx}$ is the same as the left hand side of equation (1) when R is

chosen such that RP is equal to $\dfrac{dR}{dx}$.

If $\dfrac{dR}{dx} = Rp$, separating the variables gives:

$$\frac{dR}{R} = P\,dx$$

Integrating both sides gives: $\displaystyle\int \frac{dR}{R} = \int P\,dx$

i.e. $\ln R = \int P\,dx + c$

and $R = e^{\int P\,dx\, +\, c} = e^{\int P\,dx}\, e^{c}$

i.e. $R = A e^{\int P\,dx}$ (where $A = e^{c}$, an arbitrary constant)

Substituting $R = A e^{\int P\,dx}$ in equation (1) gives:

$$A e^{\int P\,dx}\left(\frac{dy}{dx}\right) + A e^{\int P\,dx}\,Py = A e^{\int P\,dx}\,Q$$

and by cancelling the constant A:

$$e^{\int P\,dx}\left(\frac{dy}{dx}\right) + e^{\int P\,dx}\,Py = e^{\int P\,dx}\,Q \tag{2}$$

The left-hand side of equation (2) is $\dfrac{d}{dx}\left(y e^{\int P\,dx}\right)$, which may be checked by differentiating $y e^{\int P\,dx}$ with respect to x, using the product rule.

Hence from equation (2):

$$\frac{d}{dx}\left[y e^{\int P\,dx}\right] = e^{\int P\,dx}\,Q$$

Integrating both sides gives:

$$y e^{\int P\,dx} = \int e^{\int P\,dx}\,Q\,dx \tag{3}$$

Procedure to solve differential equations of the form $\dfrac{dy}{dx} + Py = Q$

(i) Rearrange the differential equation, if necessary, into the form $\dfrac{dy}{dx} + Py = Q$, where P and Q are functions of x.

(ii) Find $\int P\,dx$. Since the arbitrary constant of integration disappears (as shown in equation (2) above) it is unnecessary to include the constant for this indefinite integral.

(iii) Find the integrating factor $e^{\int P\,dx}$.

(iv) Substitute $e^{\int P\,dx}$ into equation (3).

(v) Integrate the right-hand side of equation (3) to give the general solution. If boundary conditions are given the arbitrary constant of integration may be determined and the particular solution of the differential equation may be found.

Problem 1. Solve the equation $\dfrac{1}{x}\dfrac{dy}{dx} + 2y = 1$ given that when $x = 0$, $y = 1$.

Following the above procedure:

(i) Multiplying throughout by x gives $\dfrac{dy}{dx} + 2xy = x$, which is of the form $\dfrac{dy}{dx} + Py = Q$, where $P = 2x$ and $Q = x$.

(ii) $\int P\, dx = \int 2x\, dx = x^2$.
(iii) Integrating factor $= e^{x^2}$.
(iv) Substituting into equation (3) gives: $ye^{x^2} = \int e^{x^2}(x)\, dx$
The right-hand side may be integrated by using the substitution $u = x^2$.

Hence $ye^{x^2} = \dfrac{e^{x^2}}{2} + c.$ (This is the general solution.)

When $x = 0, y = 1$, thus $(1)e^0 = \dfrac{e^0}{2} + c$

i.e. $c = \tfrac{1}{2}$

Hence the particular solution is $ye^{x^2} = \dfrac{e^{x^2}}{2} + \dfrac{1}{2}$ or $2y = 1 + e^{-x^2}$

Problem 2. Solve: $x \cos x \dfrac{dy}{dx} + (x \sin x + \cos x)y = 1.$

(i) Dividing throughout by $x \cos x$ gives:

$$\frac{dy}{dx} + \left(\frac{x \sin x + \cos x}{x \cos x} \right) y = \frac{1}{x \cos x}$$

i.e. $\dfrac{dy}{dx} + \left(\tan x + \dfrac{1}{x} \right) y = \dfrac{\sec x}{x}$

which is of the form $\dfrac{dy}{dx} + Py = Q$, where $P = \left(\tan x + \dfrac{1}{x} \right)$

and $Q = \left(\dfrac{\sec x}{x} \right)$

(ii) $\int P\, dx = \int \left(\tan x + \dfrac{1}{x} \right) dx = \ln(\sec x) + \ln x = \ln(x \sec x)$

(iii) Integrating factor $= e^{\int P\, dx} = e^{\ln(x \sec x)} = x \sec x.$

(iv) Substituting in equation (3) gives:

$$y(x \sec x) = \int (x \sec x) \left(\frac{\sec x}{x} \right) dx$$

i.e. $xy \sec x = \int \sec^2 x\, dx$

(v) Integrating gives: $\quad xy \sec x \;=\; \tan x + c,$

and dividing by $\sec x$: $\qquad xy \;=\; \dfrac{\tan x}{\sec x} + \dfrac{c}{\sec x}$

i.e. the general solution is $\;xy \;=\; \sin x + c \cos x.$

Problem 3. Find the particular solution of the differential equation $\dfrac{dy}{dx} + 2x = y$, given that $x = 0$ when $y = 2$.

(i) Rearranging gives $\dfrac{dy}{dx} - y = -2x$, which is of the form $\dfrac{dy}{dx} + Py = Q$, where $P = -1$ and $Q = -2x$.

(ii) $\int P \, dx = \int -1 \, dx = -x$

(iii) Integrating factor $= e^{\int P \, dx} = e^{-x}$.

(iv) Substituting into equation (3) gives

$$ye^{-x} \;=\; \int e^{-x}(-2x) \, dx$$

i.e. $\quad ye^{-x} \;=\; -2 \int x e^{-x} \, dx \qquad\qquad\qquad (4)$

(v) $\int x e^{-x} \, dx$ is determined using integration by parts.

Let $u = x$ and $dv = e^{-x} \, dx$.

Then $du = dx$ and $v = \int e^{-x} \, dx = -e^{-x}$.

Thus $\int x e^{-x} \, dx \;=\; x(-e^{-x}) - \int(-e^{-x}) \, dx$

$$= -x e^{-x} - e^{-x} + c.$$

Hence from equation (4):

$$ye^{-x} = -2\left(-x e^{-x} - e^{-x} + c\right)$$

which is the general solution.

When $x = 0, y = 2$, thus $2(1) = -2(0 - 1 + c)$

from which $\qquad\qquad\qquad c = 0$

Thus the particular solution is $\;ye^{-x} = 2x e^{-x} + 2e^{-x}$

or $\qquad\qquad\qquad\qquad y = 2x + 2$

or $\qquad\qquad\qquad\qquad y = 2(x + 1).$

Problem 4. Solve: $\quad (x + 2) \dfrac{dy}{dx} = 3 - \dfrac{2y}{x}$

(i) Rearranging gives: $\dfrac{dy}{dx} + \dfrac{2}{x(x + 2)}\, y = \dfrac{3}{(x + 2)}$, which is of the form $\dfrac{dy}{dx} + Py = Q$, where $P = \dfrac{2}{x(x + 2)}$ and $Q = \dfrac{3}{(x + 2)}$.

(ii) $\int P \, dx = \int \dfrac{2}{x \, (x + 2)} \, dx$ which may be integrated using partial fractions.

Let $\dfrac{2}{x \, (x + 2)} \equiv \dfrac{A}{x} + \dfrac{B}{(x + 2)} \equiv \dfrac{A(x + 2) + Bx}{x \, (x + 2)}$

Equating the numerators gives: $2 \equiv A \, (x + 2) + Bx$

When $x = 0$, $\qquad\qquad\qquad 2 = 2A$, i.e. $A = 1$

When $x = -2$, $\qquad\qquad\quad 2 = -2B$, i.e. $B = -1$.

Hence $\int \dfrac{2}{x \, (x + 2)} \, dx = \int \left\{ \dfrac{1}{x} - \dfrac{1}{(x + 2)} \right\} \, dx = \ln x - \ln (x + 2)$

$$= \ln \left(\dfrac{x}{x + 2} \right)$$

(iii) Integrating factor $= e^{\int P \, dx} = e^{\ln \, (x/x+2)} = \dfrac{x}{x + 2}$

(iv) Substituting in equation (3) gives:

$$y \left(\dfrac{x}{x + 2} \right) = \int \left(\dfrac{x}{x + 2} \right) \left(\dfrac{3}{x + 2} \right) \, dx = \int \dfrac{3x}{(x + 2)^2} \, dx$$

(v) $\int \dfrac{3x}{(x + 2)^2} \, dx$ cannot be determined 'on sight'. It is thus split into two fractions, the first fraction being specifically chosen to give a logarithmic solution.

Hence $\dfrac{xy}{x + 2} = \int \left\{ \dfrac{\frac{3}{2} \, (2x + 4)}{(x + 2)^2} - \dfrac{6}{(x + 2)^2} \right\} \, dx$

$$= \tfrac{3}{2} \, \ln (x + 2)^2 + \dfrac{6}{(x + 2)} + c$$

$\dfrac{xy}{x + 2} = \ln [(x + 2)^2]^{3/2} + \dfrac{6}{(x + 2)} + \ln k \quad$ (where $\ln k = c$)

Hence the general solution is $xy = (x + 2) \ln \, \{ k \, (x + 2)^3 \} + 6$.

Further problems on solving linear differential equations may be found in the following Section (2) (Problems 1 to 18).

2 Further problems

In Problems 1 to 10 solve the differential equations.

1. $x \dfrac{dy}{dx} + y = 2$. $[xy = 2x + c]$

2. $y = x(3 - \dfrac{dy}{dx})$ and $x = 1$ when $y = 2$. $\quad [2xy = 3x^2 + 1]$

3. $x\dfrac{dy}{dx} = y + x^3 - 2x$ and $x = 1$ when $y = 3$. $\quad [2y = x^3 + 5x - 4x \ln x]$

4. $\dfrac{dy}{dx} + 1 = \dfrac{-y}{x}$ and $x = 2$ when $y = 1$. $\quad [2xy + x^2 = 8]$

5. $\dfrac{dy}{dx} = x - \dfrac{2}{x} y$ and $x = 2$ when $y = 1$. $\quad [4y = x^2]$

6. $\cos x \dfrac{dy}{dx} = (\sin x)(1 - 2y)$. $\quad [2y = 1 + k \cos^2 x]$

7. $\tan x \dfrac{dy}{dx} + y = x^2 \tan x$. $\quad [(y - 2x) \sin x = (2 - x^2) \cos x + C]$

8. $\dfrac{dy}{dx} + x = y$ and $x = 0$ when $y = 2$. $\quad [y = x + 1 + e^x]$

9. $(x + 1)\dfrac{dy}{dx} + \dfrac{y}{x} = 2$. $\quad [xy = 2(x + 1) \ln (x + 1) + 2 + C(x + 1)]$

10. $2(x + 1)\dfrac{dy}{dx} - (1 + 2x)y = x^2 \sqrt{(1 + x)}$.

$$[2y \sqrt{(x + 1)} = C e^{-(x^2 + 2x + 2)}]$$

11. In an electrical alternating current circuit containing resistance R and inductance L the current i is given by the equation $Ri + L\dfrac{di}{dt} = E$. If $E = E_0 \sin \omega t$ and $i = 0$ when $t = 0$ show that:

$$i = \frac{E_0}{R^2 + \omega^2 L^2} (R \sin \omega t - \omega L \cos \omega t) + \frac{E_0 \omega L}{R^2 + \omega^2 L^2} e^{-Rt/L}.$$

12. Solve the following equation for θ, given that when $t = 0, \theta = \frac{1}{2}$:

$\cosh t \dfrac{d\theta}{dt} + \theta \sinh t = \sinh 2t$. $\quad [\theta = \frac{1}{2} \cosh 2t \operatorname{sech} t]$

13. An equation of motion when a particle moves in a resisting medium is given by: $\dfrac{dv}{dt} = - (kv + bt)$, where k and b are constants. Solve the equation for v given that $v = u$ when $t = 0$.

$$\left[v = \frac{b}{k^2} - \frac{bt}{k} + \left(u - \frac{b}{k^2}\right) e^{-kt}\right]$$

14. A train of total mass m is moved from rest by the engine which exerts a time-dependent force $m k (1 - e^{-t})$ on the train. The resistance to motion is mcv, where v is the speed of the train and c is a constant. Find the subsequent speed of the train if the equation of motion is given by

$$m \frac{dv}{dt} = m k (1 - e^{-t}) - mcv.$$

$$\left[v = k \left(\frac{1}{c} - \frac{e^{-t}}{(c-1)} + \frac{e^{-ct}}{c(c-1)} \right) \right]$$

15. In a unimolecular chemical reaction a substance α changes into β. If x is the number of molecules of β present at any time t, and a the initial number of molecules of α, then $\frac{dx}{dt} = k_1 (a - x) - k_2 x$, where k_1 and k_2 are constants. Show that if $x = 0$ when $t = 0$, then

$$x = \frac{k_1 a}{k_1 + k_2} \left[1 - e^{-(k_1 + k_2)t} \right]$$

16. The instantaneous current (i) passing through a solution, in a circuit of resistance R and inductance L, whose dielectric constant is to be measured is given by $\frac{di}{dt} + \frac{R}{L} i = \frac{V_0}{L} \sin pt$, where t is time and V_0 and p are constants. Show that the solution of this equation is

$$i = \left\{ \frac{V_0}{R^2 + p^2 L^2} \right\} \left\{ R \sin pt - pL \cos pt \right\} + C e^{-Rt/L}.$$

17. In the drain tank of an oil purifier in a ship the concentration (c) of impurities varies with time (t) according to the equation

$$M \frac{dc}{dt} = p + km - cm,$$

where M, p, m and k are constants.
Solve this equation given $c = c_0$ when $t = 0$.

$$\left[c = c_0 e^{-mt/M} + M \left(k + \frac{p}{m} \right) \left(1 - e^{-mt/M} \right) \right]$$

18. The angular velocity of a flywheel (ω) at time t is given by

$$I \frac{d\omega}{dt} + K \omega = A + B \sin^2 pt,$$

where I, K, A, B and p are constants. If $\omega = 0$ when $t = 0$ show that:

$$\omega = \frac{2A + B}{2K} - \frac{B(K \cos 2pt + 2 pI \sin 2pt)}{2(K^2 + 4p^2 I^2)}$$

$$- \frac{1}{K} \left(A + \frac{2 p^2 I^2 B}{K^2 + 4p^2 I^2} \right) e^{-Kt/I}.$$

Chapter 23

The solution of linear second order differential equations of the form

$$a\,\frac{d^2y}{dx^2} + b\,\frac{dy}{dx} + cy = 0\,\textbf{(CD 20)}$$

1 Introduction

An equation of the form:

$$a\frac{d^2y}{dx^2} + b\frac{dy}{dx} + cy = 0, \tag{1}$$

where a, b and c are constants, is called a **linear second order differential equation with constant coefficients**. It is termed 'linear' since y and its derivatives are all of the first degree, and it is of 'second order' since its highest derivative is $\frac{d^2y}{dx^2}$ (see Chapter 20, Section 1).

An alternative way of stating equation (1) is:

$$(a\,D^2 + b\,D + c)y = 0, \tag{2}$$

where D indicates the operation $\frac{d}{dx}$ and D^2 symbolises $\frac{d^2}{dx^2}$. When an equation is stated as in equation (2) it is said to be in **'D operator'** form. Equations of the form stated in equations (1) and (2) have important engineering applications, especially in electrical and mechanical work. Two examples include:

(i) $L\,\frac{d^2q}{dt^2} + R\,\frac{dq}{dt} + \frac{1}{c}\,q = 0$, representing an equation for charge q in an electrical circuit containing resistance R, inductance L and capacitance C in series, and

(ii) $m \dfrac{d^2 s}{dt^2} + a \dfrac{ds}{dt} + ks = 0$ defines a mechanical system, where s is the distance from a fixed point after t seconds, m is a mass, a the damping factor and k the spring stiffness.

In this chapter the solution of differential equations of the type $a \dfrac{d^2 y}{dx^2} + b \dfrac{dy}{dx} + cy = 0$ are dealt with, and in Chapter 24 the more general type, $a \dfrac{d^2 y}{dx^2} + b \dfrac{dy}{dx} + cy = f(x)$ are dealt with.

2 Types of solution of second order differential equations with constant coefficients

If in equation (1) we let $a = 0$, then $b \dfrac{dy}{dx} + cy = 0$

$$\text{i.e.} \quad \frac{dy}{dx} + \frac{c}{b} y = 0$$

Hence $\qquad \dfrac{dy}{dx} = - \dfrac{c}{b} y$

and $\qquad \dfrac{dy}{y} = - \dfrac{c}{b} \, dx$ by separating the variables

Therefore $\displaystyle\int \dfrac{dy}{y} = - \dfrac{c}{b} \int dx$

$\qquad \ln y = - \dfrac{c}{b} x + k$, where k is a constant.

Hence, $\qquad y = e^{\left(-\frac{c}{b}x + k\right)} = \left(e^{-\frac{c}{b}x}\right)(e^k)$

If we let $e^k = A = $ a constant, then

$$y = A \, e^{-\frac{c}{b} x} \quad \text{or} \quad y = A \, e^{mx}, \text{ where } m \text{ is a constant.}$$

Let us see if the solution $y = A \, e^{mx}$ also satisfies equation (1), m being a constant to be found by substituting $y = A \, e^{mx}$ into equation (1).

If $y = A \, e^{mx}$ then $\dfrac{dy}{dx} = A \, m \, e^{mx}$

$\qquad$ and $\dfrac{d^2 y}{dx^2} = A \, m^2 \, e^{mx}$

Substituting for y, $\dfrac{dy}{dx}$ and $\dfrac{d^2 y}{dx^2}$ in equation (1) gives:

$$a\frac{d^2y}{dx^2} + b\frac{dy}{dx} + cy = a\,A\,m^2\,e^{mx} + b\,A\,m\,e^{mx} + c\,A\,e^{mx} = 0$$

i.e. $A\,e^{mx}\,(am^2 + bm + c) = 0$ (3)

Equation (3) is true for all values of x. Since e^{mx} is not equal to zero and A is not zero (for this would make $y = 0$ always) then:

$am^2 + b\,m + c = 0$ (4)

which is a quadratic equation in m.

Equation (4) is called the **auxiliary equation** (or the characteristic equation).

From equation (4), m may be obtained by factorisation or by using the quadratic formula, $m = \dfrac{-b \pm \sqrt{[b^2 - 4ac]}}{2a}$

Since a, b and c are real the auxiliary equation may have either:

(i) two different real roots   (when $b^2 > 4ac$),
(ii) two real equal roots   (when $b^2 = 4ac$),
or (iii) two complex roots   (when $b^2 < 4ac$).

These three cases will now be discussed separately.

(i) Two different real roots

Let the solutions to equation (4) be $m = \alpha$ and $m = \beta$. Then either $y = A\,e^{\alpha x}$ or $y = B\,e^{\beta x}$ is a solution of equation (1). However, it is shown below that a general property of linear equations, such as equation (4), is that the sum of the two solutions is also a solution.

Hence: $y = A\,e^{\alpha x} + B\,e^{\beta x}$ (5)

is a solution of $a\dfrac{d^2y}{dx^2} + b\dfrac{dy}{dx} + cy = 0$, and since this expression contains two arbitrary constants, A and B, it is the **general solution**. If sufficient data is given the constants A and B may be determined and the **particular solution** found.

The general solution $y = A\,e^{\alpha x} + B\,e^{\beta x}$, stated in equation (5), may be derived directly as follows:

Let α and β be the roots of the auxiliary equation $am^2 + bm + c = 0$

i.e. $m^2 + \dfrac{b}{a}\,m + \dfrac{c}{a} = 0$.

From the theory of quadratic equations, $\alpha + \beta = -\dfrac{b}{a}$

$$\text{and} \quad \alpha\beta = \frac{c}{a}$$

$a\dfrac{d^2y}{dx^2} + b\dfrac{dy}{dx} + cy = 0$ may be written as

$$\frac{d^2y}{dx^2} + \frac{b}{a}\frac{dy}{dx} + \frac{c}{a}\, y = 0,$$

or $\quad \dfrac{d^2y}{dx^2} - (\alpha + \beta)\dfrac{dy}{dx} + \alpha\beta y = 0$, and by rearranging,

$$\frac{d}{dx}\left(\frac{dy}{dx} - \beta y\right) = \alpha\left(\frac{dy}{dx} - \beta y\right) \tag{6}$$

Let $\quad \dfrac{dy}{dx} - \beta y = Z$

then $\qquad \dfrac{dZ}{dx} = \alpha Z$ from equation (6)

Hence $\displaystyle\int \frac{dZ}{Z} = \int \alpha\, dx$, by separation of variables.

$\qquad\qquad \ln Z = \alpha x + k$, where k is a constant

$\qquad$ and $Z = e^{\alpha x + k} = (e^{\alpha x})(e^k) = C\,e^{\alpha x}$ where $e^k = C = $ a constant,

i.e. $Z = \dfrac{dy}{dx} - \beta y = C\,e^{\alpha x} \tag{7}$

Equation (7) is of the form of the linear first order differential equations discussed in Chapter 22 solved by introducing an integrating factor, and the solution of this equation is:

$$y\,e^{\int -\beta\,dx} = \int e^{\int -\beta\,dx}\,(C\,e^{\alpha x})\,dx$$

i.e. $\quad y\,e^{-\beta x} = \int C\,e^{(\alpha - \beta)x}\,dx \tag{8}$

$\qquad y\,e^{-\beta x} = \dfrac{C}{\alpha - \beta}e^{(\alpha - \beta)x} + B$, where B is a constant

Multiplying throughout by $e^{\beta x}$ gives:

$$y = \frac{C}{\alpha - \beta}e^{\alpha x}\,e^{-\beta x}\,e^{\beta x} + B\,e^{\beta x}$$

$$y = A\,e^{\alpha x} + B\,e^{\beta x}, \text{ where } A = \frac{C}{\alpha - \beta} = \text{ a constant.}$$

This is the same as equation (5) showing that the sum of the two solutions is also a solution.

Thus, given the differential equation $2\dfrac{d^2y}{dx^2} - \dfrac{dy}{dx} - 6y = 0$

$$\text{then } (2D^2 - D - 6)y = 0.$$

Letting $y = A\,e^{mx}$ we obtain $A\,e^{mx}\,(2m^2 - m - 6) = 0$, and since $A\,e^{mx} \neq 0$ then $2m^2 - m - 6 = 0$. This is the auxiliary equation. (Note that the auxiliary equation may be obtained from the original equation by replacing D by m.)

If $2m^2 - m - 6 = 0$, then $(2m + 3)(m - 2) = 0$,

i.e. $m = -\dfrac{3}{2}$ or $m = 2$,

i.e. two different real roots.

Hence the general solution of the differential equation

$2\dfrac{d^2y}{dx^2} - \dfrac{dy}{dx} - 6y = 0$ is $y = A\,e^{-\frac{3}{2}x} + B\,e^{2x}$.

Such a solution may always be checked:

if $y = A\,e^{-\frac{3}{2}x} + B\,e^{2x}$ then $\dfrac{dy}{dx} = -\dfrac{3}{2}\,A\,e^{-\frac{3}{2}x} + 2B\,e^{2x}$,

and $\dfrac{d^2y}{dx^2} = \dfrac{9}{4}\,A\,e^{-\frac{3}{2}x} + 4B\,e^{2x}$.

Substituting for y, $\dfrac{dy}{dx}$ and $\dfrac{d^2y}{dx^2}$ into $2\dfrac{d^2y}{dx^2} - \dfrac{dy}{dx} - 6y$ gives:

$$2\left(\dfrac{9}{4}\,A\,e^{-\frac{3}{2}x} + 4B\,e^{2x}\right) - \left(-\dfrac{3}{2}\,A\,e^{-\frac{3}{2}x} + 2B\,e^{2x}\right)$$

$$-6\left(A\,e^{-\frac{3}{2}x} + B\,e^{2x}\right) = \left(\dfrac{9}{2} + \dfrac{3}{2} - 6\right)A\,e^{-\frac{3}{2}x} + (8 - 2 - 6)B\,e^{2x} = 0.$$

Hence $y - A\,e^{-\frac{3}{2}x} + B\,e^{2x}$ is a solution of the differential equation

$2\dfrac{d^2y}{dx^2} - \dfrac{dy}{dx} - 6y = 0.$

(ii) Two real equal roots

If from the auxiliary equation $\alpha = \beta$ then the general solution is given by $y = A\,e^{\alpha x} + B\,e^{\alpha x}$.

i.e. $y = (A + B)\,e^{\alpha x} = D\,e^{\alpha x}$, where $D = A + B = $ a constant.

However this is no longer the general solution of equation (1) since the solution **must** contain **two** arbitrary constants. The analysis used in (i) can be used as far as equation (8), which becomes:

$y\,e^{-\alpha x} = \int C\,dx$, since $\alpha = \beta$,

i.e. $y\,e^{-\alpha x} = Cx + B$.

Multiplying throughout by $e^{\alpha x}$ gives: $y = Cx\,e^{\alpha x} + B\,e^{\alpha x}$

Replacing C by A gives:

$y = A\,x\,e^{\alpha x} + B\,e^{\alpha x}$

i.e. $y = (Ax + B)\,e^{\alpha x}$.

This is the general solution for the case of two real equal roots. Thus, given the differential equation $\dfrac{d^2y}{dx^2} - 6\dfrac{dy}{dx} + 9y = 0$

i.e. $(D^2 - 6D + 9)y = 0$

The auxiliary equation is $m^2 - 6m + 9 = 0$

i.e. $(m - 3)(m - 3) = 0$

Hence $m = 3$, twice.

The general solution of $\dfrac{d^2 y}{dx^2} - 6\dfrac{dy}{dx} + 9y = 0$ is $y = (Ax + B)e^{3x}$.

This solution may be checked:

If $y = (Ax + B)e^{3x}$ then $\dfrac{dy}{dx} = (Ax + B)(3e^{3x}) + (e^{3x})(A)$

and $\dfrac{d^2 y}{dx^2} = (Ax + B)(9e^{3x}) + (3e^{3x})(A) + 3A\,e^{3x}$

Substituting for y, $\dfrac{dy}{dx}$ and $\dfrac{d^2 y}{dx^2}$ into $\dfrac{d^2 y}{dx^2} - 6\dfrac{dy}{dx} + 9y$ gives:

$(9Ax\,e^{3x} + 9Be^{3x} + 3A\,e^{3x} + 3A\,e^{3x}) - 6(3Ax\,e^{3x} + 3B\,e^{3x} + A\,e^{3x})$

$+ 9(Ax\,e^{3x} + B\,e^{3x}) = 0$

Hence $y = (Ax + B)e^{3x}$ is a solution of the differential equation $\dfrac{d^2 y}{dx^2} - 6\dfrac{dy}{dx} + 9y = 0$.

(iii) Two complex roots

When solving the auxiliary equation $am^2 + bm + c = 0$, then

$$m = \frac{-b \pm \sqrt{[b^2 - 4ac]}}{2a}$$

If the coefficients of the differential equation are such that $b^2 < 4ac$ then the values of m are complex.

For example, to solve $m^2 - 6m + 13 = 0$

$$m = \frac{-(-6) \pm \sqrt{[(-6)^2 - 4(1)(13)]}}{2(1)} = \frac{6 \pm \sqrt{(-16)}}{2}$$

Now $\sqrt{(-16)} = \sqrt{[(-1)(16)]} = (\sqrt{-1})(\sqrt{16})$

By definition, $\sqrt{-1} = j$

Hence $\sqrt{(-16)} = \pm j\,4$

Thus $m = \dfrac{6 \pm j\,4}{2} = 3 \pm j\,2$

Hence the solution of the differential equation:

$$\frac{d^2 y}{dx^2} - 6\frac{dy}{dx} + 13y = 0$$

is $\quad y = A\,e^{(3+j2)x} + B\,e^{(3-j2)x}$

i.e. $\quad y = e^{3x}\left\{A\,e^{j2x} + B\,e^{-j2x}\right\}$

It may be shown that $\qquad e^{jx} = \cos x + j\sin x$
$\qquad\qquad\qquad$ and $\qquad e^{-jx} = \cos x - j\sin x$

Hence $y = e^{3x}\left\{A(\cos 2x + j\sin 2x) + B(\cos 2x - j\sin 2x)\right\}$

$\qquad y = e^{3x}\left\{(A + B)\cos 2x + j\,(A - B)\sin 2x\right\}$

It would appear at first sight that this is a complex solution. However, y (which represents a real voltage or current or displacement, etc.) cannot be equal to a complex expression. In practice it is always found that A and B are complex conjugate numbers. For example, let $A = 3 + j4$ and $B = 3 - j4$. Then $A + B = 6$, a real number and $j(A - B) = j(j8) = -8$, also a real number.

Hence, given that the solution of the auxiliary equation is $m = 3 \pm j2$ then the solution may be written down directly as $y = e^{3x}\left\{C\cos 2x + D\sin 2x\right\}$, where C and D are two real unknown constants. This solution may be checked by substituting y, $\dfrac{dy}{dx}$ and $\dfrac{d^2y}{dx^2}$ into the left hand side of the differential equation $\dfrac{d^2y}{dx^2} - 6\dfrac{dy}{dx} + 13y = 0$.

Generally, if the roots of the auxiliary equation are $\alpha \pm j\beta$, then the general solution is $y = e^{\alpha x}\left\{C\cos \beta x + D\sin \beta x\right\}$.

3 Summary of the procedure used to solve differential equations of the form $a\dfrac{d^2y}{dx^2} + b\dfrac{dy}{dx} + cy = 0$

(a) Rewrite the given differential equation $a\dfrac{d^2y}{dx^2} + b\dfrac{dy}{dx} + cy = 0$ as $(a\,D^2 + b\,D + c)y = 0$.

(b) Substitute m for D and solve the auxiliary equation $am^2 + bm + c = 0$ for m.

(c)   (i) If the roots of the auxiliary equation are **real and different**, say, $m = \alpha$ and $m = \beta$, then the general solution is:

$\qquad y = A\,e^{\alpha x} + B\,e^{\beta x}$

  (ii) If the roots of the auxiliary equation are **real and equal**, say $m = \alpha$, twice, then the general solution is:

$\qquad y = (Ax + B)\,e^{\alpha x}$.

  (iii) If the roots of the auxiliary equation are **complex**, say $m = \alpha \pm j\beta$, then the general solution is:

$\qquad y = e^{\alpha x}\left\{C\cos \beta x + D\sin \beta x\right\}$

304

(d) If the **particular solution** of a differential equation is required then substitute the given boundary conditions to find the unknown constants (i.e. to find A and B in (c, i) and (c, ii) or C and D in (c, iii)).

Worked problems on solving differential equations of the form
$$a\frac{d^2y}{dx^2} + b\frac{dy}{dx} + cy = 0$$

Problem 1. (a) Find the general solution of $6\frac{d^2y}{dx^2} + 5\frac{dy}{dx} - 4y = 0$

(b) Find the particular solution of the differential equation in (a) given that when $x = 0, y = 11$ and $\frac{dy}{dx} = 0$.

(a) In D operator form the differential equation is $(6D^2 + 5D - 4)y = 0$ where $D \equiv \frac{d}{dx}$

The auxiliary equation is $\quad 6m^2 + 5m - 4 = 0$
Factorising gives: $\qquad (3m + 4)(2m - 1) = 0$
$$\text{i.e. } m = -\frac{4}{3} \text{ or } m = \frac{1}{2}$$

Since the two roots are real and different **the general solution is**

$$y = A e^{-\frac{4}{3}x} + B e^{\frac{1}{2}x}.$$

(b) When $x = 0, y = 11$, then
$$11 = A + B \tag{1}$$

Since $y = A e^{-\frac{4}{3}x} + B e^{\frac{1}{2}x}$ then

$$\frac{dy}{dx} = -\frac{4}{3} A e^{-\frac{4}{3}x} + \frac{1}{2} B e^{\frac{1}{2}x}$$

When $x = 0$, $\frac{dy}{dx} = 0$.

Thus $0 = -\frac{4}{3} A + \frac{1}{2} B \tag{2}$

$2 \times (2)$ gives: $0 = -\frac{8}{3}A + B \tag{3}$

(1)–(3) gives: $11 = 3\frac{2}{3} A$, i.e. $A = 3$

Substituting $A = 3$ in equation (1) gives $B = 8$.

Hence the particular solution is $y = 3 e^{-\frac{4}{3}x} + 8 e^{\frac{1}{2}x}$.

Problem 2. (a) Determine the general solution of $4\dfrac{d^2y}{dt^2} - 12\dfrac{dy}{dt} + 9y = 0$.

(b) Find the particular solution of the equation given in (a)
which makes $y = 2$ at $t = 0$ and $\dfrac{dy}{dt} = 4$ at $t = 0$.

(a) In D operator from the differential equation is $(4D^2 - 12D + 9)y = 0$

where $D \equiv \dfrac{d}{dt}$.

The auxiliary equation is $\quad 4m^2 - 12m + 9 = 0$

$\quad$ i.e. $\quad (2m - 3)(2m - 3) = 0$

$\quad$ i.e. $\quad m = \dfrac{3}{2}$ twice.

Since the two roots are real and equal **the general solution is**

$y = (At + B)e^{\frac{3}{2}t}$.

(b) When $y = 2, t = 0$.

Thus $2 = (0 + B)e^0$, i.e. $B = 2$

$\dfrac{dy}{dt} = (At + B)\left(\dfrac{3}{2}\ e^{\frac{3}{2}t}\right) + (e^{\frac{3}{2}t})(A)$ by the product rule of
$\qquad\qquad\qquad\qquad\qquad\qquad\qquad\qquad$ differentiation.

When $\dfrac{dy}{dt} = 4, t = 0$.

Thus $4 = (0 + B)\left(\dfrac{3}{2}\ e^0\right) + (e^0)(A)$

i.e. $\quad 4 = \dfrac{3}{2}\ B + A$

Since $B = 2, A = 1$.
Hence the particular solution is $y = (t + 2)e^{\frac{3}{2}t}$.

Problem 3. (a) Determine the general solution of the differential equation
$\dfrac{d^2y}{dx^2} + 2\dfrac{dy}{dx} + 5y = 0$.

(b) Given the boundary conditions that when $x = 0, y = 1$ and
when $x = 0$, $\dfrac{dy}{dx} = 5$, find the particular solution of the
equation given in (a).

(a) In D operator form the differential equation is $(D^2 + 2D + 5)y = 0$

where $D \equiv \dfrac{d}{dx}$.

The auxiliary equation is $\quad m^2 + 2m + 5 = 0$.

Using the quadratic formula, $m = \dfrac{-2 \pm \sqrt{[(2)^2 - 4(1)(5)]}}{2(1)}$

$$= \frac{-2 \pm \sqrt{-16}}{2}$$

i.e. $m = \dfrac{-2 \pm j\,4}{2} = -1 \pm j\,2$

Since the roots are complex **the general solution is**
$y = e^{-x} \{\, C \cos 2x + D \sin 2x \,\}$

(b) When $x = 0, y = 1$.

Thus $1 = e^0 \{\, C \cos 0 + D \sin 0 \,\}$

Hence $1 = C$

$\dfrac{dy}{dx} = e^{-x} (-2C \sin 2x + 2D \cos 2x) - e^{-x} (C \cos 2x + D \sin 2x)$,
by the product rule,

$\qquad = e^{-x} \{\, (2D - C) \cos 2x - (2C + D) \sin 2x \,\}$

When $x = 0$, $\dfrac{dy}{dx} = 5$.

Hence $5 = e^0 \{\, (2D - C) \cos 0 - (2C + D) \sin 0 \,\}$

$\qquad\quad 5 = 2D - C$.

Since $C = 1, D = 3$.

Hence the particular solution is $y = e^{-x} \{\, \cos 2x + 3 \sin 2x \,\}$.

Since $a \cos \omega t + b \sin \omega t \equiv R \sin(\omega t + \alpha)$, where $R = \sqrt{(a^2 + b^2)}$ and

$$\alpha = \arctan \frac{a}{b},$$

$\cos 2x + 3 \sin 2x \equiv \sqrt{(1^2 + 3^2)} \sin \left(2x + \arctan \dfrac{1}{3}\right)$

$\qquad\qquad\qquad\quad \equiv \sqrt{10} \sin(2x + 18°\,26')$

$\qquad\qquad\qquad\quad \equiv \sqrt{10} \sin(2x + 0.321\,8)$

Thus the particular solution may also be expressed as

$$y = \sqrt{10}\, e^{-x} \sin(2x + 0.321\,8).$$

Problem 4. Solve the differential equations.

(a) $\dfrac{d^2 y}{dx^2} + n^2 y = 0$

(b) $\dfrac{d^2 y}{dx^2} - n^2 y = 0$, where n is a constant.

(a) $\dfrac{d^2 y}{dx^2} + n^2 y = 0$ is a differential equation representing simple harmonic motion (S.H.M.) and has many practical applications.

In D operator form the equation is $(D^2 + n^2)y = 0$
The auxiliary equation is $\qquad m^2 + n^2 = 0$

$$\text{i.e.} \qquad m^2 = -n^2$$
$$m = \sqrt{(-n^2)} = \pm j\,n$$

Since the two roots are complex the general solution is
$y = e^0 \{ C \cos nx + D \sin nx \}$, i.e. $y = C \cos nx + D \sin nx$ or

$y = R \sin (nx + \alpha)$, where $R = \sqrt{(C^2 + D^2)}$ and $\alpha = \arctan \dfrac{C}{D}$.

(b) In D operator form the differential equation $\dfrac{d^2 y}{dx^2} - n^2 y = 0$ is

$(D^2 - n^2)y = 0$.
The auxiliary equation is $m^2 - n^2 = 0$.
$$\text{Hence} \qquad m^2 = n^2$$
$$m = \pm n.$$

Since the two roots are real and different **the general solution is**
$y = A\,e^{nx} + B\,e^{-nx}$.

Since $\sinh nx = \dfrac{1}{2} (e^{nx} - e^{-nx})$ and $\cosh nx = \dfrac{1}{2} (e^{nx} + e^{-nx})$

then $\sinh nx + \cosh nx = e^{nx}$
and $\cosh nx - \sinh nx = e^{-nx}$.

Hence the general solution may also be written as:

$$y = A(\sinh nx + \cosh nx) + B(\cosh nx - \sinh nx)$$
$$\text{i.e.} \ y = (A + B) \cosh nx + (A - B) \sinh nx$$
$$\text{i.e.} \ y = C \cosh nx + D \sinh nx.$$

Problem 5. A circuit containing resistance R, inductance L and capacitance C in series has an equation for current i given by:

$$L \frac{d^2 i}{dt^2} + R \frac{di}{dt} + \frac{1}{C} i = 0.$$

If $L = 0.25$ henry, $C = 25 \times 10^{-6}$ farads and $R = 200$ ohms, solve the equation given that when time $t = 0$, $i = 0$ and $\dfrac{di}{dt} = 50$.

In D operator form the differential equation is $\left(L D^2 + R D + \dfrac{1}{C} \right) i = 0$.

The auxiliary equation is $L m^2 + R m + \dfrac{1}{C} = 0$

$$\text{Hence } m = \frac{-R \pm \sqrt{(R^2 - \frac{4L}{C})}}{2L}$$

Substituting $L = 0.25$, $C = 25 \times 10^{-6}$ and $R = 200$ gives:

$$m = \frac{-200 \pm \sqrt{\left[(200)^2 - \dfrac{4(0.25)}{25.10^{-6}}\right]}}{2(0.25)} = \frac{-200 \pm \sqrt{0}}{0.5}$$

i.e. $m = \dfrac{-200}{0.5} = -400$.

For a second order differential equation there must be two solutions. Hence $m = -400$ twice. **Hence the general solution is $i = (At + B) e^{-400t}$.**

When $t = 0$, $i = 0$.

Hence $0 = B$

$$\frac{di}{dt} = (At + B)(-400\, e^{-400t}) + (e^{-400t})(A)$$

When $t = 0$, $\dfrac{di}{dt} = 50$.

Hence $50 = -400 B + A$

Since $B = 0$, $A = 50$.

Hence the particular solution is $i = 50\, t\, e^{-400t}$.

Further problems on solving differential equations of the form
$a\dfrac{d^2y}{dx^2} + b\dfrac{dy}{dx} + cy = 0$ *may be found in the following section (4)*
(Problems 1 to 21).

4 Further problems

In Problems 1 to 6 find the general solution of the given differential equations.

1. (a) $\dfrac{d^2y}{dx^2} - 5\dfrac{dy}{dx} + 6y = 0$. $[y = A\, e^{3x} + B\, e^{2x}]$

   (b) $\dfrac{d^2y}{dx^2} - 2\dfrac{dy}{dx} + y = 0$. $[y = (Ax + B)\, e^x]$

2. (a) $\dfrac{d^2y}{dx^2} + 6\dfrac{dy}{dx} + 13y = 0$. $[y = e^{-3x}(C \cos 2x + D \sin 2x)]$

   (b) $\dfrac{d^2x}{dt^2} + 3\dfrac{dx}{dt} + 2x = 0$. $[x = A\, e^{-t} + B\, e^{-2t}]$

3. (a) $\dfrac{d^2y}{d\theta^2} - 4\dfrac{dy}{d\theta} + 4y = 0$ $[y = (A\theta + B)\, e^{2\theta}]$

   (b) $\dfrac{d^2y}{dt^2} + y = 0$. $[y = C \cos t + D \sin t]$

4. (a) $6\dfrac{d^2\theta}{dt^2} + 4\dfrac{d\theta}{dt} - 2\theta = 0$ $[\theta = A\,e^{\frac{1}{3}t} + B\,e^{-t}]$

   (b) $4\dfrac{d^2y}{dx^2} + 4\dfrac{dy}{dx} + y = 0$ $[y = (Ax + B)\,e^{-\frac{1}{2}x}]$

5. (a) $(3D^2 - 2D + 5)y = 0$ where $D \equiv \dfrac{d}{dx}$.

$$\left[y = e^{\frac{1}{3}x}\left(C\cos\frac{\sqrt{14}}{3}x + D\sin\frac{\sqrt{14}}{3}x\right)\right]$$

   (b) $(4D^2 - 7D - 15)y = 0$ where $D \equiv \dfrac{d}{dx}$.

$$\left[y = A\,e^{-\frac{5}{4}x} + B\,e^{3x} \right]$$

6. (a) $(16D^2 + 8D + 1)y = 0$ where $D \equiv \dfrac{d}{dx}$.

$$\left[y = (Ax + B)\,e^{-\frac{1}{4}x} \right]$$

   (b) $(D^2 + D + 1)\theta = 0$ where $D \equiv \dfrac{d}{dt}$.

$$\left[\theta = e^{-\frac{1}{2}t}\left(C\cos\frac{\sqrt{3}}{2}t + D\sin\frac{\sqrt{3}}{2}t\right)\right]$$

7. Show that the solution to the differential equation

$\dfrac{d^2y}{dx^2} + (2a\cos\theta)\dfrac{dy}{dx} + a^2y = 0$, where a and θ are constants, may be may be expressed as: $y = A\,e^{-ax\,\cos\theta}\cos(B + ax\sin\theta)$.

In Problems 8 to 13, find the particular solution of the given differential equations for the stated boundary conditions.

8. $12\dfrac{d^2y}{dx^2} - 3y = 0$ $y = 3$ when $x = 0$

$\dfrac{dy}{dx} = \dfrac{1}{2}$ when $x = 0$ $\left[y = 2e^{\frac{1}{2}x} + e^{-\frac{1}{2}x}\right]$

9. $9\dfrac{d^2y}{dt^2} - 12\dfrac{dy}{dt} + 4y = 0.$ $\begin{cases} y = 3 \text{ when } t = 0 \\ \\ \dfrac{dy}{dt} = 4 \text{ when } t = 0 \quad \left[y = (2t + 3)\,e^{\frac{2}{3}t} \right] \end{cases}$

10. $\dfrac{d^2y}{dx^2} + 2\dfrac{dy}{dx} + 6y = 0.$ $\begin{cases} y = 2 \text{ when } x = 0 \\ \\ \dfrac{dy}{dx} = 3 \text{ when } x = 0 \end{cases}$

$[y = e^{-x}(2\cos\sqrt{5}x + \sqrt{5}\sin\sqrt{5}x)]$

11. $(35D^2 - 11D - 6)x = 0$ where $D \equiv \dfrac{d}{dt}$
$\begin{cases} x = 5 \text{ when } t = 0 \\ \dfrac{dx}{dt} = \dfrac{12}{35} \text{ when } t = 0 \end{cases}$

$$\left[x = 3e^{-\frac{2}{7}t} + 2 e^{\frac{3}{5}t} \right]$$

12. $(2D^2 + 2D + 1)y = 0$ where $D \equiv \dfrac{d}{dx}$.
$\begin{cases} y = 4 \text{ when } x = 0 \\ \dfrac{dy}{dx} = 5 \text{ when } x = 0 \end{cases}$

$$\left[y = 2e^{-\frac{1}{2}x} \left(2 \cos \frac{x}{2} + 7 \sin \frac{x}{2} \right) \right]$$

13. $(25D^2 - 20D + 4)y = 0$ where $D \equiv \dfrac{d}{dx}$.
$\begin{cases} y = 5 \text{ when } x = 0 \\ \dfrac{dy}{dx} = 3 \text{ when } x = 0 \end{cases}$

$$\left[y = (x + 5) \, e^{\frac{2}{5}x} \right]$$

14. The oscillations of a heavily damped pendulum satisfy the differential equation $\dfrac{d^2x}{dt^2} + 7\dfrac{dx}{dt} + 12x = 0$, where x cm is the displacement of the bob at time t seconds. The initial displacement is equal to $+3$ cm and the initial velocity (i.e. $\dfrac{dx}{dt}$) is 6 cm/s. Solve the equation for x.

$$[x = 3 \, (6e^{-3t} - 5 \, e^{-4t})]$$

15. If $\dfrac{d^2 V}{dx^2} - w^2 \, V = 0$ where w is a constant show that:

$$V = V_0 \cosh wx + \frac{2}{w} \sinh wx,$$

given that when $x = 0$, $V = V_0$ and $\dfrac{dV}{dx} = 2$.

16. The charge q on a capacitor in a certain electrical circuit satisfies the differential equation $\dfrac{d^2q}{dt^2} + 3\dfrac{dq}{dt} + 4q = 0$. Initially (i.e. when $t = 0$), $q = Q$ and $\dfrac{dQ}{dt} = 0$. Show that the charge in the circuit may be expressed as:

$$q = \frac{4}{\sqrt{7}} \, Q \, e^{-\frac{3}{2}t} \sin\left(\frac{\sqrt{7}}{2} \, t + 0.723 \right).$$

17. The differential equation $I\dfrac{d^2\theta}{dt^2} + K\dfrac{d\theta}{dt} + F\theta = 0$ represents the motion

of the pointer of a galvanometer about its position of equilibrium. I is the moment of inertia of the pointer about its pivot, K is the resistance due to friction at unit angular velocity and F is the force on the spring necessary to produce unit displacement. If $I = 0.006$, $K = 0.03$ and $F = 0.187\,5$ solve the equation for θ in terms of t given that when $t = 0$,

$\theta = 0.2$ and $\dfrac{d\theta}{dt} = 0$. $\quad [\theta = e^{-2.5t} (0.2 \cos 5t + 0.1 \sin 5t)]$

18. A body moves in a straight line so that its distance S metres from the origin after time t seconds is given by $\dfrac{d^2 S}{dt^2} + n^2\, S = 0$, where n is a constant. Solve the equation for S given that $S = k$ and $\dfrac{dS}{dt} = 0$ when

$t = \dfrac{2\pi}{n}$. $\quad [S = k \cos nt]$

19. The equation $\dfrac{d^2 i}{dt^2} + \dfrac{R}{L} \dfrac{di}{dt} + \dfrac{1}{LC} i = 0$ represents a current i flowing in an electrical circuit containing resistance R, inductance L and capacitance C connected in series. Current $i = 0$ when time $t = 0$ and $\dfrac{di}{dt} = 100$ when $t = 0$. Inductance $L = 0.20$ henry and capacitance $C = 80 \times 10^{-6}$ farads. Solve the equation for i when (a) $R = 100$ ohms and (b) $R = 223.6$ ohms.

$$(a)\ \left[i = 100\, t\, e^{-250t} \right] \quad (b)\ \left[i = \frac{1}{10} \left(e^{-59t} - e^{-1059t} \right) \right]$$

20. The equation of motion of a body oscillating on the end of a spring is: $\dfrac{d^2 x}{dt^2} + 225x = 0$, where x is the displacement in metres of the body from its equilibrium position after time t seconds. Find x in terms of t given that at time $t = 0$, $x = 1$ and $\dfrac{dx}{dt} = 0$. $\quad [x = \cos 15t]$

21. The equation: $\dfrac{d^2 y}{dx^2} = k \sqrt{\left\{ 1 + \left(\dfrac{dy}{dx} \right)^2 \right\}}$ on solution will give that for the catenary. Solve it for y given that $\dfrac{dy}{dx} = 0$ when $x = 0$.

$$\left[y = \frac{1}{k} \cosh (kx) + B \right]$$

Chapter 24

The solution of linear second order differential equations of the form

$$a\,\frac{d^2y}{dx^2} + b\,\frac{dy}{dx} + cy = f(x)\ \textbf{(CD 21)}$$

1 Complementary function and particular integral

The differential equation $a\dfrac{\mathrm{d}^2y}{\mathrm{d}x^2} + b\dfrac{\mathrm{d}y}{\mathrm{d}x} + cy = f(x)$ \hfill (1)

where a, b and c are constants, is a linear, second order differential equation with constant coefficients.

In order to solve this type of equation a substitution is made as follows. Let $u + v$ be substituted for y in equation (1). The equation then becomes:

$$a\,\frac{\mathrm{d}^2}{\mathrm{d}x^2}(u+v) + b\,\frac{\mathrm{d}}{\mathrm{d}x}(u+v) + c(u+v) = f(x)$$

Hence $a\dfrac{\mathrm{d}^2u}{\mathrm{d}x^2} + a\dfrac{\mathrm{d}^2v}{\mathrm{d}x^2} + b\dfrac{\mathrm{d}u}{\mathrm{d}x} + b\dfrac{\mathrm{d}v}{\mathrm{d}x} + cu + cv = f(x)$

i.e. $\left(a\dfrac{\mathrm{d}^2u}{\mathrm{d}x^2} + b\dfrac{\mathrm{d}u}{\mathrm{d}x} + cu \right) + \left(a\dfrac{\mathrm{d}^2v}{\mathrm{d}x^2} + b\dfrac{\mathrm{d}v}{\mathrm{d}x} + cv \right) = f(x)$

Let v be any solution of the equation (1) such that:

$$a\frac{\mathrm{d}^2v}{\mathrm{d}x^2} + b\frac{\mathrm{d}v}{\mathrm{d}x} + cv = f(x) \tag{2}$$

This would mean that $\quad a\dfrac{\mathrm{d}^2u}{\mathrm{d}x^2} + b\dfrac{\mathrm{d}u}{\mathrm{d}x} + cu = 0$ \hfill (3)

The general solution, u, of equation (3) will contain two unknown constants as required for the general solution of equation (1). (The method of solution of this type of equation was discussed in the previous chapter.)

If, also, the particular solution, v, of equation (2) can be found without containing any unknown constants then $y = u + v$ will give the general solution of equation (1). Section 2 following discusses methods whereby v may be determined.

The function u is called the **complementary function (C.F.)**.

The function v is called the **particular integral (P.I.)**.

The general solution y of a linear differential equation such as equation (1) is given by the sum of the complementary function and the particular integral.

$$y = \text{C.F.} + \text{P.I.}$$
or $\quad y = u + v.$

2 Methods of finding the particular integral

The function $f(x)$ on the right-hand side of equation (1) may be any non-zero expression, the more common expressions being a constant, a polynomial (i.e. of the form $p + qx + rx^2 + \ldots$, where p, q and r are constants), an exponential function, a sine or cosine function, or a sum of product of these functions. The method of finding the particular integral differs for each expression of $f(x)$ and it is usual to assume a form of particular integral which is suggested by $f(x)$. The following examples will make this clear.

(a) $f(x) = $ a constant

(i) Straightforward case

If $f(x)$ is a constant then firstly try $v = k$, where k is a constant, as the particular integral. $v = k$ is substituted into equation (2) and k may be determined.

Problem 1. Solve $\dfrac{d^2 y}{dx^2} + 4\dfrac{dy}{dx} + 3y = 6$.

Let $v = k$ then from equation (2): $(D^2 + 4D + 3)k = 6$.
Since $D^2(k) = 0$ and $D(k) = 0$, then $3k = 6$, i.e. $k = 2$.
Hence the particular integral, $v = 2$.

The complementary function, u is obtained as in the previous chapter.

The auxiliary equation is $\quad m^2 + 4m + 3 = 0$
$$\text{i.e. } (m + 3)(m + 1) = 0$$
$$\text{i.e. } m = -3 \text{ and } m = -1.$$
Hence the complementary function $u = A e^{-3x} + B e^{-x}$.
The general solution is given by $\quad y = u + v,$
$$\text{i.e. } y = A e^{-3x} + B e^{-x} + 2.$$

(ii) 'Snag' case

When $f(x)$ is a constant the substitution of $y = k$ will not always work. If the complementary function is found to contain a constant term (as in Problem 2) then try $y = kx$ as the particular integral.

Problem 2. (a) Find the general solution of $\dfrac{d^2y}{dx^2} + 4\dfrac{dy}{dx} = 6$.

      (b) Find the particular solution for (a) given that when $x = 0$, $\overset{.}{y} = 0$ and $\dfrac{dy}{dx} = 0$.

(a) In D operator form the differential equation is $(D^2 + 4D)y = 6$.
The auxiliary equation is $\quad m^2 + 4m = 0$
$$\text{i.e. } m(m + 4) = 0$$
$$\text{Hence} \quad m = 0 \text{ or } m = -4.$$
Hence the complementary function $u = A\,e^0 + B\,e^{-4x} = A + B\,e^{-4x}$.
If we assume a particular integral of $v = k$ in this case, we obtain
$(D^2 + 4D)k = 6$ from equation (2) from which $0 = 6$ which is impossible.
This occurs because the particular integral assumed, namely a constant,
already occurs in the complementary function (i.e. as A).
In this case let $v = kx$, then $(D^2 + 4D)kx \equiv 6$,
$$\text{i.e. } D^2\,(kx) + 4D\,(kx) \equiv 6.$$
$$D^2\,(kx) = 0 \text{ and } 4D\,(kx) = 4k$$

Hence $\qquad\qquad 4k = 6, \text{i.e. } k = \dfrac{3}{2}$

Hence the particular integral, $v = \dfrac{3}{2}\,x$.

The general solution, $y = u + v = A + B\,e^{-4x} + \dfrac{3}{2}x$.

(b) When $x = 0, y = 0$ and since $y = A + B\,e^{-4x} + \dfrac{3}{2}\,x$, then

$$0 = A + B \qquad\qquad\qquad\qquad\qquad\qquad (1)$$

$$\dfrac{dy}{dx} = 0 - 4B\,e^{-4x} + \dfrac{3}{2}$$

$$\dfrac{dy}{dx} = 0 \text{ when } x = 0, \text{ thus } 0 = -4B + \dfrac{3}{2}$$

$$\text{i.e. } \quad B = \dfrac{3}{8}$$

From equation (1) $\qquad\qquad A = -\dfrac{3}{8}$

Hence the particular solution is $y = -\dfrac{3}{8} + \dfrac{3}{8}\,e^{-4x} + \dfrac{3}{2}\,x$

$$\text{i.e. } \quad y = \dfrac{3}{8}\,(e^{-4x} - 1) + \dfrac{3}{2}\,x.$$

Typical practical examples of the type of equation solved in Problems

1 and 2 are the equations for the charge q in a series-connected electrical circuit and the force equation for a 'mass, spring, damped' mechanical system, i.e.

$$L \frac{dq^2}{dt^2} + R \frac{dq}{dt} + \frac{1}{c} q = E \text{ and } m \frac{d^2 s}{dt^2} + a \frac{ds}{dt} + ks = F$$

(b) $f(x) = $ a polynomial

If $f(x) = p + qx + rx^2 + \ldots$, where p, q and r are constants, then try $v = a + bx + cx^2 + \ldots$, where a, b and c are constants, as the particular integral.

For example, if $f(x) = 4 + 3x$ then try $v = a + bx$
if $f(x) = 4x^2 + 2x - 1$ then try $v = ax^2 + bx + c$,
if $f(x) = 2x^3 - 3$ then try $v = ax^3 + bx^2 + cx + d$,

and so on.

After substitution into equation (2), the coefficients of similar terms are equated, as shown in Problem 3.

Problem 3. Solve the equation $2\frac{d^2 y}{dx^2} - \frac{dy}{dx} - 6y = 3x + 2$.

In D operator form, the equation is $(2D^2 - D - 6)y = 3x + 2$.
The auxiliary equation is $2m^2 - m - 6 = 0$
i.e. $(2m + 3)(m - 2) = 0$
i.e. $m = -1\frac{3}{2}$ or $m = 2$

The complementary function, $u = A e^{-\frac{3}{2}x} + B e^{2x}$.
Let the particular integral $v = ax + b$, then substituting into equation (2) gives:

$(2D^2 - D - 6)(ax + b) \equiv 3x + 2$
i.e. $2D^2 [ax + b] - D[ax + b] - 6[ax + b] \equiv 3x + 2$
$2D^2 (ax + b) = 0, \quad D(ax + b) = a$
Hence $0 - a - 6ax - 6b \equiv 3x + 2$

Equating the coefficients of x gives: $-6a = 3$

i.e. $a = -\frac{1}{2}$

Equating the constant terms gives: $-a - 6b = 2$

i.e. $-\left(-\frac{1}{2}\right) - 6b = 2$

$-6b = \frac{3}{2}$

$b = -\frac{1}{4}$

Hence the particular integral $v = -\dfrac{1}{2}x - \dfrac{1}{4}$

The general solution $y = u + v = A\,e^{-\frac{3}{2}x} + B\,e^{2x} - \dfrac{1}{2}x - \dfrac{1}{4}$

(c) $f(x)$ = an exponential function

(i) Straightforward case

If $f(x) = C\,e^{px}$, where C and p are constants, then try as the particular integral $v = k\,e^{px}$. The reason for such a substitution is that all the differential coefficients of e^{px} are multiples of e^{px}.

Problem 4. Find the general solution of the equation

$$3\frac{d^2y}{dx^2} + \frac{dy}{dx} - 4y = 2\,e^{-3x}$$

Given the boundary conditions that when $x = 0$, $y = \dfrac{3}{5}$ and $\dfrac{dy}{dx} = -6\dfrac{4}{5}$, find the particular solution.

In D operator form the equation is $(3D^2 + D - 4)y = 2\,e^{-3x}$.
The auxiliary equation is $3m^2 + m - 4 = 0$
$$\text{i.e. } (m-1)(3m+4) = 0$$
$$\text{i.e. } m = 1 \text{ or } m = -\frac{4}{3}$$

Hence the complementary function, $u = A\,e^x + B\,e^{-\frac{4}{3}x}$
Let the particular integral, $v = k\,e^{-3x}$ then

$(3D^2 + D - 4)\,[k\,e^{-3x}] \equiv 2\,e^{-3x}$ from equation (2)
$D(k\,e^{-3x}) = -3\,k\,e^{-3x}$
$D^2(k\,e^{-3x}) = 9\,k\,e^{-3x}$
Hence $27\,k\,e^{-3x} - 3\,k\,e^{-3x} - 4\,k\,e^{-3x} \equiv 2\,e^{-3x}$
$$\text{i.e. } 20\,k\,e^{-3x} \equiv 2\,e^{-3x}$$
$$\text{Hence } k = \frac{1}{10}$$

Hence the particular integral, $v = \dfrac{1}{10}\,e^{-3x}$

The general solution, $y = u + v = A\,e^x + B\,e^{-\frac{4}{3}x} + \dfrac{1}{10}\,e^{-3x}$

When $x = 0$, $y = \dfrac{3}{5}$, and substituting in the general solution gives:

$$\frac{3}{5} = A + B + \frac{1}{10}$$

i.e. $\dfrac{1}{2} = A + B$ (1)

$$\dfrac{dy}{dx} = A\,e^x - \dfrac{4}{3}\,B\,e^{-\frac{4}{3}x} - \dfrac{3}{10}\,e^{-3x}$$

When $x = 0$, $\dfrac{dy}{dx} = -\,6\dfrac{4}{5}$. Therefore $-6\dfrac{4}{5} = A - \dfrac{4}{3}\,B - \dfrac{3}{10}$

i.e. $\quad -\,6\dfrac{1}{2} = A - \dfrac{4}{3}\,B$ (2)

$(1) - (2)$ gives $\quad \dfrac{1}{2} - -6\,\dfrac{1}{2} = \left(1 - -\dfrac{4}{3}\right)B$

i.e. $\ 7 = \dfrac{7}{3}\,B$

i.e. $B = 3$

From (1), when $B = 3$, $A = -\,2\dfrac{1}{2}$

Hence the particular solution is $y = 3\,e^{-\frac{4}{3}x} - \dfrac{5}{2}\,e^x + \dfrac{1}{10}\,e^{-3x}$.

(ii) 'Snag' cases

If $f(x) = C\,e^{px}$, where C and p are constants, and e^{px} occurs in the complementary function the particular integral cannot be found by assuming that $v = k\,e^{px}$, for on substituting k becomes zero. In this case try $v = kx\,e^{px}$ for the particular integral (see Problem 5). If e^{px} and $x\,e^{px}$ both appear in the complementary function then try $v = k\,x^2\,e^{px}$ as the particular integral, and so on (see Problem 6).

Problem 5. Solve the differential equation $(D^2 - D - 2)y = 6\,e^{-x}$.

The auxiliary equation is $m^2 - m - 2 = 0$
i.e. $(m - 2)(m + 1) = 0$
i.e. $m = 2$ and $m = -1$
Hence the complementary function, $u = A\,e^{2x} + B\,e^{-x}$

Since e^{-x} appears in the complementary function **and** in the right-hand side of the differential equation, let the particular integral $v = kx\,e^{-x}$.

Then $(D^2 - D - 2)[kx\,e^{-x}] \equiv 6\,e^{-x}$

$D(kx\,e^{-x}) = (kx)(-e^{-x}) + (e^{-x})(k) = k\,e^{-x}\,(1 - x)$, by the product rule

$D^2\,(kx\,e^{-x}) = D(ke^{-x}\,(1 - x)) = (k\,e^{-x})(-1) + (1 - x)(-\,k\,e^{-x})$ by the

product rule

$$= -k\,e^{-x}\,(2 - x)$$

Hence $(D^2 - D - 2)[kx\,e^{-x}] = -k\,e^{-x}(2-x) - k\,e^{-x}(1-x) - 2\,kx\,e^{-x}$
$$\equiv 6\,e^{-x}$$

Thus $-2\,ke^{-x} + kx\,e^{-x} - ke^{-x} + kx\,e^{-x} - 2\,kx\,e^{-x} \equiv 6\,e^{-x}$

i.e. $\quad -3k\,e^{-x} = 6\,e^{-x}$

i.e. $\quad k = -2$

Hence the particular integral, $v = -2x\,e^{-x}$

The general solution $y = u + v = A\,e^{2x} + B\,e^{-x} - 2x\,e^{-x}$

Problem 6. Solve $\dfrac{d^2y}{dx^2} - 2\dfrac{dy}{dx} + y = 5\,e^x$

In D operator form the differential equation is $(D^2 - 2D + 1)y = 5\,e^x$
The auxiliary equation is $m^2 - 2m + 1 = 0$
i.e. $\quad (m-1)(m-1) = 0$
i.e. $\quad m = 1$ twice
Hence the complementary function, $u = (Ax + B)\,e^x$

Since e^x **and** $x\,e^x$ both appear in the complementary function, let the particular integral, $v = kx^2\,e^x$

Thus $\quad (D^2 - 2D + 1)[kx^2\,e^x] \equiv 5\,e^x$

$D(kx^2\,e^x) = (kx^2)(e^x) + (e^x)(2\,kx) \equiv k\,e^x(x^2 + 2x)$

$D^2(kx^2\,e^x) = D(ke^x(x^2 + 2x)) = (ke^x)(2x+2) + (x^2+2x)(ke^x)$
$$= ke^x(x^2 + 4x + 2)$$

Hence $(D^2 - 2D + 1)[kx^2\,e^x] = ke^x(x^2 + 4x + 2) - 2ke^x(x^2 + 2x)$
$$+ kx^2\,e^x \equiv 5e^x$$

i.e. $2\,k\,e^x = 5\,e^x$

i.e. $\quad k = \dfrac{5}{2}$

Hence the particular integral, $v = \dfrac{5}{2}\,x^2\,e^x$.

The general solution, $y = u + v = (Ax + B)\,e^x + \dfrac{5}{2}\,x^2\,e^x$

i.e. $y = e^x\left(Ax + B + \dfrac{5}{2}\,x^2\right).$

(d) $f(x) =$ a sine or cosine

(i) Straightforward case

If $f(x) = k_1 \sin px + k_2 \cos px$, where p, k_1 and k_2 are constants and either k_1 or k_2 may be zero, then try $v = A \sin px + B \cos px$ as the particular integral. The reason for this is that all the differential coefficients of either sin px or cos px are multiples of either sin px or cos px. (Note that in the suggested substitution for v it is insufficient to just let $v = A \sin pt$ or $v = B \cos pt$ even if either k_1 or k_2 are zero). Substituting for v in the differential equation and comparing coefficients of sin px and cos px on both sides produces two equations to determine A and B.

Problem 7. Solve $2\dfrac{d^2 y}{dx^2} + 5\dfrac{dy}{dx} - 3y = 4 \sin 2x$.

In D operator form the differential equation is $(2D^2 + 5D - 3)y = 4 \sin 2x$.
The auxiliary equation is $2m^2 + 5m - 3 = 0$
$$\text{i.e.} \quad (2m - 1)(m + 3) = 0$$
$$\text{i.e.} \quad m = \frac{1}{2} \text{ or } m = -3.$$
Hence the complementary function, $u = A e^{\frac{1}{2}x} + B e^{-3x}$

Let the particular integral be $v = A \sin 2x + B \cos 2x$
then $(2D^2 + 5D - 3) [A \sin 2x + B \cos 2x] \equiv 4 \sin 2x$
$D (A \sin 2x + B \cos 2x) = 2A \cos 2x - 2B \sin 2x$
$D^2(A \sin 2x + B \cos 2x) = D(2A \cos 2x - 2B \sin 2x)$
$$= -4A \sin 2x - 4B \cos 2x.$$

Hence $(2D^2 + 5D - 3) [A \sin 2x + B \cos 2x]$
$= -8A \sin 2x - 8B \cos 2x + 10A \cos 2x - 10B \sin 2x - 3A \sin 2x$
$$- 3B \cos 2x \equiv 4 \sin 2x$$

Equating coefficients of sin $2x$ gives: $-11A - 10B = 4$ ... (1)
Equating coefficients of cos $2x$ gives: $10A - 11B = 0$ ... (2)
$10 \times (1)$ gives: $-110A - 100B = 40$ ... (3)
$11 \times (2)$ gives: $110A - 121B = 0$ ... (4)
$(3) + (4)$ gives: $-221B = 40$
$$\text{i.e.} \quad B = \frac{-40}{221}$$

Substituting $B = \dfrac{-40}{221}$ into equation (1) or (2) gives:
$$A = \frac{-44}{221}$$

Hence the particular integral, $v = \dfrac{4}{221} (-11 \sin 2x - 10 \cos 2x)$

The general solution,
$$y = u + v = A e^{\frac{1}{2}x} + B e^{-3x} - \frac{4}{221} (11 \sin 2x + 10 \cos 2x)$$

(ii) 'Snag' case

If $f(x) = k_1 \sin px + k_2 \cos px$ and $\sin px$ and/or $\cos px$ occur in the complementary function then let the particular integral, $v = x (A \sin px + B \cos px)$. This case is demonstrated in Problem 8.

Problem 8. Find the general solution of $\dfrac{d^2 y}{dx^2} + 9y = 12 \cos 3x$. If $y = 2$ and $\dfrac{dy}{dx} = 3$ when $x = 0$, determine the particular solution.

In D operator form the differential equation is $(D^2 + 9)y = 12 \cos 3x$.
The auxiliary equation is $m^2 + 9 = 0$

i.e. $m = \sqrt{-9} = \pm j\,3$

Hence the complementary function $u = e^0 \ (C \cos 3x + D \sin 3x)$

$$= C \cos 3x + D \sin 3x.$$

Since $\cos 3x$ occurs in the complementary function and in the right-hand side of the differential equation, let the particular integral
$v = x \ (A \sin 3x + B \cos 3x)$.

Hence $(D^2 + 9) \ [x \ (A \sin 3x + B \cos 3x)] \equiv 12 \cos 3x$.

$D[x(A \sin 3x + B \cos 3x)] = (x)(3A \cos 3x - 3B \sin 3x)$
$\qquad\qquad\qquad\qquad + (A \sin 3x + B \cos 3x)(1)$, by the product rule

$D^2 [x(A \sin 3x + B \cos 3x)] = (x)(-9A \sin 3x - 9B \cos 3x)$
$+ (3A \cos 3x - 3B \sin 3x) (1) + (3A \cos 3x - 3B \sin 3x)$

Hence $(D^2 + 9) \ [x(A \sin 3x + B \cos 3x)] = -9Ax \sin 3x - 9Bx \cos 3x$
$+ 6A \cos 3x - 6B \sin 3x + 9Ax \sin 3x + 9Bx \cos 3x \equiv 12 \cos 3x$

i.e. $6A \cos 3x - 6B \sin 3x \equiv 12 \cos 3x$

Equating coefficients of $\cos 3x$ gives: $\quad 6A = 12$
$\qquad\qquad\qquad\qquad\qquad$ i.e. $\quad A = 2$

Equating coefficients of $\sin 3x$ gives: $\quad -6B = 0$
$\qquad\qquad\qquad\qquad\qquad$ i.e. $\quad B = 0$

Hence the particular integral, $v = x(2 \sin 3x)$
The general solution, $y = u + v = C \cos 3x + D \sin 3x + 2x \sin 3x.$

When $x = 0, y = 2$. Hence $2 = C \cos 0 + D \sin 0 + 0$
$\qquad\qquad\qquad\qquad$ i.e. $\quad C = 2$

$\dfrac{dy}{dx} = -3 \, C \sin 3x + 3 \, D \cos 3x + (2x)(3 \cos 3x) + (2 \sin 3x)$

When $x = 0, \dfrac{dy}{dx} = 3$. Hence $3 = -3 \, C \sin 0 + 3 \, D \cos 0 + 0 + 2 \sin 0$
$\qquad\qquad\qquad$ i.e. $\quad 3 = 3D$
$\qquad\qquad\qquad\qquad\quad D = 1$

Hence the particular solution is $y = 2 \cos 3x + \sin 3x + 2x \sin 3x$
    i.e. $y = 2 \cos 3x + (1 + 2x) \sin 3x$.

    Typical practical examples of the type of equation solved in Problems 7 and 8 include:

(i) the equation of motion of a body: $\dfrac{d^2 y}{dt^2} + n^2 y = k \cos \omega t$, and

(ii) the equation for variation of charge q in an alternating current circuit containing L, R and C in series:

$$L \frac{d^2 q}{dt^2} + R \frac{dq}{dt} + \frac{1}{C} q = V_0 \sin \omega t.$$

(e) $f(x) =$ a sum or product

In all of the straightforward cases of finding particular integrals it is noticed that a form of particular integral is assumed which is suggested by $f(x)$ but which contains undetermined coefficients.

    If $f(x)$ consists of a **sum** of terms the particular integral is the sum of the particular integrals corresponding to the separate terms. Thus, for example, if $f(x) = 2x + 3 \sin 4x$ then let the particular integral, $v = (ax + b) + (C \sin 2x + D \cos 2x)$.

    If $f(x)$ is a **product** of two terms then the particular integral assumed is that suggested by $f(x)$. Thus, for example, if $f(x) = e^{\alpha x} \cos \beta x$ then let the particular integral, $v = e^{\alpha x} (A \cos \beta x + B \sin \beta x)$.

Problem 9. Solve $(D^2 - 4D + 4)y = 4x + 3 \cos 2x$.

The auxiliary equation is $m^2 - 4m + 4 = 0$
          i.e. $(m - 2)(m - 2) = 0$
          i.e. $m = 2$ twice.

The complementary function, $u = (Ax + B) e^{2x}$.

Let the particular integral, $v = ax + b + C \cos 2x + D \sin 2x$,

then $(D^2 - 4D + 4) [v] = 4x + 3 \cos 2x$.

$D(v) = a - 2C \sin 2x + 2D \cos 2x$

$D^2(v) = -4C \cos 2x - 4D \sin 2x$.

Hence $(D^2 - 4D + 4) [v] = -4C \cos 2x - 4D \sin 2x - 4a + 8C \sin 2x$
                         $- 8D \cos 2x + 4ax + 4b + 4C \cos 2x$
                         $+ 4D \sin 2x \equiv 4x + 3 \cos 2x$.

Equating constant terms gives: $-4a + 4b = 0$
Equating the coefficients of x gives: $4a = 4$
              i.e. $a = 1$ and thus $b = 1$.

Equating coefficients of $\cos 2x$ gives: $-4C - 8D + 4C = 3$

i.e. $\qquad 8D = 3$

$$D = \frac{-3}{8}$$

Equating coefficients of sin 2x gives: $\qquad -4D + 8C + 4D = 0$

i.e. $\qquad 8C = 0$

$$C = 0$$

Hence the particular integral, $v = x + 1 - \dfrac{3}{8} \sin 2x.$

The general solution, $y = u + v = (Ax + B)\, e^{2x} + x + 1 - \dfrac{3}{8} \sin 2x.$

Problem 10. Solve $\dfrac{d^2 y}{dx^2} + 2\dfrac{dy}{dx} + 2y = 6\, e^x \sin 2x.$

In D operator form the differential equation is $(D^2 + 2D + 2)y = 6e^x \sin 2x.$
The auxiliary equation is $\quad m^2 + 2m + 2 = 0$

i.e. $\quad m = \dfrac{-2 \pm \sqrt{[2^2 - 4(1)(2)]}}{2} = -1 \pm j\,1$

The complementary function, $u = e^{-x}\,(C \cos x + D \sin x)$

Let the particular integral, $v = e^x\,(A \cos 2x + B \sin 2x)$

then $(D^2 + 2D + 2)\,[v] \equiv 6\, e^x \sin 2x$

$D(v) = (e^x)(-2A \sin 2x + 2B \cos 2x) + e^x\,(A \cos 2x + B \sin 2x)$
$\qquad\qquad [= e^x\,(-2A + B) \sin 2x + e^x\,(2B + A) \cos 2x]$

$D^2\,(v) = (e^x)(-4A \cos 2x - 4B \sin 2x) + (e^x)(-2A \sin 2x + 2B \cos 2x)$
$\qquad + (e^x)(-2A \sin 2x + 2B \cos 2x) + (e^x)(A \cos 2x + B \sin 2x)$
$\qquad = e^x\,[(-3A + 4B) \cos 2x + (-3B - 4A) \sin 2x]$

Hence $(D^2 + 2D + 2)\,[v] = e^x(-3A + 4B) \cos 2x + e^x(-3B - 4A) \sin 2x$
$\qquad\qquad + 2\,e^x\,(-2A + B) \sin 2x + 2\,e^x\,(2B + A) \cos 2x$
$\qquad\qquad + 2\,e^x\,A \cos 2x + 2\,e^x\,B \sin 2x \equiv 6\, e^x \sin 2x$

Equating coefficients of $e^x \sin 2x$ gives: $\quad -3B - 4A - 4A + 2B + 2B = 6$

i.e. $\quad -8A + B = 6 \qquad\qquad (1)$

Equating coefficients of $e^x \cos 2x$ gives: $\quad -3A + 4B + 4B + 2A + 2A = 0$

i.e. $\quad A + 8B = 0 \qquad\qquad (2)$

Solving equations (1) and (2) gives $\quad A = \dfrac{-48}{65}$ and $B = \dfrac{6}{65}$

Hence the particular integral, $v = e^x \left(\dfrac{-48}{65} \cos 2x + \dfrac{6}{65} \sin 2x \right)$

The general solution is

$y = u + v = e^{-x}\,(C \cos x + D \sin x) + \dfrac{e^x}{65}\,(6 \sin 2x - 48 \cos 2x)$

Further problems on solving differential equations of the type
$$a\frac{d^2y}{dx^2} + b\frac{dy}{dx} + cy = f(x) \text{ may be found in Section 4 (Problems 1 to 26)},$$
page 324.

3 Summary of procedure to solve differential equations of the type $a\dfrac{d^2y}{dx^2} + b\ \dfrac{dy}{dx} + cy = f(x)$

1. Rewrite the given differential equation as $(a D^2 + b D + c)y = f(x)$.
2. Substitute m for D and solve the **auxiliary equation** $am^2 + bm + c = 0$ for m.
3. Obtain the **complementary function, u**. (This is exactly the same procedure discussed in Section 3, Chapter 23, although it was not referred to at that stage as the complementary function.)
4. To find the **particular integral, v**, firstly assume a particular integral which is suggested by $f(x)$ but which contains undetermined coefficients. Below are a list of suggested substitutions.

Type	Straightforward cases Try as particular integral:	'Snag' cases Try as particular integral:
(a) $f(x) = $ a constant	$v = k$	$v = kx$ (used when C.F. contains a constant)
(b) $f(x) = $ polynomial ($f(x) = L + Mx + Nx^2$ $+ \ldots$ where any of the coefficients may be zero)	$v = a + bx + cx^2 + \ldots$	
(c) $f(x) = $ an exponential function. ($f(x) = A\,e^{\alpha x}$)	$v = k\,e^{\alpha x}$	$v = k\,x\,e^{\alpha x}$ (used when $e^{\alpha x}$ appears in the C.F.) $v = kx^2\,e^{\alpha x}$ (used when $e^{\alpha x}$ **and** $xe^{\alpha x}$ both appear in the C.F.), and so on
(d) $f(x) = $ a sine or cosine ($f(x) = a \sin px + b \cos px$, where a or b may be zero.)	$v = A \sin px + B \cos px$	$v = x\,(A \sin px + B \cos px)$ (used when $\sin px$ and/or $\cos px$ appears in the C.F.)

Type	Straightforward cases	'Snag' cases
	Try as particular integral:	Try as particular integral:
(e) $f(x) =$ a sum		
e.g. (i) $f(x) =$ $2x^2 + 5 \cos 3x$	(i) $v = ax^2 + bx + c$ $+ d \cos 3x$ $+ e \sin 3x$	
(ii) $f(x) = x + 1 - e^{-x}$	(ii) $v = ax + b + ce^{-x}$	
$f(x) =$ a product e.g. $f(x) = 3e^{2x} \sin 4x$	$v = e^{2x} (A \cos 4x +$ $B \sin 4x)$	

5. Substitute the suggested particular integral into the differential equation
$a\dfrac{d^2 v}{dx^2} + b\dfrac{dv}{dx} + cv = f(x)$ and equate relevant coefficients to find the
constants introduced.

6. The **general solution** is given by $y =$ complementary function + particular
integral, i.e. $y = u + v$.

7. Given sufficient **boundary conditions** the arbitrary constants in the complementary function may be determined to give the **particular solution**.

4 Further problems

In Problems 1 to 12 find the general solution of the given differential
equations.

1. (a) $\dfrac{d^2 y}{dx^2} - 5\dfrac{dy}{dx} + 6y = 3 \quad [y = A\,e^{3x} + B\,e^{2x} + \tfrac{1}{2}]$

   (b) $\dfrac{d^2 y}{dt^2} + 2\dfrac{dy}{dt} + 2y = 5 \quad [y = e^{-t}\,(C \cos t + D \sin t) + \tfrac{5}{2}]$

2. (a) $(D^2 + 9D)y = 2$ where $D \equiv \dfrac{d}{dx}$. $\quad [y = A + B\,e^{-9x} + \tfrac{2}{9}x]$

   (b) $\dfrac{d^2 x}{d\theta^2} + 7\dfrac{dx}{d\theta} = 7 \quad [x = A + B\,e^{-7\theta} + x]$

3. (a) $4\dfrac{d^2 y}{dx^2} - 12\dfrac{dy}{dx} + 9y = x \quad \left[y = (A + Bx)\,e^{\frac{3}{2}x} + \dfrac{1}{27}\,(3x + 4)\right]$

   (b) $(D^2 + 3D - 4)y = 2x + 3 \quad [y = Ae^x + Be^{-4x} - \tfrac{1}{8}\,(4x + 9)]$

4. (a) $2\dfrac{d^2 y}{dx^2} + 3\dfrac{dy}{dx} + 2y = x^2 + 2x + 1.$

$$\left[y = e^{-\frac{3}{4}x} \left(C \cos \frac{\sqrt{7}}{4} x + D \sin \frac{\sqrt{7}}{4} x \right) + \frac{1}{2} x^2 - \frac{1}{2} x + \frac{1}{4} \right]$$

(b) $(2D^2 + 5D - 3)y = 2x^3 - x + 7.$

$$\left[y = Ae^{\frac{1}{2}x} + Be^{-3x} - \frac{1}{27} (18x^3 + 90x^2 + 363x + 788) \right]$$

5. (a) $\dfrac{d^2y}{dx^2} + 3\dfrac{dy}{dx} - 10y = 2e^{3x}$ $[y = Ae^{2x} + Be^{-5x} + \frac{1}{4} e^{3x}]$

(b) $9\dfrac{d^2x}{dt^2} - 6\dfrac{dx}{dt} + x = 5e^{2t}$ $[x = (A + Bt)e^{\frac{1}{3}t} + \frac{1}{5} e^{2t}]$

6. (a) $\dfrac{d^2y}{dx^2} + \dfrac{dy}{dx} - 6y = e^{-3x}$ $[y = Ae^{2x} + Be^{-3x} - \frac{1}{6} x e^{-3x}]$

(b) $(2D^2 - D - 3)y = 5e^{-x}$ $[y = Ae^{-x} + Be^{\frac{3}{2}x} - x e^{-x}]$

7. (a) $\dfrac{d^2y}{dx^2} + 4\dfrac{dy}{dx} + 4y = 4e^{-2x}$ $[y = (A + Bx)e^{-2x} + 2x^2 e^{-2x}]$

(b) $(4D^2 + 20D + 25)y = 8e^{-\frac{5}{2}x}$ $[y = (A + Bx)e^{-\frac{5}{2}x} + x^2 e^{-\frac{5}{2}x}]$

8. (a) $2\dfrac{d^2x}{dt^2} + 5\dfrac{dx}{dt} - 3x = 2 \sin 3t.$

$$\left[x = Ae^{\frac{1}{2}t} + Be^{-3t} - \frac{1}{111} (7 \sin 3t + 5 \cos 3t) \right]$$

(b) $(9D^2 - 6D + 1)y = 40 \cos 2x.$

$$\left[y = (A + Bx)e^{\frac{1}{3}x} - (0.350\,6 \sin 2x + 1.022\,6 \cos 2x) \right]$$

9. (a) $\dfrac{d^2y}{dx^2} + 4y = 2 \sin 2x.$ $[y = C \cos 2x + D \sin 2x - \frac{1}{2} x \cos 2x]$

(b) $(D^2 + 6D + 13)y = \cos 2x.$

$$[y = e^{-3x} (C \cos 2x + D \sin 2x) + \frac{x}{26} (2 \sin 2x + 3 \cos 2x)]$$

10. (a) $15\dfrac{d^2y}{dx^2} - 2\dfrac{dy}{dx} - y = 3x + 65 \sin x.$

$$[y = Ae^{-\frac{1}{5}x} + Be^{\frac{1}{3}x} - 3x + 6 - 4 \sin x + \frac{1}{2} \cos x]$$

(b) $(4D^2 + 12D + 9)y = e^x \cos x.$

$$[y = (A + Bx)e^{-\frac{3}{2}x} + \frac{e^x}{841} (20 \sin x + 21 \cos x)]$$

11. (a) $\dfrac{d^2x}{dt^2} - 2\dfrac{dx}{dt} + 2x = e^t \sin t.$

$$[x = e^t\,(C \cos t + D \sin t) - \tfrac{1}{2}\,e^t t \cos t]$$

(b) $(2D^2 - 5D - 7)y = 3 \sin x + 4 \cos x.$

$$\left[\, y = A\,e^{\frac{7}{2}x} + Be^{-x} - \frac{1}{106}\,(47 \sin x + 21 \cos x) \right]$$

12. (a) $\dfrac{d^2y}{dx^2} - 3\dfrac{dy}{dx} + 2y = x^2 + e^x.$

$$\left[\, y = Ae^x + Be^{2x} + \frac{1}{2}\,x^2 + \frac{3}{2}\,x + \frac{7}{4} - x\,e^x \right]$$

(b) $(D^2 - 1)y = \sinh 2x.$ $\quad \left[\, y = A\,e^x + B\,e^{-x} + \dfrac{1}{3}\,\sinh 2x \right]$

In Problems 13 to 19 find the particular solution of the given differential equations for the stated boundary conditions.

13. $2\dfrac{d^2y}{dx^2} + 5\dfrac{dy}{dx} - 3y = 9$; when $x = 0, y = 0$ and $\dfrac{dy}{dx} = 2$.

$$\left[\, y = \frac{1}{7}\,\left(22e^{-\frac{1}{2}x} - e^{3x}\right) - 3 \right]$$

14. $(D^2 + 1)y = 4$; when $x = 0, y = 0$ and $D(y) = 3$. $\left(D \equiv \dfrac{d}{dx} \right).$

$$[y = 3 \sin x - 4 \cos x + 4]$$

15. $3\dfrac{d^2x}{dt^2} + \dfrac{dx}{dt} - 4x = 2e^{-t}$; when $t = 0, x = 0$ and $\dfrac{dx}{dt} = -5$.

$$\left[\, x = 3e^{-\frac{4}{3}t} - 2e^t - e^{-t} \right]$$

16. $(D^2 - 2D + 1)y = 5e^x$; when $x = 0, y = 2$ and $D(y) = 3$.

$$\left[\, y = (2 + x)e^x + \frac{5}{2}x^2\,e^x \right]$$

17. $(D^2 - 5D + 6)y = -3 \sin x$; when $x = 0, y = 0$ and $D(y) = \dfrac{2}{5}$.

$$\left[\, y = \frac{1}{10}\left\{e^{3x} + 2e^{2x} - 3(\sin x + \cos x)\right\} \right]$$

18. $\dfrac{d^2\theta}{dt^2} - 6\dfrac{d\theta}{dt} + 10\theta = 20 - e^{2t}$; when $t = 0, \theta = 4$ and $\dfrac{d\theta}{dt} = \dfrac{25}{2}$.

$$\left[\, \theta = e^{3t}\,\frac{5}{2}\,\cos t + 6 \sin t) + 2 - \frac{1}{2}\,e^{2t} \right]$$

19. $2\dfrac{d^2x}{dt^2} - \dfrac{dx}{dt} - 6x = 12\, e^t \cos t$; when $t = 0, x = \dfrac{-13}{29}$ and $\dfrac{dx}{dt} = \dfrac{-169}{29}$.

$$\left[x = 2e^{-\frac{3}{2}t} - e^{2t} + \frac{e^t}{29}(18 \sin t - 42 \cos t) \right]$$

20. The charge q in an electrical circuit of time t satisfies the differential equation $L\dfrac{d^2q}{dt^2} + R\dfrac{dq}{dt} + \dfrac{1}{C}\, q = E$, where L, R, C and E are constants. $L = 0.9, C = 40 \times 10^{-6}$ and $E = 100$. Solve the equation for q when (a) $R = 300$ and (b) R is negligible. Assume that when $t = 0, q = 0$ and $\dfrac{dq}{dt} = 0$.

$$\text{(a)} \quad \left[q = \frac{1}{250} - \left(\frac{1}{250} + \frac{2}{3}\, t \right) e^{-\frac{500}{3} t} \right]$$

$$\text{(b)} \quad \left[q = \frac{1}{250} \left(1 - \cos \frac{500}{3} t \right) \right]$$

21. Solve the following differential equation representing the motion of a body: $\dfrac{d^2y}{dt^2} + n^2 y = a \sin pt$,

where $n \neq 0$ and $p^2 \neq n^2$, given that when $t = 0, y = 0$ and $\dfrac{dy}{dt} = 0$

$$\left[y = \frac{a}{n^2 - p^2} \left(\sin pt - \frac{p}{n} \sin nt \right) \right]$$

22. In a galvanometer the deflection θ satisfies the differential equation: $\dfrac{d^2\theta}{dt^2} + 2\dfrac{d\theta}{dt} + \theta = 4$.

Solve the equation for θ given that when $t = 0, \theta = 0$ and $\dfrac{d\theta}{dt} = 0$.

$$[\theta = 4 - 4e^{-t}(1 + t)]$$

23. Solve the equation $\dfrac{d^2x}{dt^2} + n^2 x = k \cos \omega t$ representing the equation of motion of a body, given that when $t = 0, x = 0$ and $\dfrac{dx}{dt} = 0$.

$$\left[x = \frac{k}{n^2 - \omega^2} (\cos \omega t - \cos nt) \right]$$

24. The differential equation describing the variation of capacitor charge in an alternating current circuit containing inductance L, resistance R and capacitance C in series is:

$$L\frac{d^2q}{dt^2} + R\frac{dq}{dt} + \frac{1}{C}\, q = V_0 \sin \omega t.$$

Find an expression for the charge q of the capacitor in the circuit at any time t seconds, given that when $t = 0, q = 0$ and $\dfrac{dq}{dt} = -44.29$, in the following case: $R = 20$ ohms, $L = 10 \times 10^{-3}$ henry, $C = 100 \times 10^{-6}$ farads, $V_0 = 500$ volts and $\omega = 250$.

$$[q = (0.022\ 15 - 32.53t)e^{-1\ 000t} + 0.0415\ 2 \sin 250t$$
$$- 0.022\ 15 \cos 250t]$$

25. The equation of motion of a particle moving in a straight line under damping is: $2\dfrac{d^2x}{dt^2} + 2\dfrac{dx}{dt} + x = k \sin t$, where k is a constant. Show that if when $t = 0, x = n$ and $\dfrac{dx}{dt} = 0, x$ is given by:

$$x = e^{-\frac{1}{2}t}\left\{ n \cos \frac{t}{2} + (n - 2k) \sin \frac{t}{2} \right\} + k \sin t.$$

26. The motion of a mass vibrating in a given way is shown by the equation:

$$\frac{d^2y}{dt^2} + 10\frac{dy}{dt} + 81y = 4 \sin 3t$$

Solve the equation for y.

$$[y = e^{-5} (c \cos \sqrt{56}t + D \sin \sqrt{56}t) + \frac{2}{507} (12 \sin 3t - 5 \cos 3t)]$$

Chapter 25

Sampling and estimation theories (EC 14)

1 Introduction

In this chapter a brief introduction is given to certain aspects of three major statistical theories, these being:

(a) the elementary sampling theory (the word 'elementary' not meaning that it is simple, but rather a word used to distinguish it from the exact sampling theory),
(b) the statistical estimation theory for large samples, and
(c) the small sampling theory.

A full treatment of these three theories is beyond the scope of this book, as are the derivations of the formulae and coefficients introduced. However, the concepts which are introduced will be useful for those making a detailed study at a later date of inspection, control and quality control techniques.

Sampling distributions are discussed in Section 2 and parts of the elementary sampling theory for large samples are introduced in Section 3 and deal with relating the mean and standard deviation of samples to that of the population from which they are drawn. In the statistical estimation theory introduced in Section 4, the degree of confidence with which the mean and standard deviation of a population can be estimated, using large samples drawn from that population, is discussed. Finally, in Section 5, a technique is introduced for determining the confidence with which the mean of a population can be found, using a small sample drawn from the population.

The authors are aware that a full treatment of the three theories has not been given and also that some of the finer points of the theories have been omitted for the sake of clarity. However, at this stage, there is a danger of 'not seeing the wood for the trees' and only the basic concepts have been developed.

2 Sampling distributions

If, in statistics, an analysis is to be made on the numbers, say, 3, 7, 8, 4, 9, 2, 4, 7 and 11, then these numbers are said to form a **set**. The numbers in the set are called the **members** of the set, thus 3 is a member of the set of numbers 3, 7, 8, ...

The word 'population' is used in modern statistics to define the set containing **all** the data. It is a carry-over from the days when statistics were largely used to obtain data relating to the population of a country. Thus, the set containing, say, all car registration numbers is called a population, even though it does not refer to people in any way.

In statistics, it is not always possible to take into account all the members of a set and in these circumstances, **a sample**, or many samples, are drawn from a population. For example, a company manufacturing trousers for the teenage market might measure the waist and inside leg values of 250 teenagers in 30 different cities in order to find out the various sizes likely to be required. Naturally, it would be totally impractical to determine the measurements of all the several million teenagers in this country. Usually when the word sample is used, it means that a **random sample** is taken. If each member of a population has the same chance of being selected, then a sample taken from that population is called random. A sample which is not random is said to be **biased** and this usually occurs when some influence affects the selection. For example, if sampling techniques are used to make predictions on wages, and the sample is made by selecting peoples' names from telephone directories, a bias occurs, due to excluding those people who do not have a telephone installed. One way of obtaining a random sample for a relatively small population is to allocate each member of the population a number. Thus for a population of, say, 500 members, the members may be numbered from 1 to 500. If a sample of, say, 10 is required, the 500 numbers can be well mixed and 10 can be drawn, as for a raffle. Alternatively, tables of random numbers can be used to select a sample of 10. These are statistical tables giving lists of numbers and are generated in such a way that any number within a given range has an equal chance of being selected, by taking either row or column values from the table. Thus to select random numbers between 1 and 500 from the specimen table of random numbers given below, one way is to use the first three columns of the numbers of 50 000 or less, giving 240, 459, 305, 35, 156 and so on.

51 772	64 937	50 532
24 033	**15 630**	07 136
45 939	09 448	27 989
30 586	21 631	85 184
03 585	91 097	54 398

For very large populations, various artificial devices are used, and amongst these is 'Ernie' ('electronic random number indicating equipment'), used to prepare lists of winning Premium Bond numbers.

When it is necessary to make predictions about a population based on random sampling, often many samples of, say, N members are taken, before the predictions are made. If the mean value and standard deviation of each of the samples is calculated, it is found that the results vary from sample to sample, even though the samples are all taken from the same population. In the theories introduced in Sections 3, 4 and 5, it is important to know whether the differences in the values obtained are due to chance or whether the differences obtained are related in some way. If M samples of N members are drawn at random from a population, the mean values for the M samples together form a set of data. Similarly, the standard deviations of the M samples collectively form a set of data. Sets of data based on many samples drawn from a population are called **sampling distributions**. They are often used to describe the chance fluctuations of mean values and standard deviations based on random sampling.

3 The sampling distribution of the means

Suppose that it is required to obtain a sample of two items from a set containing five items. If the set is the five letters, A, B, C, D and E then the different samples which are possible are:

AB, AC, AD, AE, BC, BD, BE, CD, CE, and *DE,*

that is, ten different samples. The number of possible different samples in this case is given by $\dfrac{5 \times 4}{2 \times 1}$, i.e. 10. Similarly, the number of different ways in which a sample of three items can be drawn from a set having ten members can be shown to be $\dfrac{10 \times 9 \times 8}{3 \times 2 \times 1}$, i.e. 120. It follows that when a small sample is drawn from a large population, there are very many different combinations of members possible. With so many different samples possible, quite a large variation can occur in the mean values of various samples taken from the same population.

Usually, the greater the number of members in a sample, the closer will be the mean value of the sample to that of the population. Consider the set of numbers 3, 4, 5, 6 and 7. For a sample of 2 members, the lowest value of the mean is $\dfrac{3 + 4}{2}$, i.e. 3.5; the highest is $\dfrac{6 + 7}{2}$, i.e. 6.5, giving a range of mean values of 3. For a sample of 3 members, the range is $\dfrac{3 + 4 + 5}{3}$ to $\dfrac{5 + 6 + 7}{3}$, that is, 2. As the number in the sample increases, the range decreases until, in the limit, if the sample contains all the members of the set, the range of mean values is zero. When many samples are drawn from a population and a sampling distribution of the mean values of the samples is formed, the range of the mean values is small provided the number in the sample is large. Because the range is small it follows that the standard deviation of all the mean values will also be small, since it depends on the distance of the mean values from the distribution mean. The relationship between the standard deviation of the mean values of a sampling distribution and the number in each sample can be expressed as follows:

Theorem 1.

'If all possible samples of size N are drawn from a finite population, N_p, without replacement, and the standard deviation of the mean values of the sampling distribution of means is determined, then:

$$\sigma_{\bar{x}} = \frac{\sigma}{\sqrt{N}} \sqrt{\left(\frac{N_p - N}{N_p - 1} \right)}$$

where $\sigma_{\bar{x}}$ is the standard deviation of the sampling distribution of means and σ is the standard deviation of the population.'

The standard deviation of a sampling distribution of mean values is called the **standard error of the means,** thus

standard error of the means, $\sigma_{\bar{x}} = \dfrac{\sigma}{\sqrt{N}} \sqrt{\left(\dfrac{N_p - N}{N_p - 1} \right)}$ \hfill (1)

Equation (1) is used for a finite population of size N_p and/or for sampling without replacement. The word 'error' in the 'standard error of the means' does not mean that a mistake has been made but rather that there is a degree of uncertainty in predicting the mean value of a population based on the mean values of the samples. The formula for the standard error of the means is true for all values of the number in the sample, N. When N_p is very large compared with N or when the population is infinite (this can be considered to be the case when sampling is done with replacement), the correction factor $\sqrt{\left(\dfrac{N_p - N}{N_p - 1} \right)}$ approaches unity and equation (1) becomes

$$\sigma_{\bar{x}} = \frac{\sigma}{\sqrt{N}} \hfill (2)$$

Equation (2) is used for an infinite population and/or for sampling with replacement.

Theorem 1 can be verified by applying it to a small set of numbers as follows. Let the set be, say, 3, 4, 5, 6 and 7 and the sample size be, say, 2. The only possible different samples of size 2 which can be drawn from this set without replacement are:

$(3, 4), (3, 5), (3, 6), (3, 7), (4, 5), (4, 6), (4, 7), (5, 6), (5, 7)$ and $(6, 7)$.

The mean values of these samples form the following sampling distribution of means:

3.5, 4, 4.5, 5, 4.5, 5, 5.5, 5.5, 6 and 6.5.

The mean of the sampling distribution of means, $\mu_{\bar{x}}$, is

$$\frac{3.5 + 4 + 4.5 + \ldots + 6.5}{10} = \frac{50}{10} = 5$$

The standard deviation of the sampling distribution of means, $\sigma_{\bar{x}}$, is

$$\sqrt{\left[\frac{(3.5 - 5)^2 + (4 - 5)^2 + (4.5 - 5)^2 + \ldots + (6.5 - 5)^2}{10} \right]}$$

i.e. $\sigma_{\bar{x}} = \sqrt{\dfrac{7.5}{10}} = \pm 0.866$

Thus, the standard error of the means is 0.866.
The standard deviation of the population, σ, is

$$\sqrt{\left[\frac{(3 - 5)^2 + (4 - 5)^2 + (5 - 5)^2 + (6 - 5)^2 + (7 - 5)^2}{5} \right]}$$

i.e. $\sigma = \sqrt{2} = \pm 1.414$.

But from theorem 1:

$\sigma_{\bar{x}} = \dfrac{\sigma}{\sqrt{N}} \sqrt{\left(\dfrac{N_p - N}{N_p - 1} \right)}$, and substituting for N_p, N and σ in equation (1)

gives $\sigma_{\bar{x}} = \dfrac{\pm 1.414}{\sqrt{2}} \sqrt{\left(\dfrac{5 - 2}{5 - 1} \right)} = \sqrt{\dfrac{3}{4}} = \pm 0.866$, as obtained by con-

sidering all samples from the population. Thus theorem 1 is verified for this
particular problem.

In the example given above, it can be seen that the mean of the population,
$\left(\dfrac{3 + 4 + 5 + 6 + 7}{5} \right)$, is 5 and also that the mean of the sampling distribution
of means, $\mu_{\bar{x}}$, is 5. This result is generalised in Theorem 2.
Theorem 2.

'If all possible samples of size N are drawn from a population of size
N_p *and the mean value of the sampling distribution of means $\mu_{\bar{x}}$ is*
determined, then

$$\mu_{\bar{x}} = \mu \tag{3}$$

where μ is the mean value of the population.'

In practice, all possible samples of size N are not drawn from the population. However, if the sample size is large (usually taken as 30 or more), then
the relationship between the mean of the sampling distribution of means and
the mean of the population is very near to that shown in equation (3).
Similarly, the relationship between the standard error of the means and the
standard deviation of the population is very near to that shown in equation
(2).

Another important property of a sampling distribution is that when the
sample size, N, is large, **the sampling distribution of means approximates to a**

normal distribution, of mean value $\mu_{\bar{x}}$ and standard deviation $\sigma_{\bar{x}}$. This is true for all normally distributed populations and also for populations which are not normally distributed provided the population size is at least twice as large as the sample size. This property of normality of a sampling distribution is based on a special case of the 'central limit theorem', an important theorem relating to sampling theory. Because the sampling distribution of means and standard deviations are normally distributed, the table of the partial areas under the standardised normal curve (shown on page 350), can be used to determine the probabilities of a particular sample lying between, say, ± 1 standard deviation, and so on. This point is expanded in Worked Problem 2 following.

Worked problems on the sampling distribution of means

Problem 1. The heights of 2 500 men are normally distributed with a mean of 170 cm and a standard deviation of 7 cm. If random samples are taken of 30 men, predict the standard deviation and the mean of sampling distribution of means, if sampling is done (a) with replacement, and (b) without replacement.

For the population: number of members, N_p = 2 500;
standard deviation, σ = 7 cm; mean, $\mu = 170$ cm.
For the samples: number in each sample, $N = 30$.

(a) When sampling is done with replacement, the total number of possible samples (two or more can be the same), is infinite. Hence, from equation (2) the standard error of the mean (i.e. the standard deviation of the sampling distribution of means),

$$\sigma_{\bar{x}} = \frac{\sigma}{\sqrt{N}} = \frac{7}{\sqrt{30}} = 1.278 \text{ cm.}$$

From equation (3), the mean of the sampling distribution,

$$\mu_{\bar{x}} = \mu = 170 \text{ cm.}$$

(b) When sampling is done without replacement, the total number of possible samples is finite and hence equation (1) applies. Thus the standard error of the means, $\sigma_{\bar{x}} = \frac{\sigma}{\sqrt{N}} \sqrt{\left(\frac{N_p - N}{N_p - 1} \right)}$

$$= \frac{7}{\sqrt{30}} \sqrt{\left(\frac{2\,500 - 30}{2\,500 - 1} \right)}$$

$$= 1.278 \times 0.994\,2$$
$$= 1.271 \text{ cm.}$$

As stated, following equation (3), provided the sample size is large, the mean of the sampling distribution of means is the same for both finite and infinite populations. Hence, from equation (3),

$$\mu_{\bar{x}} = 170 \text{ cm.}$$

Problem 2. A group of 1 000 ingots of a metal have a mean mass of 7.4 kg and a standard deviation of 0.4 kg. Find the probability that a sample of 50 ingots chosen at random from the group, without replacement, will have a combined mass of (a) between 360 and 377.5 kg, and (b) more than 375 kg.

For the population: number of members, N_p = 1 000;
standard deviation, σ = 0.4 kg; mean μ = 7.4 kg.
For the sample: number in sample, N = 50.

If many samples of 50 ingots had been drawn from the group, then the mean of the sampling distribution of means, $\mu_{\bar{x}}$ would be equal to the mean of the population. Also, the standard error of means is given by

$\sigma_{\bar{x}} = \dfrac{\sigma}{\sqrt{N}} \sqrt{\left(\dfrac{N_p - N}{N_p - 1}\right)}$. In addition, the sampling distribution would

have been approximately normal. Assume that the sample given in the problem is one of many samples. For many (theoretical) samples:

the mean of the sampling distribution of means, $\mu_{\bar{x}} = \mu = 7.4$ kg.

Also, the standard error of the means, $\sigma_{\bar{x}} = \dfrac{\sigma}{\sqrt{N}} \sqrt{\left(\dfrac{N_p - N}{N_p - 1}\right)}$

$= \dfrac{0.4}{\sqrt{50}} \sqrt{\left(\dfrac{1000 - 50}{1000 - 1}\right)}$

$= 0.055\,2$ kg.

Thus, the sample under consideration is part of a normal distribution of mean value 7.4 kg and a standard error of the means of 0.055 2 kg.

(a) If the combined mass of 50 ingots is between 360 and 377.5 kg, then the mean mass of each of the 50 ingots lies between $\dfrac{360}{50}$ and $\dfrac{377.5}{50}$ kg, i.e. between 7.2 and 7.55 kg.

   Since the masses are normally distributed, it is possible to use the techniques of the normal distribution to determine the probability of the mean mass lying between 7.2 and 7.55 kg. The normal standard variate value, z, is given by $z = \dfrac{x - \bar{x}}{\sigma}$, hence for the sampling distribution of means, this becomes, $z = \dfrac{x - \mu_{\bar{x}}}{\sigma_{\bar{x}}}$. Thus, 7.2 kg corresponds to a z-value of $\dfrac{7.2 - 7.4}{0.055\,2} = -3.62$ standard deviations.

   Similarly, 7.55 kg corresponds to a z-value of $\dfrac{7.55 - 7.4}{0.055\,2} = 2.72$ standard deviations.

   Using the table given on page 350, the areas corresponding to these values of standard deviations are 0.499 9 and 0.496 7 respectively. Hence the probability of the mean mass lying between 7.2 and 7.55 kg is

0.499 9 + 0.496 7 = 0.996 6.

(This means that if 10 000 samples are drawn, 9 966 of these samples will have a combined mass of between 360 and 377.5 kg.)

(b) If the combined mass of 50 ingots is 375 kg, the mean mass of each ingot is $\frac{375}{50}$, that is, 7.5 kg.

The z-value for 7.5 kg is $\frac{7.5 - 7.4}{0.0552}$, i.e. 1.81 standard deviations. From the table given on page 350, the area corresponding to this z-value is 0.464 9. But this is the area between the ordinate $z = 0$ and ordinate $z = 1.81$. The 'more than' value required is the total area to the right of the $z = 0$ ordinate, less the value between $z = 0$ and $z = 1.81$, i.e. 0.500 0 $-$ 0.464 9. Thus, since areas are proportional to probabilities for the standardised normal curve, the probability of the mean mass being more than 7.5 kg is 0.500 0 $-$ 0.464 9, i.e. 0.035 1. (This means that only 351 samples in 10 000, for example, will have a combined mass exceeding 375 kg.)

Further problems on the sampling distribution of means may be found in *Section 7 (Problems 1 to 10), page 347.*

4 The estimation of population parameters based on a large sample size

When a population is large, it is not practical to determine its mean and standard deviation by using the basic formulae for these parameters. In fact, when a population is infinite, it is impossible to determine these values. For large and infinite populations the values of the mean and standard deviation may be estimated by using the data obtained from samples drawn from the population.

(a) Point and interval estimates

An estimate of a population parameter, such as mean or standard deviation, based on a single number is called a **point estimate.** An estimate of a population parameter given by two numbers, between which the parameter may be considered to lie is called an **interval estimate.** Thus if an estimate is made of the length of an object and the result is quoted as 200 centimetres, this is a point estimate. If the result is quoted as 200 ± 20 centimetres, this is an interval estimate and indicates that the length lies between 180 and 220 centimetres. Generally, a point estimate does not indicate how close the value is to the true value of the quantity and should be accompanied by additional information on which its merits may be judged. A statement of the error or the precision of an estimate is often called its **reliability.** In statistics, when estimates are made of population parameters based on samples, usually interval estimates are used. The word estimate does not suggest that we adopt the approach 'let's guess that the mean value is about . . .', but rather that a value is carefully selected and the degree of confidence which can be placed in the estimate is given in addition.

(b) Confidence intervals

It is stated in Section 3 that when samples are taken from a population, the mean values of these samples are approximately normally distributed, that is, the mean values forming the sampling distribution of means is approximately normally distributed. It is also true that if the standard deviations of each of the samples is found, then the standard deviations of all the samples are approximately normally distributed, that is, the standard deviations of the sampling distribution of standard deviations are approximately normally distributed. Parameters such as the mean or the standard deviation of a sampling distribution are called **sampling statistics**, S. Let μ_s be the mean value of a sampling statistic of the sampling distribution, that is, the mean value of the means of the samples or the mean value of the standard deviations of the samples. Also let σ_s be the standard deviation of a sampling statistic of the sampling distribution, that is, the standard deviation of the means of the samples or the standard deviation of the standard deviations of the samples. Because the sampling distribution of the means and of the standard deviations are normally distributed, it is possible to predict the probability of the sampling statistic lying in the intervals:

mean ± 1 standard deviation,
mean ± 2 standard deviations,
or mean ± 3 standard deviations,

by using tables of the partial areas under the standardised normal curve given in the table on page 350. From this table, the area corresponding to a z-value of + 1 standard deviation is 0.341 3, thus the area corresponding to ± 1 standard deviation is 2 × 0.341 3, that is, 0.682 6. Thus the percentage probability of a sampling statistic lying between the mean ± 1 standard deviation is 68.26 per cent. Similarly, the probability of a sampling statistic lying between the mean ± 2 standard deviations is 95.44 per cent and of lying between the mean ± 3 standard deviations is 99.74 per cent.

The values 68.26 per cent, 95.44 per cent and 99.74 per cent are called the **confidence levels** for estimating a sampling statistic. A confidence level of 68.26 per cent is associated with two distinct values, these being, S- (1 standard deviation), i.e. $S - \sigma_S$ and $S + $ (1 standard deviation), i.e. $S + \sigma_S$. These two values are called the **confidence limits** of the estimate and the distance between the confidence limits is called the **confidence interval**. A confidence interval indicates the expectation or confidence of finding an estimate of the population statistic in that interval, based on a sampling statistic. The list below is based on values given in the table on page 350, and gives some of the confidence levels used in practice and their associated z-values, some of the values given are based on interpolation. When the table is used in this context, z-values are usually indicated by 'z_c' and are called the **confidence coefficients**.

Confidence level,	%	99	98	96	95	90	80	50
Confidence coefficient,	z_c	2.58	2.33	2.05	1.96	1.645	1.28	0.674 5

Any other values of confidence levels and their associated confidence coefficients can be obtained using the table on page 350. Thus, if a confidence level of, say, 97.5 per cent is required, the corresponding confidence coefficient, z_c, can be determined as follows:

97.5 per cent is equivalent to a per unit value of 0.975 0. This indicates that the area under the standardised normal curve between $-z_c$ and $+z_c$, i.e. corresponding to 2 z_c is 0.975 0 of the total area. Hence the area between the mean value and z_c is $\dfrac{0.975\,0}{2}$, i.e. 0.487 5 of the total area. The z-value corresponding to a partial area of 0.487 5 is 2.24 standard deviations. Thus, the confidence coefficient corresponding to a confidence limit of 97.5 per cent is 2.24.

(c) Estimating the mean of a population when the standard deviation of the population is known

When a sample is drawn from a large population whose standard deviation is known, the mean value of the sample, $\bar{x}$, can be determined. This mean value can be used to make an estimate of the mean value of the population, μ. When this is done, the estimated mean value of the population is given as lying between two values, that is, lying in the confidence interval between the confidence limits. If a high level of confidence is required in the estimated value of μ, then the range of the confidence interval will be large. For example, if the required confidence level is 99 per cent, then the confidence interval is from $-z_c$ to $+z_c$, that is, $2 \times 2.58 = 5.16$ standard deviations wide. Conversely, a low level of confidence has a narrow confidence interval and a confidence level of, say, 80 per cent, has a confidence interval of 2×1.28, that is 2.56 standard deviations. The 68.26 per cent confidence level for an estimate of the population mean is given by estimating that the population mean, μ, is equal to the sample mean, $\bar{x}$, and then stating the confidence interval of the estimate. Since the 68.26 per cent confidence level is associated with '± 1 standard deviation of the means of the sampling distribution', then the 68.26 confidence level for the estimate of the population mean is given by

$$\bar{x} \pm 1 \; \sigma_{\bar{x}}.$$

In general, any particular confidence level can be obtained in the estimate, by using, $\bar{x} \pm z_c \, \sigma_{\bar{x}}$, where z_c is the confidence coefficient corresponding to the particular confidence level required. Thus for a 95 per cent confidence level, the confidence limits of the population mean are given by $\bar{x} \pm 1.96 \, \sigma_{\bar{x}}$.

Since only one sample has been drawn, the standard error of the means, $\sigma_{\bar{x}}$, is not known. However, it is shown in Section 3 that

$$\sigma_{\bar{x}} = \frac{\sigma}{\sqrt{N}} \sqrt{\left(\frac{N_p - N}{N_p - 1}\right)}.$$

Thus, the confidence limits of the mean of the population are:

$$\bar{x} \pm \frac{z_c \sigma}{\sqrt{N}} \sqrt{\left(\frac{N_p - N}{N_p - 1}\right)} \tag{4}$$

for a finite population of size N_p.

The confidence limits for the mean of the population are:

$$\bar{x} \pm \frac{z_c \sigma}{\sqrt{N}} \qquad (5)$$

for an infinite population.

Thus for a sample of size N and mean $\bar{x}$, drawn from an infinite population having a standard deviation of σ, the mean value of the population is estimated to be, for example, $\bar{x} \pm \dfrac{2.58\,\sigma}{\sqrt{N}}$ for a confidence level of 99 per cent. This indicates that the mean value of the population lies between $\bar{x} - \dfrac{2.58\,\sigma}{\sqrt{N}}$ and $\bar{x} + \dfrac{2.58\,\sigma}{\sqrt{N}}$, with 99 per cent confidence in this prediction.

(d) Estimating the mean and standard deviation of a population from sample data

The standard deviation of a large population is not known and, in this case, several samples are drawn from the population. The mean of the sampling distribution of means, $\mu_{\bar{x}}$ and the standard deviation of the sampling distribution of means (i.e. the standard error of the means), $\sigma_{\bar{x}}$, may be determined. The confidence limits of the mean value of the population, μ, are given by

$$\mu_{\bar{x}} \pm z_c\, \sigma_{\bar{x}} \qquad (6)$$

where z_c is the confidence coefficient corresponding to the confidence level required.

To make an estimate of the standard deviation, σ, of a normally distributed population:

(i) a sampling distribution of the standard deviations of the samples is formed, and

(ii) the standard deviation of the sampling distribution is determined by using the basic standard deviation formula.

This standard deviation is called the standard error of the standard deviations and is usually signified by σ_s. If s is the standard deviation of a sample, then the confidence limits of the standard deviation of the population, are given by:

$$s \pm z_c\, \sigma_s \qquad (7)$$

where z_c is the confidence coefficient corresponding to the required confidence level.

Worked problems on the estimation of population parameters based on a large sample size

Problem 1. Over a long period of time, it is found that the standard deviation of the diameters of rivets produced by a certain machine is 0.021 cm. The diameters of a random sample of 100 rivets produced by this machine in a day have a mean value of 0.493 cm. If the machine produces 2 700 rivets a day, determine (a) the 90 per cent confidence limits, and (b) the 97 per cent confidence limits for an estimate of the mean diameter of all the rivets produced by the machine in a day.

For the population: standard deviation, $\sigma = 0.021$ cm
    number in the population, $N_p = 2\,700$.
For the sample: number in the sample, $N = 100$
    mean, $\bar{x} = 0.493$ cm.

    There is a finite population and the standard deviation of the population is known, hence expression (4) is used for determining an estimate of the confidence limits of the population mean, i.e.

$$\bar{x} \pm z_c \; \frac{\sigma}{\sqrt{N}} \; \sqrt{\left(\frac{N_p - N}{N_p - 1} \right)}$$

(a) For a 90 per cent confidence level, the value of z_c, the confidence coefficient, is given on page 337, and is $z_c = 1.645$. Hence, the estimate of the confidence limits of the population mean, μ, is

$$0.493 \pm \frac{1.645 \times 0.021}{\sqrt{100}} \sqrt{\left(\frac{2\,700 - 100}{2\,700 - 1} \right)}$$

that is, $0.493 \pm 0.003\,45 \times 0.981\,5$,
i.e. $0.493 \pm 0.003\,4$ cm.
Thus, the 90 per cent confidence limits are 0.490 cm and 0.496 cm.

This indicates that if the mean diameter of a sample of 100 rivets is 0.493 cm, then it is predicted that the mean diameter of all the rivets will be between 0.490 cm and 0.496 cm and this prediction is made with confidence that it will be correct nine times out of ten.

(b) For a 97 per cent confidence level, the value of z_c has to be obtained from the table of partial areas under the standardised normal curve given on page 350, as it is not one of the values given in the table on page 337. The total area between ordinates drawn at $-z_c$ and $+z_c$ has to be 0.970 0. Because the standardised normal curve is symmetrical, the area between $z_c = 0$ and z_c, is $\dfrac{0.970\,0}{2}$, i.e. 0.485 0. From the table, an area of 0.485 0 corresponds to a z_c -value of 2.17. Hence, the estimated value of the confidence limits of the population mean is between $\bar{x} \pm z_c \dfrac{\sigma}{\sqrt{N}} \sqrt{\left(\dfrac{N_p - N}{N_p - 1} \right)}$

i.e. between $0.493 \pm \dfrac{2.17 \times 0.021}{\sqrt{100}} \sqrt{\left(\dfrac{2\,700 - 100}{2\,700 - 1} \right)}$, that is, between
$0.493 \pm 0.004\,5$ cm.
Thus the 97 per cent confidence limits are 0.489 cm and 0.498 cm.

    It can be seen that the higher value of confidence level required in part (b) results in a larger confidence interval.

Problem 2. It is required to determine the mean diameter of a long length of wire. The diameter of the wire is measured in 16 places selected at random throughout its length and the mean of these values is 0.314 mm. If the standard deviation of the diameter of the wire is given by the manufacturers as 0.025 mm, determine (a) the 80 per cent confidence interval of the estimated mean diameter of the wire, and (b) with what degree of confidence it can be said that 'the mean diameter is 0.314 ± 0.01 mm'.

For the population: $\sigma = 0.025$ mm.
For the sample: $N = 16$, $\bar{x} = 0.314$ mm.

Because an infinite number of measurements can be obtained for the diameter of the wire, the population is infinite and the estimated value of the confidence interval of the population mean is given by expression (5).

(a) For an 80 per cent confidence level, the value of z_c is obtained from the table on page 337, and is 1.28.

The 80 per cent confidence level estimate of the confidence interval

$$\text{of } \mu = \bar{x} \pm \frac{z_c \, \sigma}{\sqrt{N}}$$

$$= 0.314 \pm \frac{1.28 \times 0.025}{\sqrt{16}}$$

$$= 0.314 \pm 0.008 \text{ mm}.$$

That is, **the 80 per cent confidence interval is from 0.306 mm to 0.322 mm.** This indicates that the estimated mean diameter of the wire is between 0.306 and 0.322 mm and that this prediction is likely to be correct 80 times out of 100.

(b) To determine the confidence level, the given data is equated to expression (5), giving

$$0.314 \pm 0.01 = \bar{x} \pm z_c \, \frac{\sigma}{\sqrt{N}}.$$

But $\bar{x} = 0.314$, therefore

$$\pm z_c \, \frac{\sigma}{\sqrt{N}} = \pm 0.01$$

$$\text{i.e. } z_c = \frac{0.01 \sqrt{N}}{\sigma} = \frac{\pm 0.01 \times 4}{0.025}$$

$$= \pm 1.6.$$

Using the table of partial areas under the standardised normal curve given on page 350, a z_c-value of 1.6 standard deviations corresponds to an area of 0.445 2 between the mean value, $(z_c = 0)$, and $+1.6$ standard deviations. Because the standardised normal curve is symmetrical, the area between the mean and ± 1.6 standard deviations is $0.445\ 2 \times 2$, i.e. $0.890\ 4$. Thus **the confidence level corresponding to 0.314 ± 0.01 mm is 89.04 per cent.**

Problem 3. Several samples of thirty fuses selected at random from a large batch are tested when operating at a 20 per cent overload current and the mean time of the sampling distribution before the fuses failed is 12.56 minutes. The standard error of the means is 2.3 minutes. Determine the estimated mean time to failure of the batch of fuses for a confidence level of 80 per cent.

For the sampling distribution: the mean, $\mu_{\bar{x}} = 12.56$,
the standard error of the means, $\sigma_{\bar{x}} = 2.3$.

The estimated mean of the population is based on sampling distribution data only and so expression (6) is used, i.e. the confidence limits of the estimated mean of the population are: $\mu_{\bar{x}} \pm z_c\, \sigma_{\bar{x}}$.

For an 80 per cent confidence level, $z_c = 1.28$, thus
$$\mu_{\bar{x}} \pm z_c\, \sigma_{\bar{x}} = 12.56 \pm 1.28 \times 2.3$$
$$= 12.56 \pm 2.94 \text{ minutes.}$$

Thus, **the 80 per cent confidence level of the mean time to failure is from 9.62 minutes to 15.50 minutes.**

Problem 4. The sampling distribution of random samples of capacitors drawn from a large batch is found to have a standard error of the standard deviations of 0.27 μF. Determine the 88 per cent confidence interval for the estimate of the standard deviation of the whole batch, if in a particular sample, the standard deviation is 0.73 μF. It can be assumed that the values of capacitance of the batch are normally distributed.

For the sample: the standard deviation, $s = 0.73$ μF.
For the sampling distribution: the standard error of the standard deviations,
$$\sigma_s = 0.27 \ \mu\text{F.}$$

When the confidence level is 88 per cent, then by using the table of partial areas under the standardised normal curve, area $= \dfrac{0.880\,0}{2} = 0.440\,0$, giving z_c as ± 1.555 standard deviations, since an area of 0.440 0 lies half way between z-values of 1.55 and 1.56. Because the population is normally distributed, the confidence limits of the standard deviation of the population may be estimated by using expression (7), i.e. $s \pm z_c\, \sigma_s$.
$$s \pm z_c\, \sigma_s = 0.73 \pm 1.555 \times 0.27$$
$$= 0.73 \pm 0.42 \ \mu\text{F.}$$

Thus, **the 88 per cent confidence interval for the estimate of the standard deviation for the batch is from 0.31 μF to 1.15 μF.**

Further problems on the estimation of population parameters based on a large sample size may be found in Section 7 (Problems 11 to 20), page 348.

5 Estimating the mean of a population based on a small sample size

The methods used in Section 4 to estimate the population mean and standard deviation rely on a relatively large sample size, usually taken as 30 or more. This is because when the sample size is large the sampling distribution of a parameter is approximately normally distributed. When the sample size is small, usually taken as less than 30, the techniques used for estimating the population parameters in Section 4 become more and more inaccurate as the sample size becomes smaller, since the sampling distribution no longer approximates to a normal distribution. Investigations were carried out into the effect of small sample sizes on the estimation theory by W. S. Gosset in the early twentieth century and, as a result of his work, tables are available which enable a realistic estimate to be made, when sample sizes are small. In these tables, the t-value is determined from the relationship

$t = \dfrac{(\bar{x} - \mu)}{s} \sqrt{(N - 1)}$ where $\bar{x}$ is the mean value of a sample, μ is the mean

value of the population from which the sample is drawn, s is the standard deviation of the sample and N is the number of independent observations in the sample. He published his findings under the pen name of 'Student', and these tables are often referred to as the **'Student's t distribution'**.

The confidence limits of the mean value of a population based on a small sample drawn at random from the population is given by

$$\bar{x} \pm \frac{t_c s}{\sqrt{(N - 1)}} \tag{8}$$

In this estimate, t_c is called the confidence coefficient for small samples, analogous to z_c for large samples, s is the standard deviation of the sample, $\bar{x}$ is the mean value of the sample and N is the number of members in the sample.

The table given on page 351 is called 'percentile values for Student's t distribution'. The columns are headed t_p, where p is equal to 0.995, 0.99, . . ., 0.55. For a confidence level of, say, 95 per cent, the column headed $t_{.95}$ is selected and so on. The rows are headed with the Greek letter 'nu', v, and are numbered from 1 to 30 in steps of 1, together with the numbers 40, 60, 120 and ∞. These numbers represent a quantity called the **degrees of freedom**, which is defined as follows:

'the sample number, N, minus the number of population parameters which must be estimated for the sample'.

When determining the t-value, given by $t = \dfrac{(\bar{x} - \mu)}{s} \sqrt{(N - 1)}$, it is

necessary to know the sample parameters $\bar{x}$ and s and the population parameter μ. $\bar{x}$ and s can be calculated for the sample, but usually an estimate has to be made of the population mean μ, based on the sample mean value. The number of degrees of freedom, v, are given by the number of independent observations in the sample, N, minus the number of population parameters which have to

be estimated, k, i.e. $v = N - k$. For the equation $t = \dfrac{(\bar{x} - \mu)}{s} \sqrt{(N - 1)}$,

only μ has to be estimated, hence $k = 1$, and $v = N - 1$.

For all the work to be done in this section, only one population parameter is to be estimated, and hence v can always be taken as $(N - 1)$. The method used to estimate the mean of a population based on a small sample is shown in the worked problems following.

Worked problems on estimating the mean of a population based on a small sample size

---

Problem 1. A sample of 8 measurements of the diameter of a bar are made and the mean of the sample is 2.470 cm. The standard deviation of the samples is 0.21 mm. Determine (a) the 95 per cent confidence limits and (b) the 80 per cent confidence limits for an estimate of the actual diameter of the bar.

For the sample: the sample size, $N = 8$; mean, $\bar{x} = 2.470$ cm; standard
deviation, $s = 0.021$ cm.

Because the sample number is less than 30, the small sample estimate as given in expression (8) must be used. The number of degrees of freedom, i.e., sample size minus the number of estimations of population parameters to be made, is $8 - 1$, i.e. 7.

(a) The percentile value corresponding to a confidence coefficient value of $t_{.95}$ and a degree of freedom value of $v = 7$ can be found by using the table on page 351, and is 1.90, that is, $t_c = 1.90$. The estimated value of the mean of the population is given by

$$\bar{x} \pm \frac{t_c \, s}{\sqrt{(N - 1)}} \, .$$

Substituting the numerical values gives:

$$\bar{x} \pm \frac{t_c \, s}{\sqrt{(N - 1)}} = 2.470 \pm \frac{1.90 \times 0.021}{\sqrt{7}}$$

$$= 2.470 \pm 0.015 \text{ cm.}$$

Thus, the 95 per cent confidence limits are 2.455 cm and 2.485 cm.
This indicates that the actual diameter is likely to lie between 2.455 and 2.485 centimetres and that this prediction stands a 95 per cent chance of being correct.

(b) The percentile value corresponding to $t_{.80}$ and to $v = 7$ is obtained from the table on page 351, and is 0.896, that is, $t_c = 0.896$.
The estimated value of the 80 per cent confidence limits is given by:

$$\bar{x} \pm \frac{t_c \, s}{\sqrt{(N - 1)}} = 2.470 \pm \frac{0.896 \times 0.021}{\sqrt{7}}$$

$$= 2.470 \pm 0.007 \text{ cm.}$$

Thus, **the 80 per cent confidence limits are 2.463 cm and 2.477 cm,** that is, the actual diameter of the bar is between 2.463 and 2.477 centimetres and this result has an 80 per cent probability of being correct.

Problem 2. A sample of 15 electric lamps are selected randomly from a large batch and are tested until they fail. The mean and standard deviations of the time to failure are 1 177 hours and 25 hours respectively. Determine the confidence level based on an estimated failure time of 1 177 ± 5.8 hours.

For the sample: sample size, $N = 15$; standard deviation, $s = 25$ hours; mean, $\bar{x} = 1\,177$ hours.
The confidence limits are given by:

$$\bar{x} \pm \frac{t_c\, s}{\sqrt{(N-1)}}\text{, and these are equal to } 1\,177 \pm 5.8.$$

Because $\bar{x} = 1\,177$ hours, then $\pm \dfrac{t_c\, s}{\sqrt{(N-1)}} = \pm 5.8$

$$\text{i.e. } t_c = \frac{5.8\sqrt{(N-1)}}{s}$$

$$= \frac{5.8\sqrt{14}}{25} = \pm 0.868.$$

From the table given on page 351, a t_c value of 0.868, having a v-value of 14, gives a t_p value of $t_{.80}$. That is, **the confidence level of an estimated failure time of 1177 ± 5.8 hours is 80 per cent,** that is, it is likely that 80 per cent of all of the lamps will fail between 1 171.2 and 1 182.8 hours.

Further problems on estimating the mean of a population based on a small sample size may be found in Section 7 (Problems 21 to 25), page 349.

6 Summary of symbols and main points

Symbols
Population: N_p number of members in the population
                     μ mean of the population
                     σ standard deviation of the population
Sample: N number of members in the sample
                     $\bar{x}$ mean of the sample
                     s standard deviation of the sample
Sampling distribution: $\mu_{\bar{x}}$ mean of the sampling distribution of means
                     $\sigma_{\bar{x}}$ standard error of the means
                     σ_s standard error of the standard deviations.

Main points

Standard error of the means

Standard error of the means of a sampling distribution, that is, the standard deviation of the means of samples, is:

(1) $\sigma_{\bar{x}} = \dfrac{\sigma}{\sqrt{N}} \sqrt{\left(\dfrac{N_p - N}{N_p - 1} \right)}$, for a finite population and/or for sampling

without replacement, and

(2) $\sigma_{\bar{x}} = \dfrac{\sigma}{\sqrt{N}}$, for an infinite population and/or for sampling with replace-

ment.

The relationship between sample mean and population mean

(3) $\mu_{\bar{x}} = \mu$ when all possible samples of size N are drawn from a population of size N_p.

Estimating the mean of a population (σ known)

The confidence coefficient for a large sample size, $(N \geqslant 30)$, is z_c, see tables on pages 337 and 350. The confidence limits of a population mean based on sample data are given by:

(4) $\bar{x} \pm \dfrac{z_c \, \sigma}{\sqrt{N}} \sqrt{\left(\dfrac{N_p - N}{N_p - 1} \right)}$ for a finite population of size N_p, and by

(5) $\bar{x} \pm \dfrac{z_c \, \sigma}{\sqrt{N}}$ for an infinite population.

Estimating the mean of a population (σ not known).

The confidence limits of a population mean based on sample data are given by:

(6) $\mu_{\bar{x}} \pm z_c \, \sigma_{\bar{x}}$

Estimating the standard deviation of a population

The confidence limits of the standard deviation of a population based on sample data are given by:

(7) $s \pm z_c \, \sigma_s$

Estimating the mean of a population based on a small sample size

The confidence coefficient for a small sample size, $(N < 30)$, is t_c which can be determined using the table on page 351.

The confidence limits of a population mean based on sample data are given by:

(8) $\bar{x} \pm \dfrac{t_c s}{\sqrt{(N - 1)}}$

7 Further problems

The sampling distribution of means

1. Determine the mean and standard deviation of the set of numbers 1, 2, 4, 5 and 6, correct to three decimal places. By selecting all possible different samples of size 2 which can be drawn with replacement (25 pairs) determine
   (a) the mean of the sampling distribution of means, and
   (b) the standard error of the means, correct to three decimal places.
   $[\mu = 3.600, \sigma = 1.855,$ (a) $\mu_{\bar{x}} = 3.600,$ (b) $\sigma_{\bar{x}} = 1.312]$

2. Determine the standard error of the means for Problem 1, if sampling is without replacement, correct to three significant figures. $[\sigma_{\bar{x}} = 1.136]$

3. The lengths of 1 500 bolts are normally distributed with a mean of 22.4 cm and a standard deviation of 0.048 cm. If 30 samples are drawn at random from this population, each of size 36 bolts, determine the mean of the sampling distribution and standard error of the means when sampling is done with replacement. $[\mu_{\bar{x}} = 22.4$ cm, $\sigma_{\bar{x}} = 0.008$ cm$]$

4. Determine the standard error of the means in Problem 3, if sampling is done without replacement, correct to four decimal places.
   $$[\sigma_{\bar{x}} = 0.0079 \text{ cm}]$$

5. An automatic machine produces 10 000 components in a seven-hour day, having a mean value of diameter of 0.300 cm and a standard deviation of 0.020 mm. A random sample of 25 components is taken every 15 minutes without replacement. Determine the mean of the sampling distribution of means and the standard error of the means, correct to four significant figures, for a day's output from the machine.
   $$[0.300\,0 \text{ cm}, 3.995 \times 10^{-3} \text{ cm}]$$

6. A power punch produces 1 800 washers per hour. The mean inside diameter of the washers is 1.70 cm and the standard deviation is 0.013 mm. Random samples of 20 washers are drawn every 5 minutes. Determine the mean of the sampling distribution of means and the standard error of the means for one hour's output from the punch, (a) with replacement and (b) without replacement, correct to three significant figures.
   (a) $[\mu_{\bar{x}} = 1.70$ cm, $\sigma_{\bar{x}} = 2.91 \times 10^{-3}]$
   (b) $[\mu_{\bar{x}} = 1.70$ cm, $\sigma_{\bar{x}} = 2.89 \times 10^{-3}]$

A large batch of lamps have a mean time to failure of 800 hours and the standard deviation of the batch is 60 hours. Use this data and also the table on page 350 to solve Problems 7 to 9.

7. If a random sample of 64 lamps is drawn from the batch, determine the probability that the mean time to failure will be less than 785 hours, correct to three decimal places. [0.023]

8. Determine the probability that the mean time to failure of a random sample of 16 lamps will be between 790 hours and 810 hours, correct to three decimal places. [0.497]

9. For a random sample of 64 lamps, determine the probability that the mean time to failure will exceed 820 hours, correct to two significant figures. [0.003 8]

10. The contents of a consignment of 1 200 tins of a product have a mean mass of 0.504 kg and a standard deviation of 2.3 g. Determine the probability that a random sample of 40 tins drawn from the consignment will have a combined mass of (a) less than 20.13 kg, (b) between 20.13 kg and 20.17 kg, and (c) more than 20.17 kg, correct to three significant figures. (a) [0.017 9] (b) [0.740] (c) [0.242]

The estimation of population parameters based on a large sample size

11. Measurements are made on a random sample of 100 components drawn from a population of size 1 546 and having a standard deviation of 2.93 mm. The mean measurement of the components in the sample is 67.45 mm. Determine the 95 per cent and 99 per cent confidence limits for an estimate of the mean of the population.
[66.89 and 68.01 mm, 66.72 and 68.18 mm]

12. The standard deviation of the masses of 500 blocks is 150 kg. A random sample of 40 blocks have a mean mass of 2.40 Mg.
  (a) Determine the 95 per cent and 99 per cent confidence intervals for estimating the mean mass of the remaining 460 blocks, and
  (b) with what degree of confidence can it be said that the mean mass of the remaining 460 blocks is 2.40 ± 0.035 Mg?
      (a) [2.355 to 2.445 Mg; 2.341 to 2.459 Mg,]   (b) [86 per cent]

13. The contents of a random sample of 50 bottles being filled on an automatic machine are measured and the mean of the contents is found to be 1.01 dm^3. If the standard deviation of the contents of bottles filled by the machine is 0.05 dm^3, determine, correct to three decimal places,
  (a) the 95 per cent confidence interval, and
  (b) the 79.6 per cent confidence limits for estimating the mean content value for the output of this machine.
  (a) [0.996 to 1.024 dm^3]   (b) [1.001 and 1.019 dm^3]

14. In Problem 13, determine the confidence level if the confidence limits of the mean value of the contents for the machine are 1.01 ± 0.003 dm^3.
[32.56 per cent]

15. In order to estimate the thermal expansion of a metal, measurements of the change of length for a known change of temperature are taken by a group of students. The sampling distribution of the results has a mean of 12.81×10^{-4} m and a standard error of the means of 0.04×10^{-4} $m°C^{-1}$. Determine the 95 per cent confidence interval for an estimate of the true value of the thermal expansion of the metal, correct to two decimal places.
[12.73×10^{-4} to 12.89×10^{-4} $m°C^{-1}$]

16. The standard deviation of the time to failure of an electronic component is estimated as 100 hours. Determine how large a sample of these components must be, in order to be 90 per cent confident that the error in the estimated time to failure will not exceed (a) 20 hours and (b) 10 hours.   (a) [at least 68]   (b) [at least 271]

17. A sample of 60 slings of a certain diameter, used for lifting purposes, are tested to destruction (that is, loaded until they snapped). The mean and standard deviation of the breaking loads are 11.09 tonnes and 0.73

tonnes respectively. Find the 95 per cent confidence interval for the mean of the snapping loads of all the slings of this diameter produced by this company.   [10.91 t to 11.27 t]

18. The time taken to assemble a servo-mechanism is measured for 40 operatives and the mean time is 14.63 minutes with a standard deviation of 2.45 minutes. Determine the maximum error in estimating the true mean time to assemble the servo-mechanism for all operatives, based on a 95 per cent confidence level.   [45.6 seconds]

19. The resistances of 200 similar resistors, selected at random with replacement from a large batch of resistors, were measured and the standard deviation of the value of resistance obtained was found to be 100 milliohms. If the standard error of the standard deviation is 5 mΩ, determine the 99 per cent confidence interval of the standard deviation of all the resistors in the batch.   [87.1 to 112.9 mΩ]

20. Use the data given in Problem 19 above to determine the 95 per cent confidence interval of the standard deviation of all the resistors in the batch.   [90.2 to 109.8 mΩ]

Estimating the mean of a population based on a small sample size

21. The value of the ultimate tensile strength of a material is determined by measurements on 10 samples of the materials. The mean and standard deviation of the results are found to be 4.38 MPa and 0.06 MPa respectively. Determine the 95 per cent confidence interval for the mean of the ultimate tensile strength of the material.   [4.343 to 4.417 MPa]

22. Use the data given in Problem 21 above to determine the 99 per cent confidence interval for the mean of the ultimate tensile strength of the material.   [4.324 to 4.436 MPa]

23. The time taken for a chemical reaction to take place is measured 5 times and is found to be:
0.28 hours, 0.30 hours, 0.27 hours, 0.33 hours and 0.31 hours. Determine the 95 per cent and 99 per cent confidence intervals for the estimated true reaction time.   [0.275 to 0.321 hours; 0.258 to 0.338 hours]

24. The specific resistance of a reel of German silver wire of nominal diameter 0.5 mm is estimated by determining the resistance of 7 samples of the wire. These were found to have resistance values (in ohms per metre) of:
1.12   1.15   1.10   1.14   1.15   1.10 and 1.11.
Determine the 95 per cent confidence interval for the true specific resistance of the reel of wire.   [1.11 to 1.14 Ω m^{-1}]

25. In determining the melting point of a metal, 5 determinations of the melting point are made. The mean and standard deviation of the five results are 232.27°C and 0.742°C. Calculate the confidence with which the prediction 'the melting point of the metal is between 231.48°C and 233.06°C' can be made.   [95 per cent]

Partial areas under the standardised normal curve.

$z = \dfrac{x - \bar{x}}{\sigma}$	0	1	2	3	4	5	6	7	8	9
0.0	0.0000	0.0040	0.0080	0.0120	0.0159	0.0199	0.0239	0.0279	0.0319	0.0359
0.1	0.0398	0.0438	0.0478	0.0517	0.0557	0.0596	0.0636	0.0678	0.0714	0.0753
0.2	0.0793	0.0832	0.0871	0.0910	0.0948	0.0987	0.1026	0.1064	0.1103	0.1141
0.3	0.1179	0.1217	0.1255	0.1293	0.1331	0.1388	0.1406	0.1443	0.1480	0.1517
0.4	0.1554	0.1891	0.1628	0.1664	0.1700	0.1736	0.1772	0.1808	0.1844	0.1879
0.5	0.1915	0.1950	0.1985	0.2019	0.2054	0.2086	0.2123	0.2157	0.2190	0.2224
0.6	0.2257	0.2291	0.2324	0.2357	0.2389	0.2422	0.2454	0.2486	0.2517	0.2549
0.7	0.2580	0.2611	0.2642	0.2673	0.2704	0.2734	0.2760	0.2794	0.2823	0.2852
0.8	0.2881	0.2910	0.2939	0.2967	0.2995	0.3023	0.3051	0.3078	0.3106	0.3133
0.9	0.3159	0.3186	0.3212	0.3238	0.3264	0.3289	0.3215	0.3340	0.3365	0.3389
1.0	0.3413	0.3438	0.3451	0.3485	0.3508	0.3531	0.3554	0.3577	0.3599	0.3621
1.1	0.3643	0.3665	0.3686	0.3708	0.3729	0.3749	0.3770	0.3790	0.3810	0.3830
1.2	0.3849	0.3869	0.3888	0.3907	0.3925	0.3944	0.3962	0.3980	0.3997	0.4015
1.3	0.4032	0.4049	0.4066	0.4082	0.4099	0.4115	0.4131	0.4147	0.4162	0.4177
1.4	0.4192	0.4207	0.4222	0.4236	0.4251	0.4265	0.4279	0.4292	0.4306	0.4319
1.5	0.4332	0.4345	0.4357	0.4370	0.4382	0.4394	0.4406	0.4418	0.4430	0.4441
1.6	0.4452	0.4463	0.4474	0.4484	0.4495	0.4505	0.4515	0.4525	0.4535	0.4545
1.7	0.4554	0.4564	0.4573	0.4582	0.4591	0.4599	0.4608	0.4616	0.4625	0.4633
1.8	0.4641	0.4649	0.4656	0.4664	0.4671	0.4678	0.4686	0.4693	0.4699	0.4706
1.9	0.4713	0.4719	0.4726	0.4732	0.4738	0.4744	0.4750	0.4756	0.4762	0.4767
2.0	0.4772	0.4778	0.4783	0.4785	0.4793	0.4798	0.4803	0.4808	0.4812	0.4817
2.1	0.4821	0.4826	0.4830	0.4834	0.4838	0.4842	0.4846	0.4850	0.4854	0.4857
2.2	0.4861	0.4864	0.4868	0.4871	0.4875	0.4878	0.4881	0.4884	0.4882	0.4890
2.3	0.4893	0.4896	0.4898	0.4901	0.4904	0.4906	0.4909	0.4911	0.4913	0.4916
2.4	0.4918	0.4920	0.4922	0.4925	0.4927	0.4929	0.4931	0.4932	0.4934	0.4936
2.5	0.4938	0.4940	0.4941	0.4943	0.4945	0.4946	0.4948	0.4949	0.4951	0.4952
2.6	0.4953	0.4955	0.4956	0.4957	0.4959	0.4960	0.4961	0.4962	0.4963	0.4964
2.7	0.4965	0.4966	0.4967	0.4968	0.4969	0.4970	0.4971	0.4972	0.4973	0.4974
2.8	0.4974	0.4975	0.4976	0.4977	0.4977	0.4978	0.4979	0.4980	0.4980	0.4981
2.9	0.4981	0.4982	0.4982	0.4983	0.4984	0.4984	0.4985	0.4985	0.4986	0.4986
3.0	0.4987	0.4987	0.4987	0.4988	0.4988	0.4989	0.4989	0.4989	0.4990	0.4990
3.1	0.4990	0.4991	0.4991	0.4991	0.4992	0.4992	0.4992	0.4992	0.4993	0.4993
3.2	0.4993	0.4993	0.4994	0.4994	0.4994	0.4994	0.4994	0.4995	0.4995	0.4995
3.3	0.4995	0.4995	0.4995	0.4996	0.4996	0.4996	0.4996	0.4996	0.4996	0.4997
3.4	0.4997	0.4997	0.4997	0.4997	0.4997	0.4997	0.4997	0.4997	0.4997	0.4998
3.5	0.4998	0.4998	0.4998	0.4998	0.4998	0.4998	0.4998	0.4998	0.4998	0.4998
3.6	0.4998	0.4998	0.4999	0.4999	0.4999	0.4999	0.4999	0.4999	0.4999	0.4999
3.7	0.4999	0.4999	0.4999	0.4999	0.4999	0.4999	0.4999	0.4999	0.4999	0.4999
3.8	0.4999	0.4999	0.4999	0.4999	0.4999	0.4999	0.4999	0.4999	0.4999	0.4999
3.9	0.5000	0.5000	0.5000	0.5000	0.5000	0.5000	0.5000	0.5000	0.5000	0.5000

Percentile values (t_p) for Student's t distribution with v degrees of freedom
(shaded area $= p$)

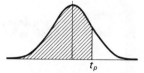

v	$t_{0.995}$	$t_{0.99}$	$t_{0.975}$	$t_{0.95}$	$t_{0.90}$	$t_{0.80}$	$t_{0.75}$	$t_{0.70}$	$t_{0.60}$	$t_{0.55}$
1	63.66	31.82	12.71	6.31	3.08	1.376	1.000	0.727	0.325	0.158
2	9.92	6.96	4.30	2.92	1.89	1.061	0.816	0.617	0.289	0.142
3	5.84	4.54	3.18	2.35	1.64	0.978	0.765	0.584	0.277	0.137
4	4.60	3.75	2.78	2.13	1.53	0.941	0.741	0.569	0.271	0.134
5	4.03	3.36	2.57	2.02	1.48	0.920	0.727	0.559	0.267	0.132
6	3.71	3.14	2.45	1.94	1.44	0.906	0.718	0.553	0.265	0.131
7	3.50	3.00	2.36	1.90	1.42	0.896	0.711	0.549	0.263	0.130
8	3.36	2.90	2.31	1.86	1.40	0.889	0.706	0.546	0.262	0.130
9	3.25	2.82	2.26	1.83	1.38	0.883	0.703	0.543	0.261	0.129
10	3.17	2.76	2.23	1.81	1.37	0.879	0.700	0.542	0.260	0.129
11	3.11	2.72	2.20	1.80	1.36	0.876	0.697	0.540	0.260	0.129
12	3.06	2.68	2.18	1.78	1.36	0.873	0.695	0.539	0.259	0.128
13	3.01	2.65	2.16	1.77	1.35	0.870	0.694	0.538	0.259	0.128
14	2.98	2.62	2.14	1.76	1.34	0.868	0.692	0.537	0.258	0.128
15	2.95	2.60	2.13	1.75	1.34	0.866	0.691	0.536	0.258	0.128
16	2.92	2.58	2.12	1.75	1.34	0.865	0.690	0.535	0.258	0.128
17	2.90	2.57	2.11	1.74	1.33	0.863	0.689	0.534	0.257	0.128
18	2.88	2.55	2.10	1.73	1.33	0.862	0.688	0.534	0.257	0.127
19	2.86	2.54	2.09	1.73	1.33	0.861	0.688	0.533	0.257	0.127
20	2.84	2.53	2.09	1.72	1.32	0.860	0.687	0.533	0.257	0.127
21	2.83	2.52	2.08	1.72	1.32	0.859	0.686	0.532	0.257	0.127
22	2.82	2.51	2.07	1.72	1.32	0.858	0.686	0.532	0.256	0.127
23	2.81	2.50	2.07	1.71	1.32	0.858	0.685	0.532	0.256	0.127
24	2.80	2.49	2.06	1.71	1.32	0.857	0.685	0.531	0.256	0.127
25	2.79	2.48	2.06	1.71	1.32	0.856	0.684	0.531	0.256	0.127
26	2.78	2.48	2.06	1.71	1.32	0.856	0.684	0.531	0.256	0.127
27	2.77	2.47	2.05	1.70	1.31	0.855	0.684	0.531	0.256	0.127
28	2.76	2.47	2.05	1.70	1.31	0.855	0.683	0.530	0.256	0.127
29	2.76	2.46	2.04	1.70	1.31	0.854	0.683	0.530	0.256	0.127
30	2.75	2.46	2.04	1.70	1.31	0.854	0.683	0.530	0.256	0.127
40	2.70	2.42	2.02	1.68	1.30	0.851	0.681	0.529	0.255	0.126
60	2.66	2.39	2.00	1.67	1.30	0.848	0.679	0.527	0.254	0.126
120	2.62	2.36	1.98	1.66	1.29	0.845	0.677	0.526	0.254	0.126
∞	2.58	2.33	1.96	1.645	1.28	0.842	0.674	0.524	0.253	0.126

Chapter 26

Significance testing (EC 15)

1 Hypotheses

The practical application of statistics in industry is often concerned with making decisions about populations and population parameters. For example, decisions about which is the better of two processes or decisions about whether to discontinue production on a particular machine because it is producing an economically unacceptable number of defective components are often based on deciding the mean or standard deviation of a population, calculated using sample data drawn from the population. In reaching these decisions, certain assumptions are made, which may or may not be true. The assumptions made are called **statistical hypotheses** or just **hypotheses** and are usually concerned with statements about probability distributions of populations.

In order to decide whether a dice is fair, that is, unbiased, a hypothesis can be made that a particular number, say 5, should occur with a probability of one in six, since there are six numbers on a dice. Such a hypothesis is called a **null hypothesis** and is an initial statement. The symbol H_0 is used to indicate a null hypothesis. Thus, if p is the probability of throwing a 5, then

$$H_0 : p = \frac{1}{6}$$

means, 'the null hypothesis that the probability of throwing a 5 is $\frac{1}{6}$'. Any hypothesis which differs from a given hypothesis is called an **alternative hypothesis**, and is indicated by the symbol H_1. Thus, if after many trials, it is found that the dice is biased and that a 5 only occurs, on average, one in every seven throws, then several alternative hypotheses may be formulated. For example:

$$H_1 : p = \frac{1}{7} \; ; \qquad \text{or } H_1 : p < \frac{1}{6}$$

$$\text{or } \quad H_1 : p > \frac{1}{8} \qquad \text{or } H_1 : p \neq \frac{1}{6}$$

are all possible alternative hypotheses to the null hypothesis that $p = \frac{1}{6}$.

Hypotheses may also be used when comparisons are being made. If we wish to compare, say, the strength of two metals, a null hypothesis may be formulated that there is **no difference** between the strengths of the two metals. If the forces that the two metals can withstand are F_1 and F_2, then the null hypothesis is

$$H_0 : F_1 = F_2.$$

If it is found that the null hypothesis has to be rejected, that is, that the strengths of the two metals are not the same, then the alternative hypotheses could be of several forms. For example,

$$H_1 : F_1 > F_2$$
$$\text{or } \quad H_1 . F_2 > F_1$$
$$\text{or } \quad H_1 : F_1 \neq F_2$$

These are all alternative hypotheses to the original null hypothesis.

2 Type I and type II errors

To illustrate what is meant by type I and type II errors, let us consider an automatic machine producing, say, small bolts. These are stamped out of a length of metal and various faults may occur. For example, the heads or the threads may be incorrectly formed, the length might be incorrect, and so on. Assume that, say, 3 bolts out of every 100 produced are defective in some way. If a sample of 200 bolts is drawn at random, then the manufacturer might be satisfied that his defect rate is still 3 per cent provided there are 6 defective bolts in the sample. Also, the manufacturer might be satisfied that his defect rate is 3 per cent or less provided that there are 6 or less bolts defective in the sample. He might then formulate the following hypotheses:

$H_0 : p = 0.03$ (the null hypothesis that the defect rate is 3 per cent)

The null hypothesis indicates that a 3 per cent defect rate is acceptable to the manufacturer. Suppose that he also makes a decision that should the defect rate rise to 5 per cent or more, he will take some action. Then the alternative hypothesis is:

$H_1 . p \geqslant 0.05$ (the alternative hypothesis that the defect rate is equal to or greater than 5 per cent)

The manufacturer's decisions, which are related to these hypotheses, might well be:

(a) a null hypothesis that a 3 per cent defect rate is acceptable, on the assumption that the associated number of defective bolts are insufficient to endanger his firm's good name;
(b) if the null hypothesis is rejected and the defect rate rises to 5 per cent or over, stop the machine and adjust or renew parts as necessary. Since the machine is not then producing bolts, this will reduce his profit.

These decisions may seem logical at first sight, but by applying the statistical concepts introduced in previous chapters it can be shown that the manufacturer is not necessarily making very sound decisions. This is shown as follows.

When drawing a random sample of 200 bolts from the machine with a defect rate of 3 per cent, by the laws of probability, some samples will contain no defective bolts, some samples will contain one defective bolt and so on. A binomial distribution can be used to determine the probabilities of getting 0, 1, 2, ..., 9 defective bolts in the sample. Thus the probability of getting 10 or more defective bolts in a sample, **even with a 3 per cent defect rate**, is given by

$1 -$ (the sum of probabilities of getting 0, 1, 2, . . . 9 defective bolts)

This is an extremely large calculation, given by:

$$1 - (0.97^{200} + 200 \times 0.97^{199} \times 0.03 + \frac{200 \times 199}{2} \times 0.97^{198} \times 0.03^2 \text{ to}$$
$$10 \text{ terms})$$

An alternative way of calculating the required probability is to use the normal approximation to the binomial distribution. This may be stated as follows:

'if the probability of a defective item is p and a non-defective item is q, then if a sample of N items is drawn at random from a large population, provided both Np and Nq are greater than 5, the binomial distribution approximates to a normal distribution of mean Np and standard deviation $\sqrt{(Npq)}$'.

The defect rate is 3 per cent, thus $p = 0.03$. Since $q = 1 - p$, $q = 0.97$.
Sample size $N = 200$. Since Np and Nq are greater than 5, a normal approximation to the binomial distribution can be used.
The mean of the normal distribution, $\bar{x} = Np = 200 \times 0.03 = 6$.
The standard deviation of the normal distribution, $\sigma = \sqrt{(Npq)}$
$$= \sqrt{(200 \times 0.03 \times 0.97)}$$
$$= 2.41$$

The normal standard variate for 10 bolts is, $z = \dfrac{\text{variate} - \text{mean}}{\text{standard deviation}}$

$$= \frac{10 - 6}{2.41} = 1.66.$$

The table on page 350 is used to determine the area between the mean and a z-value of 1.66, and is 0.451 5.

The probability of having 10 or more defective bolts is the total area under the standardised normal curve minus the area to the left of the $z = 1.66$ ordinate, i.e.

$1 - (0.5 + 0.451\ 5)$, that is, $1 - 0.951.5 = 0.048\ 5 \simeq 5$per cent.

Thus the probability of getting 10 or more defective bolts in a sample of 200 bolts, **even though the defect rate is still 3 per cent,** is 5 per cent. It follows that as a result of the manufacturer's decisions, for 5 times in every 100 the number of defects in the sample will exceed 10, the alternative hypothesis will be adopted and the machine will be stopped (and profit lost) unnecessarily. In general terms:

'a hypothesis has been rejected when it should have been accepted'.

When this occurs, it is called a **type I error**, and, in this example, the type I error is 5 per cent.

Assume now that the defect rate has risen to 5 per cent, i.e. the expectancy of a defective bolt is now 10. A second error resulting from his decisions occurs, due to the probability of getting less than 10 defective bolts in a random sample, even though the defect rate has risen to 5 per cent. Using the normal approximation to a binomial distribution:

$N = 200, \quad p = 0.05, \quad q = 0.95$

Np and Nq are greater than 5, hence a normal approximation to a binomial distribution is a satisfactory method. The normal distribution has:

$$\text{mean, } \bar{x} = Np \qquad = 200 \times 0.05 = 10$$
$$\text{standard deviation, } \sigma = \sqrt{(Npq)} \quad = \sqrt{(200 \times 0.05 \times 0.95)}$$
$$= 3.08$$

The normal standard variate for 9 defective bolts, $z = \dfrac{\text{variate} - \text{mean}}{\text{standard deviation}}$

$$= \frac{9 - 10}{3.05} = -0.32$$

Using the table of partial areas under the standardised normal curve given on page 350, a z-value of -0.32 corresponds to an area between the mean and the ordinate at $z = -0.32$ to 0.125 5. Thus, the probability of there being 9 or less defective bolts in the sample is given by the area to the left of the $z = -0.32$ ordinate, i.e. $0.500\ 0 - 0.125\ 5$, that is, 0.374 5. Thus, the probability of getting 9 or less defective bolts in a sample of 200 bolts, **even though the defect rate has risen to 5 per cent,** is 37 per cent. It follows that as a result of the manufacturer's decisions, for 37 samples in every 100, the machine will be left running even though the defect rate has risen to 5 per cent. In general terms:

'a hypothesis has been accepted when it should have been rejected'.

When this occurs, it is called a **type II error**, and, in this example, the type II error is 37 per cent.

Tests of hypotheses and rules of decisions should be designed to minimise the errors of decision. This is achieved largely by trial and error for a particular set of circumstances. Type I errors can be reduced by increasing the number of defective items allowable in a sample, but this is at the expense of allowing a larger percentage of defective items to leave the factory, increasing the criticism from customers. Type II errors can be reduced by increasing the percentage defect rate in the alternative hypothesis. If a higher percentage defect rate is given in the alternative hypothesis, the type II errors are reduced very effectively, as shown in the second of the two tables below, relating the decision rule to the magnitude of the type II errors. Some examples of the magnitude of type I errors are given below, for a sample of 1 000 components being produced by a machine with a mean defect rate of 5 per cent.

Decision rule	Type I error
Stop production if the number of defective components is equal to or greater than	(Per cent)
52	38.6
56	19.2
60	7.35
64	2.12
68	0.45

The magnitude of the type II errors for the output of the same machine, again based on a random sample of 1 000 components and a mean defect rate of 5 per cent, is given below.

Decision rule	Type II error
Stop production when the number of defective components is 60, when the defect rate is (per cent):	(Per cent)
$5\frac{1}{2}$	75.49
7	10.75
$8\frac{1}{2}$	0.23
10	0.00

When testing a hypothesis, the largest value of probability which is acceptable for a type I error is called the **level of significance** of the test. The level of significance is indicated by the symbol α (alpha) and the levels commonly adopted are 0.1, 0.05, 0.01, 0.005 and 0.002. A level of significance of, say, 0.05 means that 5 times in 100 the hypothesis has been rejected when it should have been accepted.

In significance tests, the following terminology is frequently adopted:

(i) if the level of significance is 0.01 or less, i.e. the confidence level is 99 per cent or more, the results are considered to be **highly significant**, that is, the results are considered likely to be correct,

(ii) if the level of significance is 0.05 or between 0.05 and 0.01, i.e. the confidence level is 95 per cent or between 95 per cent and 99 per cent, the results are considered to be **probably significant**, that is, the results are probably correct,

(iii) if the level of significance is greater than 0.05, i.e. the confidence level is less than 95 per cent, the results are considered to be **not significant**, that is, there are doubts about the correctness of the results obtained. This terminology indicates that the use of a level of significance of 0.05 for 'probably significant' is, in effect, a rule of thumb. Situations can arise when the probability changes with the nature of the test being done and the use being made of the results.

The example of a machine producing bolts, used to illustrate type I and type II errors, is based on a single random sample being drawn from the output of the machine. In practice, sampling is a continuous process and using the data obtained from several samples, sampling distributions are formed. From the concepts introduced in Chapter 25, the means and standard deviations of samples are normally distributed, thus for a particular sample its mean and standard deviation are part of a normal distribution. For a set of results to be probably significant a confidence level of 95 per cent is required for a particular hypothesis being probably correct. This is equivalent to the hypothesis being rejected when the level of significance is greater than 0.05. For this to occur, the z-value of the mean of the samples will lie between -1.96 and $+1.96$ (since the area under the standardised normal distribution curve between these z-values is 95 per cent). The shaded area in Fig. 1 is

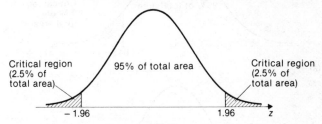

Fig. 1 Type I error, level of significance of 0.05. Two-tailed test.

based on results which are probably significant, i.e. having a level of significance of 0.05, and represents the probability of rejecting a hypothesis when it is correct. The z-values of less than -1.96 and more than 1.96 are called **critical values** and the shaded areas in Fig. 1 are called the **critical regions** or regions for which the hypothesis is rejected. Having formulated hypotheses, the rules of decision and a level of significance, the magnitude of the type I error is given. Nothing can now be done about type II errors and in most cases they are accepted in the hope that they are not too large.

When critical regions occur on both sides of the mean of a normal distribution, as shown in Fig. 1, they are as a result of **two-tailed** or **two-sided** tests. In such tests, consideration has to be given to values on both sides of the mean. For example, if it is required to show that the percentage of metal,

p, in a particular alloy is x per cent, then a two-tailed test is used, since the null hypothesis is incorrect if the percentage of metal is either less than p or more than p. The hypothesis is then of the form:

$H_0: p = x$ per cent
$H_1: p \neq x$ per cent

However, for the machine producing bolts, the manufacturer's decision is not affected by the fact that a sample contains say 1 or 2 defective bolts. He is only concerned with the sample containing, say, 10 **or more** defective bolts. Thus a 'tail' on the left of the mean is not required. In this case a **one-tailed** test or a **one-sided** test is really required. If the defect rate is, say, d and the per unit values economically acceptable to the manufacturer are u_1 and u_2, where u_1 is an acceptable defect rate and u_2 is the maximum acceptable defect rate, then the hypotheses in this case are of the form:

$H_0: d = u_1$
$H_1: d > u_2$

and the critical region lies on the right-hand side of the mean, as shown in Fig. 2(a). A one-tailed test can have its critical region either on the right-hand

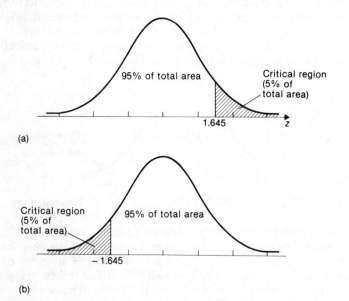

(a)

(b)

Fig. 2   (a) Type I error, level of significance 0.05.
$H_0 : d = u_1 ; H_1 : d > u_2$, i.e. a one-tailed test.

(b) Type I error, level of significance 0.05.
$H_0 : l = h ; H_1 : l < h$, i.e. a one-tailed test.

side or on the left-hand side of the mean. For example, if lamps are being tested and the manufacturer is only interested in those lamps whose life length does not meet a certain minimum requirement, then the hypotheses are of the form:

$H_0: \quad l = h$
$H_1 = l < h,$

where l is the life length and h are the number of hours to failure. In this case the critical region lies on the left-hand side of the mean, as shown in Fig. 2(b).

In Chapter 25, the z-values for various levels of confidence are given. The corresponding levels of significance (a confidence level of 95 per cent is equivalent to a level of significance of 0.05 in a two-tailed test) and their z-values for both one-tailed and two-tailed tests are given in the table below. It can be seen that two values of z are given for one-tailed tests, the negative value for critical regions lying to the left of the mean and a positive value for critical regions lying to the right of the mean.

Level of Significance, α	0.1	0.05	0.01	0.005	0.002
z-value, one-tailed test	-1.28 or 1.28	-1.645 or 1.645	-2.33 or 2.33	-2.58 or 2.58	-2.88 or 2.88
z-value, two-tailed test	-1.645 and 1.645	-1.96 and 1.96	-2.58 and 2.58	-2.81 and 2.81	-3.08 and 3.08

The problem of the machine producing 3 per cent defective bolts can now be reconsidered from a significance testing point of view. A random sample of 200 bolts is drawn, and the manufacturer is interested in a change in the defect rate in a **specified direction** (i.e. an increase), hence the hypotheses tests are designed accordingly. If the manufacturer is willing to accept a defect rate of 3 per cent, but wants adjustments made to the machine if the defect rate exceeds 3 per cent, then the hypotheses will be:

(i) a null hypothesis such that the defect rate, p, is equal to 3 per cent, i.e. $H_0: p = 0.03$, and
(ii) an alternative hypothesis such that the defect rate is greater than 3 per cent, i.e. $H_1: p > 0.03$.

The first rule of decision is as follows: let the level of significance, α, be 0.05; this will limit the type I error, that is, the error due to rejecting the hypothesis when it should be accepted, to 5 per cent, which means that the results are probably correct. The second rule of decision is to decide the number of defective bolts in a sample for which the machine is stopped and adjustments are made. For a one-tailed test, a level of significance of 0.05 and the critical region lying to the right of the mean of the standardised normal distribution, the z-value from the table above is 1.645. If the defect rate p is 0.03 per unit, the mean of the normal distribution is given by Np, that is, 200×0.03, i.e. 6 and the standard deviation is $\sqrt{(Npq)}$, that is $\sqrt{(200 \times 0.03 \times 0.97)}$, i.e. 2.41, using the normal approximation to a binomial distribution. Since the z-value is $\dfrac{\text{variate} - \text{mean}}{\text{standard deviation}}$, then $1.645 = \dfrac{\text{variate} - 6}{2.41}$, giving a variate value of 9.96. This variate is the number of defective bolts in a sample such that when this number is reached or exceeded the null hypothesis is rejected. For 95 times out of 100 this will be the correct thing to do. The second rule of decision will thus be 'reject H_0 if the number of defective bolts in a random sample is equal to or exceeds 10, otherwise accept H_0'. That is, the machine is adjusted when the number of defective bolts in a random sample reaches 10 and this will be the correct decision for 95 per cent of the time. The type II error can now be calculated, but there is little point, since having fixed the sample number and the level of significance, there is nothing that can be done about it.

A two-tailed test is used when it is required to test for changes in an **unspecified direction**. For example, if the manufacturer of bolts, used in the previous example, is inspecting the diameter of the bolts, he will want to know whether the diameters are too large or too small. Let the nominal diameter of the bolts be 2 millimetres. In this case the hypotheses will be:

$H_0 : d = 2.00$ mm
$H_1 : d \neq 2.00$ mm,

where d is the mean diameter of the bolts. His first decision is to set the level of significance, to limit his type I error. A two-tailed test is used, since adjustments must be made to the machine if the diameter does not lie **within** specified limits. The method of using such a significance test is given in Section 3.

When determining the magnitude of type I and type II errors, it is often possible to reduce the amount of work involved by using a normal or a Poisson distribution rather than binomial distribution. A summary of the criteria for the use of these distributions and their form is given below, for a sample of size N, a probability of defective components p and a probability of non-defective components q.

Binomial distribution

The probability of having $0, 1, 2, 3, \ldots$ defective components in a random sample of N components is given by the successive terms of the expansion of $(q + p)^N$, taken from the left. Thus:

number of defective components	0	1	2	3	...
Probability	q^N	$Nq^{N-1}p$	$\dfrac{N(N-1)}{2!}q^{N-2}p^2$	$\dfrac{N(N-1)(N-2)}{3!}q^{N-3}p^3$	...

Poisson approximation to a binomial distribution

When $N \geqslant 50$ and $Np < 5$, the Poisson distribution is approximately the same as the binomial distribution. In the Poisson distribution, the expectation $\lambda = Np$ and the probability of $0, 1, 2, 3, \ldots$ defective components in a random sample of N components is given by the successive terms of

$$e^{-\lambda}\left(1 + \lambda + \frac{\lambda^2}{2!} + \frac{\lambda^3}{3!} + \ldots\right)$$ taken from the left. Thus

number of defective components	0	1	2	3	...
Probability	$e^{-\lambda}$	$\lambda e^{-\lambda}$	$\dfrac{\lambda^2 e^{-\lambda}}{2!}$	$\dfrac{\lambda^3 e^{-\lambda}}{3!}$	...

Normal approximation to a binomial distribution

When both Np and Nq are greater than 5, the normal distribution is approximately the same as the binomial distribution. The normal distribution has a mean of Np and a standard deviation of $\sqrt{(Npq)}$.

Worked problems on type I and type II errors

Problem 1. Wood screws are produced by an automatic machine and it is found over a period of time that 7 per cent of all the screws produced are defective. Random samples of 80 screws are drawn periodically from the output of the machine. If a decision is made that production continues until a sample contains more than 7 defective screws, determine the type I error based on this decision for a defect rate of 7 per cent. Also determine the magnitude of the type II error when the defect rate has risen to 10 per cent.

$N = 80$, $p = 0.07$, $q = 0.93$.

Since both Np and Nq are greater than 5, a normal approximation to the binomial distribution is used.
Mean of the normal distribution, $Np = 80 \times 0.07 = 5.6$.
Standard deviation of the normal distribution,
$\sqrt{(Npq)} = \sqrt{(80 \times 0.07 \times 0.93)}$
$= 2.28$

A type I error is the probability of rejecting a hypothesis when it is correct, hence, the type I error in this problem is the probability of stopping the machine, that is, the probability of getting more than 7 defective screws in a sample, even though the defect rate is still 7 per cent. The z-value corresponding to 7 defective screws is given by:

$\dfrac{\text{variate} - \text{mean}}{\text{standard deviation}}$, i.e. $\dfrac{7 - 5.6}{2.28}$, that is, 0.61.

Using the table of partial areas under the standardised normal curve given on page 350, the area between the mean and a z-value of 0.61 is 0.229 1. Thus, the probability of more than 7 defective screws is the area to the right of the z-ordinate at 0.61, that is,

(total area $-$ (area to the left of mean $+$ area between mean and $z = 0.61$)),

i.e. $1 - (0.5 + 0.229\ 1)$. This gives a probability of 0.270 9. It is usual to express type I errors as a percentage, giving

type I error = 27.1 per cent.

A type II error is the probability of accepting a hypothesis when it should be rejected. The type II error in this problem is the probability of a sample containing less than 7 defective screws, even though the defect rate has risen to 10 per cent. The values are now:

$N = 80, \quad p = 0.1, \quad q = 0.9.$

As Np and Nq are both greater than 5, a normal approximation to a binomial distribution is used, in which the mean Np is $80 \times 0.1 = 8$ and the standard deviation $\sqrt{(Npq)}$ is $\sqrt{(80 \times 0.1 \times 0.9)} = 2.68$.

The z-value for a variate of 7 defective screws is $\dfrac{7-8}{2.68}$, i.e. -0.37.

Using the table of partial area given on page 350, the area between the mean and $z = -0.37$ is 0.144 3. Hence, the probability of getting less than 7 defective screws, even though the defect rate is 10 per cent is (area to the left of mean $-$ area between mean and a z-value of -0.37), i.e. $0.5 - 0.144\ 3$, that is, 0.355 7.

It is usual to express type II errors as a percentage, giving

type II error = 35.6 per cent.

Problem 2. The sample size in Problem 1 is reduced to 50. Determine the type I error if the defect rate remains at 7 per cent and the type II error when the defect rate rises to 9 per cent. The decision is now to stop the machine for adjustment if a sample contains 4 or more defective screws.

$N = 50, \quad p = 0.07$

When $N \geqslant 50$ and $Np < 5$, the Poisson approximation to a binomial distribution is used. The expectation, $\lambda = Np = 3.5$. The probabilties of $0, 1, 2, 3, \ldots$ defective screws are given by

$$e^{-\lambda}, \lambda e^{-\lambda}, \frac{\lambda^2 e^{-\lambda}}{2!}, \frac{\lambda^3 e^{-\lambda}}{3!} \ldots \quad \text{Thus,}$$

probability of a sample containing no defective screws, $e^{-\lambda}$ = 0.030 2
probability of a sample containing 1 defective screw, $\lambda e^{-\lambda}$ = 0.105 7
probability of a sample containing 2 defective screws, $\dfrac{\lambda^2 e^{-\lambda}}{2!}$ = 0.185 0
probability of a sample containing 3 defective screws, $\dfrac{\lambda^3 e^{-\lambda}}{3!}$ = 0.215 8

probability of a sample containing 0, 1, 2, or 3 defective screws
is 0.536 7

Hence, probability of a sample containing 4 or more defective screws is $1 - 0.536\ 7$, i.e. 0.463 3. Thus the type I error, that is, rejecting the hypothesis when it should be accepted or stopping the machine for adjustment when it should continue running, is **46 per cent.**

When the defect rate has risen to 9 per cent, $p = 0.09$ and $Np = \lambda = 4.5$. Because $N \geqslant 50$ and $Np < 5$, the Poisson approximation to a binomial distribution can still be used. Thus,

probability of a sample containing no defective screws, $e^{-\lambda}$ = 0.011 1
probability of a sample containing 1 defective screw, $\lambda e^{-\lambda}$ = 0.050 0

probability of a sample containing 2 defective screws, $\dfrac{\lambda^2 e^{-\lambda}}{2!} = \underline{0.112\,5}$

probability of a sample containing 3 defective screws, $\dfrac{\lambda^3 e^{-\lambda}}{3!} = \underline{0.168\,7}$

probability of a sample containing 0, 1, 2, or 3 defective screws

is $0.342\,3$

That is, the probability of a sample containing less than 4 defective screws is 0.342 3. Thus, the type II error, that is, accepting the hypothesis when it should have been rejected or leaving the machine running when is should be stopped, is **34 per cent**.

Problem 3. The sample size in Problem 1 is further reduced to 25. Determine the type I error if the defect rate remains at 7 per cent and the type II error when the defect rate rises to 10 per cent. The decision is now to stop the machine for adjustment if a sample contains 3 or more defective screws.

$N = 25, \quad p = 0.07, \quad q = 0.93.$

The criteria for a normal approximation to a binomial distribution and for a Poisson approximation to a binomial distribution are not met, hence the binomial distribution is applied.

Probability of no defective screws in a sample,
$$q^N \quad = 0.93^{25} \qquad\qquad = 0.163\,0$$
Probability of 1 defective screw in a sample,
$$Nq^{N-1}p = 25 \times 0.93^{24} \times 0.07 \qquad = 0.306\,6$$
Probability of 2 defective screws in a sample,
$$\frac{N(N-1)}{2!} q^{N-2}p^2 = \frac{25 \times 24}{2} \times 0.93^{23} \times 0.07^2 = \underline{0.277\,0}$$
Probability of 0, 1, or 2 defective screws in a sample is $\qquad 0.746\,6$

Thus, the probability of a type I error, that is, stopping the machine even though the defect rate is still 7 per cent, is $1 - 0.746\,6$, i.e. $0.253\,4$.
Hence, **type I error $= 25$ per cent**.

When the defect rate has risen to 10 per cent; $N = 25, p = 0.1, q = 0.9$.

Probability of no defective screws in a sample,
$$q^N \quad = 0.9^{25} \qquad\qquad = 0.071\,8$$
Probability of 1 defective screw in a sample,
$$Nq^{N-1}p = 25 \times 0.9^{24} \times 0.1 \qquad = 0.199\,4$$
Probability of 2 defective screws in a sample,
$$\frac{N(N-1)}{2!} q^{N-2}p^2 = \frac{25 \times 24}{2} \times 0.9^{23} \times 0.1^2 \quad = \underline{0.265\,9}$$
Probability of 0, 1, or 2 defective screws in a sample is $\qquad 0.537\,1,$

i.e. the probability of a type II error, that is, leaving the machine running even though the defect rate has risen to 10 per cent, is **54 per cent**.

Further problems on type I and type II errors may be found in Section 5 (Problems 1 to 6), page 375.

3 Significance tests for population means

When carrying out tests or measurements, it is often possible to form a hypothesis as a result of these tests. For example, the boiling point of water is found to be:

101.3°C, 99.8°C, 100.4°C, 100.7°C, 99.5°C and 98.9°C,

as a result of six tests. The mean of these six results is 100.1°C. Based on these results, how confidently can it be predicted, that at this particular height above sea level and at this particular barometric pressure, water boils at 100.1°C? In other words, are the results based on sampling **significantly different** from the true result? There are a variety of ways of testing significance, but only one or two of these in common use are introduced in this section. Usually, in significance tests, some predictions about population parameters, based on sample data, are required. In significance tests for population means, a random sample is drawn from the population and the mean value of the sample, $\bar{x}$, is determined. The testing procedure depends on whether or not the standard deviation of the population is known.

(a) When the standard deviation of the population is known

A null hypothesis is made that there is no difference between the value of a sample mean $\bar{x}$ and that of the population mean, μ, i.e.

$$H_0 : \bar{x} = \mu$$

If many samples had been drawn from a population and a sampling distribution of means had been formed, then, provided N is large (usually taken as $N \geqslant 30$) the mean value would form a normal distribution, having a mean value of $\mu_{\bar{x}}$ and a standard deviation or standard error of the means $\sigma_{\bar{x}}$ (see Chapter 25, Section 2).

The particular value of $\bar{x}$ of a large sample drawn for a significance test is therefore part of a normal distribution and it is possible to determine by how much $\bar{x}$ is likely to differ from $\mu_{\bar{x}}$ in terms of the normal standard variate z. The relationship is

$$z = \frac{\bar{x} - \mu_{\bar{x}}}{\sigma_{\bar{x}}}.$$

However, with reference to Chapter 25, equations (1), (2) and (3),

$$\sigma_{\bar{x}} = \frac{\sigma}{\sqrt{N}} \sqrt{\left(\frac{Np - N}{Np - 1}\right)}, \text{ for finite populations,}$$

$$= \frac{\sigma}{\sqrt{N}}, \text{ for infinite populations.}$$

and $\mu_{\bar{x}} = \mu$, where N is the sample size, Np is the size of the population, μ is the mean of the population and σ the standard deviation of the population. Substituting for $\mu_{\bar{x}}$ and $\sigma_{\bar{x}}$ in the equation for z, gives:

$$z = \frac{\bar{x} - \mu}{\dfrac{\sigma}{\sqrt{N}}} \text{ for infinite populations,} \tag{1}$$

$$z = \frac{\overline{x} - \mu}{\frac{\sigma}{\sqrt{N}}\sqrt{\left(\frac{Np - N}{Np - 1}\right)}} \quad \text{for populations of size } Np \qquad (2)$$

In Section 2 of this chapter, the relationship between z-values and levels of significance for both one-tailed and two-tailed tests are given (see page 359). It can be seen from this table for a level of significance of, say, 0.05 and a two-tailed test, the z-value is ± 1.96, and z-values outside of this range are not significant. Thus, for a given level of significance (i.e. a known value of z), the mean of the population, μ, can be predicted by using equations (1) and (2) above, based on the mean of a sample x. Alternatively, if the mean of the population is known, the significance of a particular value of z, based on sample data, can be established. If the z-value based on the mean of a random sample for a two-tailed test is found to be, say, 2.01, then at a level of significance of 0.05, that is, the results being probably significant, the mean of the sampling distribution is said to differ significantly from what would be expected as a result of the null hypothesis (i.e. that $\overline{x} = \mu$), due to the result of the test being classed as 'not significant' (see page 357). The hypothesis would then be rejected and an alternative hypothesis formed, (i.e $H_1 : \overline{x} \neq \mu$). The rules of decision for such a test would be:

(i) reject the hypothesis at a 0.05 level of significance, i.e. if the z-value of the sample mean is outside of the range -1.96 to $+1.96$.

(ii) accept the hypothesis otherwise.

For small sample sizes (usually taken as $N < 30$), the sampling distribution is not normally distributed, but approximates to Student's t-distributions (see Chapter 25, Section 5). In this case, t-values rather than z-values are used and the equations analogous to equations (1) and (2) are:

$$|t| = \frac{\overline{x} - \mu}{\frac{\sigma}{\sqrt{N}}} \quad \text{for infinite populations} \qquad (3)$$

$$|t| = \frac{\overline{x} - \mu}{\frac{\sigma}{\sqrt{N}}\sqrt{\left(\frac{Np - N}{Np - 1}\right)}} \quad \text{for populations of size } Np \qquad (4)$$

where $|t|$ means the modulus of t, i.e. the positive value of t.

(b) When the standard deviation of the population is not known

It is found, in practice, that if the standard deviation of a sample is determined, its value is less than the value of the standard deviation of the population from which it is drawn. This is as expected, since the range of a sample is likely to be less than the range of the population. The difference between the two standard deviations becomes more pronounced when the sample size is small. Investigations have shown that the variance, s^2, of a sample of N items is approximately related to the variance, σ^2, of the population from which it is drawn by

$$s^2 = \left(\frac{N - 1}{N}\right)\sigma^2.$$

The factor $\left(\dfrac{N-1}{N}\right)$ is known as **Bessel's correction**. This relationship may be used to find the relationship between the standard deviation of a sample, s, and an estimate of the standard deviation of a population, $\hat{\sigma}$, and is:

$$\hat{\sigma}^2 = s^2 \left(\frac{N}{N-1}\right) \text{, i.e. } \hat{\sigma} = s \sqrt{\left(\frac{N}{N-1}\right)}.$$

For large samples, say, a minimum of N being 30, the factor $\sqrt{\left(\dfrac{N}{N-1}\right)}$ is

$\sqrt{\dfrac{30}{29}}$ which is approximately equal to 1.017. Thus, for large samples s is

very nearly equal to $\hat{\sigma}$ and the factor $\sqrt{\left(\dfrac{N}{N-1}\right)}$ can be omitted without

introducing any appreciable error. In equations (1) and (2), s can be written for σ, giving:

$$z = \frac{\bar{x} - \mu}{\dfrac{s}{\sqrt{N}}} \text{ for infinite populations} \tag{5}$$

and $$z = \frac{\bar{x} - \mu}{\dfrac{s}{\sqrt{N}} \sqrt{\left(\dfrac{Np - N}{Np - 1}\right)}} \quad \text{for populations of size } Np \tag{6}$$

For small samples, the factor $\sqrt{\left(\dfrac{N}{N-1}\right)}$ cannot be disregarded and

substituting $\sigma = s \sqrt{\left(\dfrac{N}{N-1}\right)}$ in equations (3) and (4) gives:

$$|t| = \bar{x} - \mu \left/ \frac{s\sqrt{\left(\dfrac{N}{N-1}\right)}}{\sqrt{N}} \right. = \frac{(\bar{x} - \mu)\sqrt{(N-1)}}{s} \tag{7}$$

for infinite populations, and

$$|t| = \frac{\bar{x} - \mu}{\dfrac{s\sqrt{\left(\dfrac{N}{N-1}\right)}\sqrt{\left(\dfrac{Np - N}{Np - 1}\right)}}{\sqrt{N}}} = \frac{(\bar{x} - \mu)\sqrt{(N-1)}}{s\sqrt{\left(\dfrac{Np - N}{Np - 1}\right)}} \tag{8}$$

for populations of size Np.

The equations given in this section are parts of tests which are applied to determine population means. The way in which some of them are used is shown in the worked problems following.

Worked problems on significance tests for population means

Problem 1. Sugar is packed in bags by an automatic machine. The mean mass of the contents of a bag is 1.000 kg. Random samples of 36 bags are selected throughout the day and the mean mass of a particular sample is found to be

1.003 kg. If the manufacturer is willing to accept a standard deviation on all bags packed of 0.01 kg and a level of significance of 0.05, above which values the machine must be stopped and adjustments made, determine if, as a result of the sample under test, the machine should be adjusted.

Population mean, $\mu = 1.000$ kg, sample mean, $\bar{x} = 1.003$ kg, population standard deviation, $\sigma = 0.01$ kg.

A null hypothesis for this problem is that the sample mean and the mean of the population are equal, i.e.

$H_0 : \bar{x} = \mu$

Because the manufacturer is interested in deviations on both sides of the mean, the alternative hypothesis is that the sample mean is not equal to the population mean, i.e.

$H_1 : \bar{x} \neq \mu$.

The decision rules associated with these hypotheses are:

(i) reject H_0 if the z-value of the sample mean is outside of the range of the z-values corresponding to a level of significance of 0.05 for a two-tailed test, i.e. stop the machine and adjust, and

(ii) accept H_0 otherwise, i.e. keep the machine running.

The sample size is over 30 so this is a 'large sample' problem and the population can be considered to be infinite. Because values of $\bar{x}, \mu, \sigma$ and N are all known, equation (1) can be used to determine the z-value of the sample mean, i.e.

$$z = \frac{\bar{x} - \mu}{\dfrac{\sigma}{\sqrt{N}}}$$

$$= \frac{1.003 - 1.000}{\dfrac{0.01}{\sqrt{36}}} = \frac{\pm 0.003}{0.0016} = \pm 1.8.$$

The z-value corresponding to a level of significance of 0.05 for a two-tailed test is given on page 359 and is ± 1.96. Because the z-value of the sample is within this range, the null hypothesis is accepted and the machine should not be adjusted.

Problem 2. The mean lifetime of a random sample of 50 similar torch bulbs drawn from a batch of 500 bulbs is 72 hours. The standard deviation of the lifetime of the sample is 10.4 hours. The batch is classed as inferior if the mean lifetime of the batch is less than the population mean of 75 hours. Determine whether, as a result of the sample data, the batch is considered to be inferior at a level of significance of (a) 0.05 and (b) 0.01.

Population size, $Np = 500$; population mean, $\mu = 75$ hours;
mean of sample, $\bar{x} = 72$ hours; standard deviation of sample, $s = 10.4$ hours;
size of sample, $N = 50$.

The null hypothesis is that the mean of the sample is equal to the mean of the population, i.e.

$$H_0: \bar{x} = \mu$$

The alternative hypothesis is that the mean of the sample is less than the mean of the population, i.e.

$$H_1: \bar{x} < \mu$$

(The fact that $\bar{x} = 72$ should not lead to the conclusion that the batch is necessarily inferior. At a level of significance of 0.05, the result is 'probably significant', but since this corresponds to a confidence level of 95 per cent, there are still 5 times in every 100 when the result can be significantly different, that is, be outside of the range of z-values for this data. This particular sample result may be one of these 5 times).

The decision rules associated with the hypotheses are:

(i) reject H_0 if the z-value (or t-value) of the sample mean is less than the z-value (or t-value) corresponding to a level of significance of (a) 0.05 and (b) 0.01, i.e. the batch is inferior,

(ii) accept H_0 otherwise, i.e. the batch is not inferior.

The data given is N, N_p, $\bar{x}$, s and μ. The alternative hypothesis indicates a one-tailed distribution and since $N > 30$ the 'large sample' theory applies. From equation (6)

$$z = \frac{\bar{x} - \mu}{\dfrac{s}{\sqrt{N}} \sqrt{\left(\dfrac{N_p - N}{N_p - 1} \right)}}$$

$$= \frac{72 - 75}{\dfrac{10.4}{\sqrt{50}} \sqrt{\left(\dfrac{500 - 50}{500 - 1} \right)}} = \frac{-3}{1.471 \times 0.9496}$$

$$= -2.15$$

(a) For a level of significance of 0.05 and a one-tailed test, all values to the left of the z-ordinate at -1.645 (see page 359) indicate that the results are 'not significant', that is, they differ significantly from the null hypothesis. Since the z-value of the sample mean is $-2.15.$, i.e. less than -1.645, **the batch is considered to be inferior at a level of significance of 0.05.**

(b) The z-value for a level of significance of 0.01 for a one-tailed test is -2.33 and in this case, z-values of sample means lying to the left of the z-ordinate at -2.33 are 'not significant'. Since the z-value of the sample lies to the right of this ordinate, it does not differ significantly from the null hypothesis and **the batch is not considered to be inferior at a level of significance of 0.01.** (At first sight, for a mean value to be significant at a level of significance of 0.05, but not at 0.01, appears to be incorrect. However, it is stated earlier in the chapter that for a result to be probably significant, i.e. at a level of significance of between 0.01 and 0.05, the range of z-values is less than the range for the result to be highly significant, that is, having a level of significance of 0.01 or better. Hence the results of the problem are logical.)

Problem 3. An analysis of the mass of carbon in six similar specimens of cast iron, each of mass 425.0 g, yielded the following results:

17.1 g,   17.3 g,   16.8 g,   16.9 g,   17.8 g,   and 17.4 g.

Test the hypothesis that the percentage of carbon is 4.00 per cent assuming an arbitrary level of significance of (a) 0.02 and (b) 0.1.

The sample mean, $\bar{x} = \dfrac{17.1 + 17.3 + \ldots + 17.4}{6} = 17.22$,

The sample standard deviation

$$s = \sqrt{\left\{ \frac{(17.1 - 17.22)^2 + (17.3 - 17.22)^2 + \ldots + (17.4 - 17.22)^2}{6} \right\}}$$

$$= 0.334$$

The null hypothesis is that the sample and population means are equal, i.e.

$H_0 : \bar{x} = \mu$

The alternative hypothesis is that the sample and population means are not equal, i.e. $H_1 : \bar{x} \neq \mu$
The decision rules are:

(i) reject H_0 if the z- or t-value of the sample mean is outside of the range of the z- or t-value corresponding to a level of significance of (a) 0.2 and (b) 0.1, i.e. the mass of carbon is not 4.00 per cent,

(ii) accept H_0 otherwise, i.e. the mass of carbon is 4.00 per cent.

The number of tests taken, N, is 6 and an infinite number of tests could have been taken, hence the population is considered to be infinite. Because $N < 30$, a t-distribution is used. If the mean mass of carbon in the bulk of the metal is 4.00 per cent, the mean mass of carbon in a specimen is 4.00 per cent of 425.0, i.e. 17.00 g, thus $\mu = 17.00$.

From equation 7, $|t| = \dfrac{(\bar{x} - \mu)(\sqrt{(N-1)})}{s}$

$$= \frac{(17.22 - 17.00)(\sqrt{(6-1)})}{0.334}$$

$$= 1.473.$$

In general, for any two-tailed distribution there is a critical region both to the left and to the right of the mean of the distribution. For a level of significance of 0.2, 0.1 of the percentile value of a t-distribution lies to the left of the mean and 0.1 of the percentile value lies to the right of the mean. Thus, for a level of significance of α, a value of $t_{(1-\alpha/2)}$ is required for a two-tailed distribution when using the table on page 351. This conversion is necessary because the t-distribution is given in terms of levels of confidence and for a one-tailed distribution. The row t-value for a value of α of 0.2 is $t_{(1-\frac{0.2}{2})}$, i.e. $t_{0.90}$. The degrees of freedom, v are $N-1$, that is 5. From the t-distribution table on page 351, the percentile value corresponding to $(t_{0.90}, v = 5)$ is 1.48, and for a two-tailed test, ± 1.48. Because the mean value of the sample is within this range, the hypothesis is accepted at a level of significance of 0.2.

The t-value for $\alpha = 0.1$ is $t_{(1-\frac{0.1}{2})}$, i.e. $t_{0.95}$. The percentile value corresponding to $t_{0.95}, v = 5$ is 2.02 and since the mean value of the sample is within the range ± 2.02, the hypothesis is also accepted at this level of significance. **Thus, it is probable that the mass of metal contains 4 per cent carbon at levels of significance of 0.2 and 0.1.**

Further problems on significance tests for population means may be found in Section 5 (Problems 7 to 15), page 376.

4 Comparing two sample means

The techniques introduced in Section 3 can be used for comparison purposes. For example, it may be necessary to compare the performance of, say, two similar lamps produced by different manufacturers or different operators carrying out a test, or tests on the same items using different equipment. The null hypothesis adopted for tests involving two different populations is that there is **no difference** between the mean values of the populations.

The technique is based on the following theorem:

'If $\bar{x}_1$ and $\bar{x}_2$ are the means of random samples of size N_1 and N_2 drawn from populations having means of μ_1 and μ_2 and standard deviations of σ_1 and σ_2, then the sampling distribution of the differences of the means, $(\bar{x}_1 - \bar{x}_2)$, is a close approximation to a normal distribution, having a mean of zero and a standard deviation of

$$\sqrt{\left(\frac{\sigma_1^2}{N_1} + \frac{\sigma_2^2}{N_2}\right)}.$$

For large samples, when comparing the mean values of two samples, the variate is the difference in the means of the two samples, $\bar{x}_1 - \bar{x}_2$; the mean of the sampling distribution (and hence the difference in population means) is zero and the standard error of the sampling distribution $\sigma_{\bar{x}}$ is $\sqrt{\left(\frac{\sigma_1^2}{N_1} + \frac{\sigma_2^2}{N_2}\right)}$.

Hence, the z-value is
$$\frac{(\bar{x}_1 - \bar{x}_2) - 0}{\sqrt{\left(\frac{\sigma_1^2}{N_1} + \frac{\sigma_2^2}{N_2}\right)}} \tag{9}$$

For small samples, Student's t-distribution values are used and in this case:

$$|t| = \frac{\bar{x}_1 - \bar{x}_2}{\sqrt{\left(\dfrac{\sigma_1^2}{N_1} + \dfrac{\sigma_2^2}{N_2}\right)}} \tag{10}$$

where $|t|$ means the modulus of t, i.e. the positive value of t.

When the standard deviation of the population is not known, then Bessel's correction is applied to estimate it from the sample standard deviation (i.e. the estimate of the population variance, $\sigma^2 = s^2 \left(\dfrac{N}{N-1}\right)$, (see page 366). For large populations, the factor $\dfrac{N}{N-1}$ is small and may be neglected. However, when $N < 30$, this correction factor should be included. Also, since estimates of both σ_1 and σ_2 are being made, the k factor in the degrees of freedom in Student's t-distribution tables becomes 2 and v is given by $(N_1 + N_2 - 2)$. With these factors taken into account, when testing the hypotheses that samples come from the same population, or that there is no difference between the mean values of two populations, the t-value is given by:

$$|t| = \frac{\bar{x}_1 - \bar{x}_2}{\sigma \sqrt{\left(\dfrac{1}{N_1} + \dfrac{1}{N_2}\right)}} \tag{11}$$

An estimate of the standard deviation σ is based on a concept called 'pooling'. This states that if one estimate of the variance of a population is based on a sample, giving a result of $\sigma_1^2 = \dfrac{N_1 s_1^2}{N_1 - 1}$ and another estimate is based on a second sample, giving $\sigma_2^2 = \dfrac{N_2 s_2^2}{N_2 - 1}$, then a better estimate of the population variance, σ^2, is given by:

$$\sigma^2 = \frac{N_1 s_1^2 + N_2 s_2^2}{(N_1 - 1) + (N_2 - 1)}$$

i.e. $\sigma = \sqrt{\left(\dfrac{N_1 s_1^2 + N_2 s_2^2}{N_1 + N_2 - 2}\right)}$ (12)

Worked problems on comparing two sample means

Problem 1. An automatic machine is producing components, and as a result of many tests the standard deviation of their size is 0.02 cm. Two samples of 40 components are taken, the mean size of the first sample being 1.51 cm and the second 1.52 cm. Determine whether the size has altered appreciably if a level of significance of 0.05 is adopted, i.e. that the results are probably significant.

Because both samples are drawn from the same population, $\sigma_1 = \sigma_2 = \sigma = 0.02$ cm. Also $N_1 = N_2 = 40$ and $\bar{x}_1 = 1.51$ cm, $\bar{x}_2 = 1.52$ cm. The level of significance, $\alpha = 0.05$.

The null hypothesis is that the size of the component has not altered, i.e. $\bar{x}_1 = \bar{x}_2$, hence it is

$$H_0 : \bar{x}_1 - \bar{x}_2 = 0.$$

The alternative hypothesis is that the size of the components has altered. i.e. that $\bar{x}_1 \neq \bar{x}_2$, hence it is

$$H_1 : \bar{x}_1 - \bar{x}_2 \neq 0$$

For a large sample having a known standard deviation of the population, the z-value of the difference of means of two samples is given by equation (9), i.e.

$$z = \frac{\bar{x}_1 - \bar{x}_2}{\sqrt{\left(\dfrac{\sigma_1^2}{N_1} + \dfrac{\sigma_2^2}{N_2}\right)}}$$

Because $N_1 = N_2 =$ say, N and $\sigma_1 = \sigma_2 = \sigma$, this equation becomes

$$z = \frac{\bar{x}_1 - \bar{x}_2}{\sigma \sqrt{\left(\dfrac{2}{N}\right)}}$$

$$= \frac{1.51 - 1.52}{0.02 \sqrt{\left(\dfrac{2}{40}\right)}} = -2.236$$

Because the difference between $\bar{x}_1$ and $\bar{x}_2$ has no specified direction, a two-tailed test is indicated. The z-value corresponding to a level of significance of 0.05 and a two-tailed test is ± 1.96 (see page 359). The result for the z-value for the difference of means is outside of the range ± 1.96, that is, it is probable that the size has altered appreciably at a level of significance of 0.05.

Problem 2. The electrical resistance of two products are being compared. The parameters of product 1 are:

sample size 40, mean value of sample 74 ohms,
standard deviation of whole of product 1 batch is 8 ohms.

Those of product 2 are:

sample size 50, mean value of sample 78 ohms,
standard deviation of whole of product 2 batch is 7 ohms.

Determine if there is any significant difference between the two products at a level of significance of (a) 0.05 and (b) 0.01.

Let the mean of the batch of product 1 be μ_1, and that of product 2 be μ_2. The null hypothesis is that the means are the same, i.e.

$H_0: \mu_1 - \mu_2 = 0$

The alternative hypothesis is that the means are not the same, i.e.

$H_1: \mu_1 - \mu_2 \neq 0$.

The population standard deviations are known, i.e. $\sigma_1 = 8$ ohms and $\sigma_2 = 7$ ohms, the sample means are known, i.e. $\bar{x}_1 = 74$ ohms and $\bar{x}_2 = 78$ ohms. Also the sample sizes are known, i.e. $N_1 = 40$ and $N_2 = 50$. Hence, equation (9) can be used to determine the z-value of the difference of the sample means. From equation (9),

$$z = \frac{\bar{x}_1 - \bar{x}_2}{\sqrt{\left(\dfrac{\sigma_1^2}{N_1} + \dfrac{\sigma_2^2}{N_2}\right)}} = \frac{74 - 78}{\sqrt{\left(\dfrac{8^2}{40} + \dfrac{7^2}{50}\right)}}$$

$$= \frac{-4}{1.606} = -2.49$$

(a) For a two-tailed test, the results are probably significant at a 0.05 level of significance when z lies between -1.96 and $+1.96$. Hence the z-value of the difference of means shows there is 'no significance', i.e. that **product 1 is significantly different from product 2 at a level of significance of 0.05.**

(b) For a two-tailed test, the results are highly significant at a 0.01 level of significance when z lies between -2.58 and 2.58. Hence there is **no significant difference between product 1 and product 2 at a level of significance of 0.01.**

Problem 3. The reaction time in seconds of two people, A and B, are measured by electrodermal responses and the results of the tests are as shown below.

Person A (s)   0.243 0.243 0.239 0.232 0.229 0.241
Person B (s)   0.238 0.239 0.225 0.236 0.235 0.234

Find if there is any significant difference between the reaction times of the two people at a level of significance of 0.1.

The mean, $\bar{x}$ and standard deviation, s, of the response times of the two people are determined.

$$\bar{x}_A = \frac{0.243 + 0.243 + \ldots + 0.241}{6} = 0.237\,8 \text{ seconds}$$

$$\bar{x}_B = \frac{0.238 + 0.239 + \ldots + 0.234}{6} = 0.234\,5 \text{ seconds}$$

$$s_A = \sqrt{\left[\frac{(0.243 - 0.237\,8)^2 + \ldots + (0.241 - 0.237\,8)^2}{6}\right]} = 0.005\,43s$$

$$s_B = \sqrt{\left[\frac{(0.238 - 0.234\,5)^2 + \ldots + (0.234 - 0.234\,5)^2}{6}\right]} = 0.004\,57s$$

The null hypothesis is that there is no difference between the reaction times of the two people, i.e.

$$H_0 : \bar{x}_A - \bar{x}_B = 0$$

The alternative hypothesis is that the reaction times are different, i.e.

$$H_1 : \bar{x}_A - \bar{x}_B \neq 0$$

indicating a two-tailed test.

The sample numbers (combined) are less than 30 and a t-distribution is used. The standard deviation of all the reaction times of the two people is not known, so an estimate based on the standard deviations of the samples is used. Applying Bessel's correction, the estimate of the standard deviation of

the population, $\sigma^2 = s^2 \dfrac{N}{N-1}$, gives $\sigma_A = \sqrt{\left(\dfrac{0.005\,43^2 \times 6}{5} \right)} =$

$0.005\,95$ and $\sigma_B = \sqrt{\left(\dfrac{0.004\,57^2 \times 6}{5} \right)} = 0.005\,01$

From equation (10), the t-value of the difference of the means is given by:

$$|t| = \frac{\bar{x}_A - \bar{x}_B}{\sqrt{\left(\dfrac{\sigma_A^2}{N_A} + \dfrac{\sigma_B^2}{N_B} \right)}} = \frac{0.237\,8 - 0.234\,5}{\sqrt{\left(\dfrac{0.005\,95^2}{6} + \dfrac{0.005\,01^2}{6} \right)}}$$

$$|t| = 1.039$$

For a two-tailed test and a level of significance of 0.1, the column heading in the t-distribution table (see page 351) is $t_{0.95}$ (refer to worked problem 3, Section 3). The degrees of freedom due to k being 2 is $v = N_1 + N_2 - 2$, i.e. $6 + 6 - 2 = 10$. The corresponding t-value from the table is 1.81. Since the t-value of the difference of the means is within the range ± 1.81, there is **no significance difference between the reaction times at a level of significance of 0.1**.

Problem 4. An analyst carries out 10 analyses on equal masses of a substance which is found to contain a mean of 49.20 g of a metal, with a standard deviation of 0.41 g. A trainee operator carries out 12 analyses on equal masses of the same substance which is found to contain a mean of 49.30 g, with a standard deviation of 0.32 g. Is there any significance between the results of the operators?

Let μ_1 and μ_2 be the mean values of the amounts of metal found by the two operators. The null hypothesis is that there is no difference between the results obtained by the two operators, i.e.

$$H_0 : \mu_1 = \mu_2$$

The alternative hypothesis is that there is a difference between the results of the two operators, i.e.

$$H_1 : \mu_1 \neq \mu_2$$

Under the hypothesis H_0 the standard deviations of the amount of metal, σ, will be the same, and from equation (12)

$$\sigma = \sqrt{\left(\frac{N_1 s_1^2 + N_2 s_2^2}{N_1 + N_2 - 2}\right)}$$

$$= \sqrt{\left(\frac{(10 \times 0.41^2 + 12 \times 0.32^2}{10 + 12 - 2}\right)}$$

$$= 0.381\,4.$$

The t-value of the results obtained is given by equation (11), i.e.

$$|t| = \frac{\bar{x}_1 - \bar{x}_2}{\sigma \sqrt{\left(\dfrac{1}{N_1} + \dfrac{1}{N_2}\right)}}$$

$$= \frac{49.20 - 49.30}{0.381\,4 \sqrt{\left(\dfrac{1}{10} + \dfrac{1}{12}\right)}} = -0.612$$

For the results to be probably significant, a two-tailed test and a level of significance of 0.05 is taken. H_0 is rejected outside of the range $t_{-.975}$ and $t_{.975}$. The number of degrees of freedom are $N_1 + N_2 - 2$. For $t_{.975}$, $v = 20$, from the table on page 351, the range is from -2.09 to 2.09. Since the t-value based on the sample data is within this range, **there is no significant difference between the results of the two operators at a level of significance of 0.05.**

Further problems on comparing two sample means may be found in the following Section (5) (Problems 16 to 24), page 378.

5 Further problems

Type I and type II errors

Problems 1 and 2 refer to an automatic machine producing piston rings for car engines. Random samples of 1 000 rings are drawn from the output of the machine periodically for inspection purposes. A defect rate of 5 per cent is acceptable to the manufacturer, but if the defect rate is believed to have exceeded this value, the machine producing the rings is stopped and adjusted.

In Problem 1, determine the type I errors which occur for the decision rules stated.

1. Stop production and adjust the machine if a sample contains (a) 54, (b) 62 and (c) 70 or more defective rings.
   (a) [28.1 per cent] (b) [4.09 per cent] (c) [0.19 per cent]

In Problem 2, determine the type II errors which are made if the decision rule is to stop production if there are more than 60 defective components in the sample.

2. When the actual defect rate has risen to (a) 6 per cent, (b) 7.5 per cent and (c) 9 per cent.
   (a) [55.2 per cent]   (b) [4.65 per cent]   (c) [0.07 per cent]
3. A random sample of 100 components is drawn from the output of a machine whose defect rate is 3 per cent. Determine the type I error if the decision rule is to stop production when the sample contains: (a) 4 or more defective components, (b) 5 or more defective components, and (c) 6 or more defective components.
   (a) [35.3 per cent]   (b) [18.5 per cent]   (c) [8.4 per cent]
4. If there are 4 or more defective components in a sample drawn from the machine given in Problem 3 above, determine the type II error when the actual defect rate is: (a) 5 per cent, (b) 6 per cent and (c) 7 per cent.
   (a) [26.5 per cent]   (b) [15.1 per cent]   (c) [8.2 per cent]
5. A lecturer tests a class of students by setting 10 questions and asking his students whether the answer given is true or false. He adopts the decision rule based on an assumption that if 7 or more questions are answered correctly, the student is not guessing. Assuming the probability of guessing the correct answer is 0.5, determine the type I error based on this decision rule.   [17.2 per cent]
6. Determine the type II error for Problem 5 above if the probability of guessing the correct answer is taken as (a) 0.6 and (b) 0.4.
   (a) [61.8 per cent]   (b) [94.5 per cent]

Significance tests for population means

7. A batch of cables produced by a manufacturer have a mean breaking strength of 2 000 kN and a standard deviation of 100 kN. A sample of 50 cables are found to have a mean breaking strength of 2 050 kN. Test the hypothesis that the breaking strength of the sample is greater than the breaking strength of the population from which it is drawn at a level of significance of 0.01.
   [z (sample) = 3.54, z_α = 2.33, hence hypothesis is rejected, where z_α is the z-value corresponding to a level of significance of α.]
8. Fruit is prepacked in bags having an average mass of 1.00 kg and standard deviation of 0.1 kg. In a random sample of 100 bags the average mass of each bag is 1.03 kg. Determine whether the mean of all the bags can be considered to have altered significantly from 1.00 kg, at a level of significance of 0.01.
   [z (sample) = 3.00, z_α = ± 2.58, hence hypothesis rejected]
9. The strength of cotton thread is 343 N, with a standard deviation of 49.5 N. A random sample is cut from 36 reels of cotton selected from a large batch of reels and the mean strength of these samples is 315 N. Determine if the sample indicates that the batch is inferior at a level of significance of (a) 0.05 and (b) 0.01.
   [z (sample) = −3.39, (a) z_α = ± 1.645, (b) z_α = ± 2.33, hence batch inferior at both levels of significance.]

10. Nine estimations of the percentage of copper in a bronze alloy have a mean of 80.8 per cent and standard deviation of 1.2 per cent. Assuming that the percentage of copper in samples is normally distributed, test the null hypothesis that the true percentage of copper is 80 per cent against an alternative hypothesis that it exceeds 80 per cent, at a level of significance of 0.05.

[$t_{.95}, v_8 = 1.86, |t| = 1.88$, hence null hypothesis rejected]

11. A random sample of 100 similar tyres produced by a particular process had an effective life of 20 300 miles with a standard deviation of 1 500 miles. Test the hypothesis that the true mean effective life of all the similar tyres produced by this process is 20 000 miles at a level of significance of 0.05.

[z (sample) = 2.00, $z_\alpha = \pm 1.96$, hence hypothesis is rejected]

12. The internal diameter of a pipe has a mean diameter of 3.000 0 cm with a standard deviation of 0.015 cm. A random sample of 30 measurements are taken and the mean of the samples is 3.007 8 cm. Test the hypothesis that the mean diameter of the pipe is 3.000 0 cm at a level of significance of 0.01.

[z (sample) = 2.85, $z_\alpha = \pm 2.58$, hence hypothesis is rejected]

13. A fishing line has a mean breaking strength of 10.25 kN. Following a special treatment on the line, the following results are obtained for 20 specimens taken from the line.

Breaking strength (kN)	9.8	10	10.1	10.2	10.5	10.7	10.8	10.9	11.0
Frequency	1	1	4	5	3	2	2	1	1

Test the hypothesis that the special treatment has improved the breaking strength at a level of significance of 0.05.

[$\bar{x} = 10.38, s = 0.33, t_{0.95} \, v_{19} = 1.73, |t| = 1.72$, hence hypothesis is accepted.]

14. A machine produces ball bearings having a mean diameter of 0.50 cm. A sample of 10 ball bearings is drawn at random and the sample mean is 0.53 cm with a standard deviation of 0.03 mm. Test the hypothesis that the mean diameter is 0.50 cm at a level of significance of (a) 0.05 and (b) 0.01.

$|t| = 3.00$, (a) $t_{.975} \, v_9 = 2.26$, hence hypothesis rejected, (b) $t_{.995} \, v_9 = 3.25$, hence hypothesis is accepted]

15. Six similar switches are tested to destruction at an overload of 20 per cent of their normal maximum current rating. The mean number of operations before failure are 8 200 with a standard deviation of 145. The manufacturer of the switches claims that they can be operated at least 8 000 times at a 20 per cent overload current. Can the manufacturers claim be supported at a level of significance of (a) 0.05 and (b) 0.01?

[$|t| = 3.08$, (a) $t_{.95} \, v_5 = 2.02$, hence claim supported, (b) $t_{.99} \, v_5 = 3.36$, hence claim not supported]

Comparing two sample means

16. The mean heights of 50 soldiers is 173.2 cm with a standard deviation of 6.35 cm, whilst the mean height of 50 sailors is 171.45 cm with a standard deviation of 7.11 cm. Test the hypothesis that soldiers are taller than sailors using a level of significance of (a) 0.05 and (b) 0.01.
    [$z = 1.30$, (a) $z_{0.05}$, one-tailed test $= \pm 1.645$, hence hypothesis accepted, (b) $z_{0.1}$, one-tailed test $= \pm 1.28$, hence hypothesis rejected]

17. A comparison is being made between batteries used in calculators. Batteries of type A have a mean lifetime of 24 hours with a standard deviation of 4 hours, this data being calculated from a sample of 100 of the batteries. A sample of 80 of the type B batteries have a mean lifetime of 40 hours with a standard deviation of 6 hours. Test the hypothesis that the type B batteries have a mean lifetime of at least 15 hours more than those of type A, at a level of significance of 0.05.
    [Take $\bar{x}$ as $24 + 15$, i.e. 39 hours, $z = 1.28$, $z_{0.05}$, one-tailed test $= 1.645$, hence hypothesis is accepted]

18. Two randomly selected groups of 50 operatives in a factory are timed during an assembly operation. The first group take a mean time of 112 minutes with a standard deviation of 12 minutes. The second group take a mean time of 117 minutes with a standard deviation of 9 minutes. Test the hypothesis that the mean time for the assembly operation is the same for both groups of employees at a level of significance of 0.05.
    [$z = 2.357$, $z_{0.05}$, two-tailed test $= \pm 1.96$, hence hypothesis is rejected]

19. Capacitors having a nominal capacitance of 24 μF but produced by two different companies are tested. The values of actual capacitance are:

Company X	21.4	23.6	24.8	22.4	26.3
Company Y	22.4	27.7	23.5	29.1	25.8

    Test the hypothesis that the mean capacitance of capacitors produced by company Y are higher than those produced by company X at a level of significance of 0.01.

    $$\left[\text{Bessel's correction is } \hat{\sigma}^2 = \frac{s^2 N}{N-1} \, . \right]$$

    [$\bar{x}_1 = 23.7$, $s_1 = 1.72$, $\sigma_1 = 1.58$, $\bar{x}_2 = 25.7$, $s_2 = 2.50$, $\sigma_2 = 2.28$
    $|t| = 1.61$, $t_{.995}$ $v_8 = 3.36$, hence hypothesis is accepted.]

20. A sample of 100 relays produced by manufacturer A operated on average 1 190 times before failure occurred, with a standard deviation of 90. 75 relays produced by manufacturer B, operated on average 1 220 times before failure with a standard deviation of 120. Determine if the number of operations before failure are significantly different for the two manufacturers at a level of significance of (a) 0.05 and (b) 0.1.
    [z (sample) $= 1.82$, (a) $z_{0.05}$, two-tailed test $= \pm 1.96$, no significance, (b) $z_{0.1}$, two-tailed test $= \pm 1.645$, significant difference]

21. The hardening time for two different epoxy resins in hours at $30°C$ are as shown:

Type A 1.75   2.5   2.25   2.0   2.75   2.5
Type B 2.5   3.0   3.25   3.5   2.75   4.0   3.75   3.5

Test the hypothesis that there is no difference in the hardening times of the different resins at a level of significance of 0.01.

(Bessel's correction is $\hat{\sigma}^2 = \dfrac{s^2 N}{N-1}$.)

$[\bar{x}_A = 2.292, s_A = 0.0336, \sigma_A = 0.311, \bar{x}_B = 3.281, s_B = 0.475,$
$\sigma_B = 0.448, |t| = 4.87, t_{.995}, v_{12} = 3.06,$ hence hypothesis is rejected.]

22. A sample of 12 car engines produced by manufacturer A showed that the mean petrol consumption over a measured distance was 4.8 litres with a standard deviation of 0.40 litres. Twelve similar engines for manufacturer B were tested over the same distance and the mean petrol consumption was 5.1 litres with a standard deviation of 0.36 litres. Test the hypothesis that the engines produced by manufacturer A are more economical than those produced by manufacturer B at a level of significance of (a) 0.01 and (b) 0.05.
[Assuming null hypothesis of no difference, $\sigma = 0.397, |t| = 1.85,$
(a) $t_{.99}, v_{22} = 2.51,$ hypothesis rejected, (b) $t_{.95}, v_{22} = 1.72,$ hypothesis accepted]

23. Two-star and four-star petrol is tested in 5 similar cars under identical conditions. For two-star petrol, the cars covered a mean distance of 21.4 kilometres with a standard deviation of 0.54 kilometres for a given mass of petrol. For the same mass of four-star petrol, the mean distance covered was 22.6 kilometres with a standard deviation of 0.48 km. Test the hypothesis that four-star petrol gives more kilometres per litre than two-star petrol at a level of significance of 0.05.
[$\sigma = 0.571, |t| = 3.32, t_{.95}, v_8 = 1.86,$ hence hypothesis is rejected]

24. The ATP–ASE activity of mitochondria of 12 rats and 6 guinea pigs is given below.

Rats   25 23 22 27 28 21 19 27 26 24 21 23
Guinea pigs   16 18 15 14 17 18

Is there a significant difference between the mitochondria of rats and guinea pigs?
[$\bar{x}_1 = 23.83, \bar{x}_2 = 16.33, s_1 = 2.70, s_2 = 1.49, |t| = 5.98, t_{.975} \, v_{16} = \pm 2.12,$ hence there is a significant difference]

Chapter 27

Correlation (ED 15)

1 Linear correlation

The techniques associated with correlation are essentially those of predicting trends or of showing that no trend exists. For example, a company spends different amounts of money each month on advertising its products and wishes to assess whether it is spending its money wisely. It will be spending its money on advertising effectively provided that an increase in the cost of advertising results in a corresponding increase in the volume of sales. One way to determine that this is so is to plot a graph of expenditure on advertising and the volume of sales, as shown in Fig. 1. An examination of the graph shows that there is a trend upwards, as shown by the broken line. When using

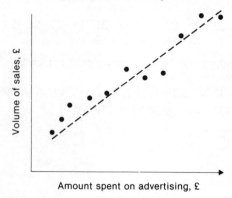

Fig. 1

correlation techniques, a value can be put on how closely the points plotted in this way represent a trend. Thus linear correlation is an investigation into the degree of the relationship between two variables. The technique used is to determine how well a linear equation of the form $y = mx + c$ describes the relationship between the variables. If the graph of the equation, say, $y = 2x + 3$ is correctly drawn, all the co-ordinates of x and y lie exactly on a straight line, and **perfect correlation** is said to exist between x and y. In statistics, when two variables are being investigated, the location of the co-ordinates on a rectangular co-ordinate system is called a **scatter diagram**.

When determining experimental values, say, measuring load F and extension x to verify Hooke's law, it is unlikely that all the values of F and x will give perfect correlation, due to measurement and other errors. The scatter diagram of the co-ordinates of F and x would probably be similar to that shown in Fig. 2(a), if many measurements are taken. In general, if Y increases as X increases, **positive or direct linear correlation** is said to exist between the values of X and Y, hence Fig. 2(a) shows positive linear correlation.

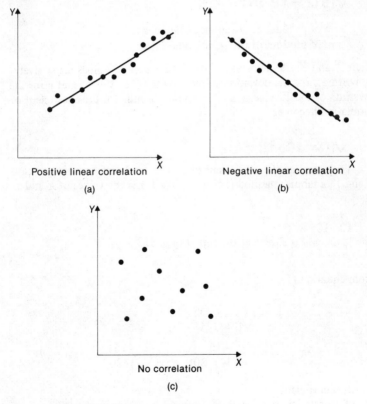

Fig. 2 (a) Positive linear correlation.
       (b) Negative linear correlation.
       (c) No correlation.

When Y decreases as X increases, then **inverse** or **negative linear correlation** is said to exist between the values of X and Y. When there is no apparent relationship between the values of X and Y, then **no correlation** exists between the values of X and Y. Negative linear correlation and no correlation are shown in Figs 2(b) and 2(c) respectively.

2 The product-moment formula for determining the linear correlation coefficient

One of the measures of the degree of linear correlation between two variables is called the **coefficient of correlation**, denoted by the symbol r. The derivation of the product-moment formula for determining the value of r for a given set of co-ordinates is based on techniques not covered by this text. However, the formula for two variables, say X and Y, is given by:

$$r = \frac{\Sigma (X - \bar{X})(Y - \bar{Y})}{\sqrt{[\Sigma (X - \bar{X})^2 \cdot \Sigma (Y - \bar{Y})^2]}}$$

$$= \frac{\text{sum of product of deviation of the means}}{\text{sum of product of the standard deviations}}$$

where $\bar{X}$ and $\bar{Y}$ are the means of the X-values and Y-values respectively. By writing x for the deviation of the X-values (i.e. $X - \bar{X}$) and y for the deviation of the Y-values (i.e. $Y - \bar{Y}$) the formula for the coefficient of correlation becomes:

$$r = \frac{\Sigma xy}{\sqrt{[(\Sigma x^2)(\Sigma y^2)]}} \tag{1}$$

To determine the coefficient of correlation between variables, say, X and Y, a tabular method is normally used. Let the values of X and Y be

X 4   5   6   9
Y 12   10   8   6

The table used is shown at the top of page 383.

From equation (1),

$$r = \frac{\Sigma xy}{\sqrt{[(\Sigma x^2)(\Sigma y^2)]}}$$

$$= \frac{-16}{\sqrt{[(14)(20)]}} = \frac{-16}{\sqrt{[280]}}$$

i.e. the coefficient
of correlation $r = -0.956\ 2$

Table 1

1	2	3	4	5	6	7
X	Y	$x = X - \bar{X}$	$y = Y - \bar{Y}$	xy	x^2	y^2
4	12	−2	3	−6	4	9
5	10	−1	1	−1	1	1
6	8	0	−1	0	0	1
9	6	3	−3	−9	9	9

$\Sigma X = 24$ $\Sigma Y = 36$ $\Sigma xy = -16$ $\Sigma x^2 = 14$ $\Sigma y^2 = 20$

$$\bar{X} = \frac{24}{4} = 6 \qquad \bar{Y} = \frac{36}{4} = 9$$

A result for the value of the coefficient of correlation has been obtained but what does the value obtained mean? It can be shown by the techniques of partial differentiation that values of r always lie between −1 and +1. If the correlation between two sets of values is perfect and direct, then the coefficient of correlation is + 1. When the correlation between two sets of numbers is perfect and inverse, the coefficient of correlation is −1. Between these extreme values, the smaller the numerical value of the coefficient the less the degree of correlation between the variables. A correlation coefficient lying between −0.5 and +0.5 indicates little or no correlation. When it is about +0.7 (and −0.7) a fair amount of correlation exists, the amount of correlation increasing to good as the value of the correlation coefficient approaches +1 (and −1). Because the correlation coefficient is only a guide to the amount of correlation between variables, it is not usual to express it to an accuracy of more than two significant figures. The sign indicates the nature of the correlation (direct or inverse). Thus the result of the correlation problem worked above (i.e. −0.96), indicates a good inverse correlation.

Care must be taken in interpreting the results obtained by applying the product-moment formula. The following points should be considered before coming to any firm conclusions.

(i) Even though a high degree of correlation may exist between two sets of data, they could be entirely unrelated. For example, a strong correlation may be found to exist between the number of car accidents in city A in some given periods of time and the number of times a particular machine tool breaks down in city B in the same periods of time. Obviously no 'cause and effect' exists between these relationships.

(ii) It is possible that both variables are changing due to the action of a third variable. For example, a relationship exists between the pressure of a gas and its volume. However, each of these quantities is affected by the thermodynamic temperature of the gas and wrong conclusions on the correlation between pressure and volume can easily be drawn if the temperature of the gas is disregarded.

(iii) A low correlation coefficient does not necessarily mean that no relationship exists between the variables. The coefficient of correlation shows a

linear relationship only and, for example, the parabola $y = x^2$ would have a very poor coefficient of correlation. However, perfect non-linear correlation exists between the values of y and x and techniques are available for checking non-linear correlation, which are outside the scope of this book.

(iv) The relationship between two values of correlation coefficients is approximately a square law relationship. Thus, when comparing correlation coefficients of, say, 0.8 and 0.4, this does not mean that the first value is twice as good as the second. The true comparison is between 0.8^2 and 0.4^2, i.e. between 0.64 and 0.16, showing that a coefficient of correlation of 0.8 is four times better than a correlation coefficient of 0.4.

Worked problem on the product-moment formula for determining the linear correlation coefficient

Problem 1. The data given below gives the experimental values obtained for the torque output from an electric motor, X, against the current taken from the supply, Y. Determine the value, degree and nature of the coefficient of linear correlation between the variables X and Y (if there is one).

X	0	1	2	3	4	5	6	7	8	9
Y	4	6	6	6	8	10	10	10	14	12

Using the tabular method:

Table 2

X	Y	$x = X - \bar{X}$	$y = Y - \bar{Y}$	xy	x^2	y^2
0	4	−4.5	−4.6	20.7	20.25	21.16
1	6	−3.5	−2.6	9.1	12.25	6.76
2	6	−2.5	−2.6	6.5	6.25	6.76
3	6	−1.5	−2.6	3.9	2.25	6.76
4	8	−0.5	−0.6	0.3	0.25	0.36
5	10	0.5	1.4	0.7	0.25	1.96
6	10	1.5	1.4	2.1	2.25	1.96
7	10	2.5	1.4	3.5	6.25	1.96
8	14	3.5	5.4	18.9	12.25	29.16
9	12	4.5	3.4	15.3	20.25	11.56
$\Sigma X = 45$ $\bar{X} = \dfrac{45}{10}$ $= 4.5$	$\Sigma Y = 86$ $\bar{Y} = \dfrac{86}{10}$ $= 8.6$			$\Sigma xy = 81.0$	$\Sigma x^2 = 82.5$	$\Sigma y^2 = 88.4$

From equation (1), the coefficient of correlation,
$$r = \frac{\Sigma xy}{\sqrt{[(\Sigma x^2)(\Sigma y^2)]}}$$
$$= \frac{81}{\sqrt{[(82.5)(88.4)]}}$$
$$= 0.95.$$

Thus, a good direct correlation exists between the values of X and Y.

Further problems on the product-moment formula for determining the linear coefficient of correlation may be found in Section 4 (Problems 1 to 8), page 387.

3 The significance of a coefficient of correlation

The values of X and Y used to determine a coefficient of correlation can be considered to be a sample from the population of all possible values of X and Y. For example, various values of force and the corresponding extension are obtained experimentally when verifying Hooke's law, but very many more values could be obtained. For this reason, values of X and the corresponding values of Y are often called **sample pairs**, and the assumption made when determining the significance of a coefficient of correlation is that sample pairs are part of a normal distribution of all sample pairs forming the population.

The population of all the possible values of X and Y has a theoretical coefficient of correlation, denoted by ρ, and by using the significance testing theory introduced in Chapter 26 an estimate of ρ can be made, based on the sample coefficient of correlation, r. It is usual to adopt a null hypothesis that there is no correlation between the population co-ordinates, i.e. $\rho = 0$. However, any null hypothesis can be adopted, say, $\rho = 0.7$ or $\rho = 1$, and the significance of the hypothesis can be tested. When the null hypothesis adopted is $\rho = 0$, then a "sample pairs value" of r may be tested by using Student's t-distribution tables and the expression shown below, based on the standard error for a t-distribution,

$$|t| = \frac{r\sqrt{(N-2)}}{\sqrt{(1-r^2)}} \tag{2}$$

where $|t|$ is the modulus or positive t-value, N is the number of co-ordinates, $N = 2$ is the number of degrees of freedom, v, and r is the sample pairs coefficient of correlation. The method used to determine the significance of an r value is shown in the worked problems below.

Worked problems on the significance of a coefficient of correlation

Problem 1. The correlation between hours of sunshine and the growth rate of a plant is being examined and a coefficient of correlation based on 20 sample pair values is calculated and found to be 0.50. Test the hypothesis that the coefficient of correlation of the population from which the sample pairs are drawn is zero at a level of significance of 0.01.

The null hypothesis is that the coefficient of correlation of the population is zero, i.e. $H_0 : \rho = 0$. The alternative hypothesis is that ρ is greater than zero, i.e. $H_1 : \rho > 0$.

The t-value is obtained by using equation (2), i.e.

$$|t| = \frac{r\sqrt{(N-2)}}{\sqrt{(1-r^2)}} = \frac{0.5\sqrt{(20-2)}}{\sqrt{(1-0.5^2)}}$$

i.e. $|t| = 2.45.$

Because the alternative hypothesis is that ρ is greater than zero, a one-tailed test is indicated. Using the table given on page 351, the $t_{0.99}$ v_{18} value is 2.55. The $|t|$ sample value is less than 2.55, hence the hypothesis that $\rho = 0$ is acceptable at a level of significance of 0.01.

Problem 2. The correlation between the value of goods produced by a company and the value of goods it exports is being determined. Find the minimum number of sample pairs at which the null hypothesis $H_0 : \rho = 0$ differs significantly from zero, for a coefficient of correlation of the sample pairs of 0.5 and a level of significance of 0.05.

The null hypothesis is $H_0 : \rho = 0$ and the alternative hypothesis is $H_1 : \rho > 0$. The t-value for the sample pairs is given by:

$$|t| = \frac{r\sqrt{(N-2)}}{\sqrt{(1-r^2)}}.$$

The value of N is determined in the following way. A first approximation of the value of $|t|$ is found by assuming that the number of degrees of freedom is infinite. This will usually give a value for N which is three or four less than the true value.

For an infinite value of the number of degrees of freedom, v and for $t_{.95}$, since the alternative hypothesis indicates a one-tailed test, the t-value is 1.645 (see page 351). Substituting this value in equation (2) gives:

$$1.645 = \frac{0.5\sqrt{(N-2)}}{\sqrt{(1-0.5^2)}} = \frac{0.5\sqrt{(N-2)}}{0.866}$$

i.e., $N = \dfrac{(1.645 \times 0.866)^2}{0.25} + 2$

$= 10.12.$

Thus the first approximation of the required value of N is about 10. For $N = 10$: the $t_{.95}$ v_8 value is 1.86 from the table on page 351, and

$$|t| = \frac{0.5\sqrt{8}}{0.866} = 1.63.$$

Since $|t| <$ the $t_{.95}$ v_8 value, H_0 is accepted.
For $N = 11$: the $t_{.95}$ v_9 value is 1.83 from the table on page 351, and

$$|t| = \frac{0.5\sqrt{9}}{0.866} = 1.73$$

Since $|t| <$ the $t_{.95}$ v_9 value, H_0 is accepted.

For $N = 12$: the $t_{.95}$ v_{10} value is 1.81 from the table on page 351, and

$$|t| = \frac{0.5\sqrt{10}}{0.866} = 1.83$$

Since $|t| >$ the $t_{.95}$ v_{10} value, H_0 is rejected.

Thus, the minimum number of sample pairs at which the null hypothesis differs significantly from zero is 12.

Further problems on the significance of a coefficient of correlation may be found in the following Section (4) (Problems 9 to 16), page 388.

4 Further problems

The product-moment formula for determining the linear correlation coefficient

1. The co-ordinates given below refer to an experiment to verify Newton's law of cooling over a limited range of values. Determine the value, degree and nature of the coefficient of correlation.

Time (min)	4	8	10	12	16	22
Temperature (°C)	46	34	30	26	24	20

   [−0.92, good, inverse]

2. The following results were obtained experimentally when verifying Hooke's law:

Load (N)	2	5	8	11	15
Extension (mm)	2	23	62	119	223

   Determine the value, degree and nature of the coefficient of correlation.
   [0.97, good, direct]

3. The thickness of case-hardening achieved varies with temperature and some co-ordinate obtained by experiment are as shown.

Temperature (°C)	400	420	350	320	400	480	440	370
Thickness (μm)	3.7	3.4	3.7	3.8	3.6	3.3	3.4	3.7

   Determine the coefficient of correlation based on these values.   [0.93]

4. The data given below refers to the relationship between man-hours worked and production achieved in a factory. Determine the coefficient of correlation.

Index of production man-hour basis	100	97	100	101	93	103	91	89	110	86
Index of production, actual basis	94	91	100	105	84	112	83	80	123	78

   [0.97]

5. The number of man-days lost per week due to sickness in two similar departments of a factory are as shown for a 12-week period.

   Department A: 20 18 19 21 17 18 12 16 14 17 13 15
   Department B: 18 21 18 20 17 19 16 15 15 18 16 18

   Determine the coefficient of correlation and comment on its degree and nature.   [0.70, fair, direct]

6. The masses and heights of ten people were measured and the results are as shown.

Mass (kg)   38   38   38   44   44   51   32   51   77   32
Height (cm) 135 140 137 141 147 145 132 149 164 130

Calculate the coefficient of correlation for this data.   [0.97]

7. The relationship between the pressure and volume of a gas was measured and the following results were obtained:

Pressure (kPa) 58   62   67   73   81   81   86   92   104
Volume (m^3) 0.36 0.97 0.43 0.52 0.48 0.29 0.31 0.75 0.27

Determine the coefficient of correlation and comment on the result obtained.   [−0.31. It is probable that the measurements were made at different temperatures.]

8. The caloric intake of rats varies with body mass as shown below.

Body mass (g)   2.0 3.1 3.6 4.6 5.0 6.0 7.0 8.0 8.5 9.0 10.0
Caloric intake
   (cal h^{-1})   1.5 2.1 3.2 3.6 3.6 3.9 4.1 4.2 4.5 4.6 5.9

Is there a linear correlation between these results?
                    [$r = 0.94$, hence there is a good, direct correlation]

The significance of a coefficient of correlation

9. The significance of the value of r is required for experimental values of the variation of the pressure of a gas with volume. The coefficient of correlation, r, of 12 sample pairs is found to have a value of −0.87. Determine if the hypothesis $\rho = 0$ is accepted, at a level of significance of 0.01.   [$t_{.099} v_{10} = 2.76$, $|t| = 5.58$, hypothesis is rejected, i.e. ρ differs significantly from zero]

10. Determine the coefficient of correlation for the data given below and test the null hypothesis that $\rho = 0$ at a level of significance of 0.1. The data given relates the number of hours of sunshine per week to the hours lost due to sickness.

Hours of sunshine/week: 10 13 15 17 18 20 22 23 24
Hours lost due to sickness: 90 75 75 65 55 45 55 45 35
[$r = -0.95$, $t_{.99} v_7 = 1.42$, $|t| = 8.05$, hypothesis is rejected]

11. Determine whether the data given in problem 2 is significant at a level of significance of 0.01.
[$r = 0.97$, $t_{.99} v_3 = 4.54$, $|t| = 6.91$, hence null hypothesis that $\rho = 0$ is rejected, i.e. ρ is significantly different from zero at a level of significance of 0.01.]

12. The amount of power taken by an electrical circuit as the applied voltage is varied is measured experimentally. Seventeen sample pairs have a correlation coefficient of −0.82. Test the significance of the sample against the null hypothesis that the coefficient of correlation of the

population from which the sample is drawn is zero, at a level of significance of 0.005.

$[t_{.995} \, v_{15} = 2.95, |t| = 5.55$, hence the hypothesis is rejected]

13. The coefficient of correlation of N sample pairs is 0.32. Determine the minimum value of N at which the null hypothesis, $H_0 : \rho = 0$ is significant at a level of significance of 0.05. $[N = 28]$

14. It is required to determine the modulus of elasticity of a sample of wire by finding values for the stress and strain at various loads. Determine the minimum number of readings for which the null hypothesis $H_0 : \rho = 0$ is significant at a level of significance of 0.01, if the coefficient of correlation of the sample pairs is –0.55. $[N = 18]$

15. The data below refers to the number of subjects taken in an examination and the average marks awarded.

No. of subjects	2	1	3	4	2	3	3	1	4	2
Average marks	62	35	60	71	44	65	59	45	76	50

Test the null hypothesis that the population has no correlation at a level of significance of 0.05.

$[r = 0.91, t_{.95} \, v_8 = 1.86, |t| = 6.20$, hence hypothesis is rejected]

16. The relationship between the percentage of tin in an alloy and the melting point of the alloy is as shown below:

Tin (%)	44.1	44.9	44.4	44.7	45.1	45.0	44.7	44.6
Melting point (°C)	513	512	511	510	513	514	521	514

Tin (%)	46.3	44.9	45.1	44.5	45.1	43.0	44.8	44.2
Melting point (°C)	526	525	522	521	513	537	513	519

Tin (%)	45.2	45.5
Melting point (°C)	512	514

Determine the coefficient of correlation correct to two decimal places. Test the null hypothesis that the percentage of tin and melting point are uncorrelated at a level of significance of 0.05.

$[r = -0.30, t_{.95} \, v_{16} = 1.75, |t| = 1.26$, hence hypothesis is rejected]

Chapter 28

Linear regression (ED 16)

1 Independent and dependent variables

In Chapter 27 it is shown that correlation is concerned with the degree or amount of association between variables. If a scatter diagram is drawn and the line which most nearly indicates the trend of the co-ordinates is then drawn, this line is called the **regression line**. A study of the techniques of regression enables us to draw the lines of best fit for a given set of co-ordinates and to represent the set of co-ordinates by a mathematical equation. Thus **regression analysis** (or just **regression**) is concerned with the nature of association between variables. If a law exists connecting the variables, the nature of the association is stated as a mathematical equation. The equation can then be used to predict values of one variable for given values of the other. The variable which is being predicted is called the dependent variable and is usually associated with the vertical axis of a graph. The variable being used as the basis of the prediction is called the **independent variable** and is usually associated with the horizontal axis of a graph. For example, when plotting the graph of $y = 2x^2$, one way is to assign x certain values, say, $-2, -1, 0, 1, 2$ and the x-values are then the independent variables. The corresponding values of y, that is, 8, 2, 0, 2 and 8, are the values of the dependent variables. When performing a regression analysis, it is necessary to decide which is the dependent and which is the independent variable, and the decision is normally based on a 'cause and effect' relationship. For example, when verifying Hooke's law experimentally, the applied force is the 'cause' and is the independent variable and the extension is the 'effect' and is the dependent variable.

2 Regression lines

One method of obtaining an equation (or regression model) associated with a regression analysis is to draw a line of 'best fit' on a scatter diagram. When a relationship exists between the co-ordinate on a scatter diagram, it can either be represented by the equation of a straight line, called a linear equation, or by the equation of a curve, called a curvilinear equation. At this stage only linear relationships between variables will be examined, giving a regression model of the form $Y = a_0 + a_1 X$. This 'statistical' equation is analogous to the 'algebraic' equation $y = mx + c$ and a_0 is the Y-axis intercept value when $X = 0$ and a_1 is the gradient of the line.

Several methods are commonly used to draw the line of best fit and amongst these are:

(i) the visual method, i.e. drawing the line which 'looks right', and
(ii) the 'as many points below the line as above it' method.

However, both these methods are 'hit and miss' methods, and in statistics the line of best fit is the one which makes the total deviation of the points from the line the smallest. This is achieved by finding what is called the **least-squares regression line.**

Let us assume that various co-ordinates, $(X_1, Y_1), (X_2, Y_2), \ldots, (X_N, Y_N)$, are given, where the X-values are the independent variables and the Y-values the dependent variables, and it is required to draw the line of best fit for this co-ordinate system. With reference to Fig. 1, the co-ordinates are plotted on a scatter diagram and any line LM is drawn in the general direction of the co-ordinates. If D_N is the vertical distance of co-ordinate (X_N, Y_N) from the line, then the line of best fit is defined as:

'that line which makes the value of $D_1^2 + D_2^2 + \ldots + D_N^2$ a minimum value'. This line is the least-squares regression line. Two other methods of defining the least-squares line can be used, these being:

(i) by taking the D-values as the perpendicular distances from the co-ordinates to the line, this method being rarely used, and
(ii) by taking the D-values as the horizontal distances from the co-ordinates to the line, which is used later in this chapter.

Let the equation of the least-squares line be of the form $Y = a_0 + a_1 X$. The values of a_0 and a_1 to make the sum of the squares of the D-values a minimum are found from two equations called the **normal equations** of the least-squares line.

The normal equations are:

$$\Sigma Y = a_0 N + a_1 \Sigma X \tag{1}$$
$$\Sigma XY = a_0 \Sigma X + a_1 \Sigma X^2, \tag{2}$$

where N is the number of co-ordinates. (These stated equations can be derived by applying the techniques of partial differentiation and the maximum and minimum theory to an equation representing the total deviation of the co-ordinates from the line.)

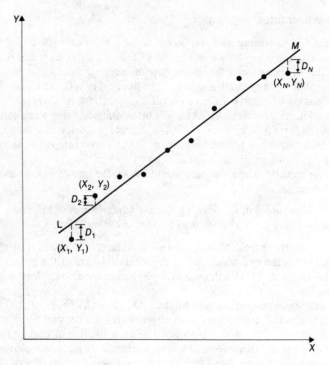

Fig. 1

The way in which these equations are used to solve a_0 and a_1 is shown below. Let X be the independent variable and let the co-ordinate values be:

X	1	3	5	7	9
Y	3	5	5	6	8

Usually, a tabular approach is used as shown in Table 1.

Table 1

1	2	3	4	5
X	Y	X^2	XY	Y^2
1	3	1	3	9
3	5	9	15	25
5	5	25	25	25
7	6	49	42	36
9	8	81	72	64
$\Sigma X = 25$	$\Sigma Y = 27$	$\Sigma X^2 = 165$	$\Sigma XY = 157$	$\Sigma Y^2 = 159$

Substituting for ΣY (the sum of column 2) and ΣX (the sum of column 1), in equation (1), where N is 5, gives:

$$27 = 5 a_0 + 25 a_1 \tag{3}$$

Substituting for ΣXY (the sum of column 4), ΣX and ΣX^2 (the sum of column 3), in equation (2), gives:

$$157 = 25 a_0 + 165 a_1 \tag{4}$$

Solving equations (3) and (4) simultaneously gives $a_0 = 2.65$ and $a_1 = 0.55$. Thus the equation of the least-squares or regression line is

$$Y = 2.65 + 0.55 X$$

and this is the line of best fit for the co-ordinate system given. The values of a_0 and a_1 are called the **regression coefficients**. The regression line can be represented graphically as shown in Fig. 2 by the continuous line and is called

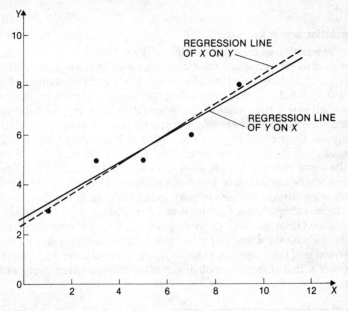

Fig. 2

the regression line of Y on X. It can be used to estimate the values of Y for given values of X.

If Y is the independent variable, a different regression line is obtained, based on the D-values being horizontal. In this case equations (1) and (2) have the X and Y values interchanged and the regression coefficients are given the symbols b_0 and b_1. Thus the normal equations become:

$$\Sigma X = b_0 N + b_1 \Sigma Y \tag{5}$$
$$\Sigma YX = b_0 \Sigma Y + b_1 \Sigma Y^2 \tag{6}$$

For the co-ordinate values given in this example, by using the values in the sum row of Table 1, equation (5) becomes:

$$25 = 5 b_0 + 27 b_1$$

and equation (6) becomes:

$$157 = 27 b_0 + 159 b_1.$$

Solving these equations simultaneously gives $b_0 = -4$ and $b_1 = 1.6\dot{6}$.
    Hence, $X = -4 + 1.6\dot{6}\ Y$.

When plotted on a graph, this equation represents the regression line of X on Y as shown by the broken line in Fig. 2. It can be used to estimate the values of X for given values of Y. The lines in Fig. 2 show that for any set of co-ordinate values, two regression lines can be drawn, depending on whether X is the independent or the dependent variable. The two regression lines cross at $(\bar{X}, \bar{Y})$, the mean values of X and Y respectively. In this case $\bar{X} = 5$ and $\bar{Y} = 5.4$.

Interpolation and extrapolation

When the equation of the regression line of Y on X has been determined, it can be used to determine any value of Y corresponding to a given value of X. If X is, say, 6.2, then since $Y = 2.65 + 0.55\ X$ in the above example, $Y = 2.65 + 0.55 \times 6.2 = 6.06$. When finding the value of a dependent variable corresponding to a given value of the independent variable lying **between** two values of the independent variable, the process is called **linear interpolation**. If the value of X corresponding to a given value of Y is required, say $Y = 10.3$, the equation of the regression line of X on Y is used. Because $X = -4 + 1.6\dot{6}$ Y in the above example, then $X = -4 + 1.6\dot{6} \times 10.3 = 13.1\dot{6}$. When finding the value of a dependent variable corresponding to a given value of independent variable lying **outside** of values of given independent variables, the process is called **linear extrapolation**. Care must be taken when making estimates by extrapolation since although the co-ordinates may have a 'good fit' to a straight line within the range of values given, the relationship may be non-linear outside of this range. For example, an exponential curve is approximately linear over a limited range, but obviously it is not linear over a fairly wide range.

Worked problems on regression lines

Problem 1. The data given below refers to the results obtained during a tensile test on a steel specimen. Assuming the regression of the extension on the force to be linear, determine the equation of the regression line of extension on force.

Tensile force (kN)	4.5	9.0	13.0	18.0	22.0	27.0
Extension (mm)	3.8	8.9	10.4	16.0	19.6	21.3

The force is the independent variable, X, and the extension is the dependent variable, Y, thus the equation of the regression line of Y on X is of the form

$Y = a_0 + a_1 X.$

The normal equations are:

$$\Sigma Y = a_0 N + a_1 \Sigma X \tag{1}$$
$$\text{and } \Sigma XY = a_0 \Sigma X + a_1 \Sigma X^2 \tag{2}$$

Using a tabular approach, Table 2 is constructed.

Table 2

Force, X	Extension, Y	X^2	XY	Y^2
4.5	3.8	20.25	17.1	14.44
9.0	8.9	81	80.1	79.21
13.0	10.4	169	135.2	108.16
18.0	16.0	324	288.0	256.00
22.0	19.6	484	431.2	384.16
27.0	21.3	729	575.1	453.69

$\Sigma X = 93.5$ $\Sigma Y = 80.00$ $\Sigma X^2 = 1\,807.25$ $\Sigma XY = 1\,526.7$ $\Sigma Y^2 = 1\,295.66$

The number of co-ordinates, N is 6. Substituting for ΣY, ΣX and N in equation (1) gives:

$$80.0 = 6\,a_0 + 93.5\,a_1 \tag{3}$$

Substituting for ΣXY, ΣX and ΣX^2 in equation (2) gives:

$$1\,526.7 = 93.5\,a_0 + 1807.25\,a_1 \tag{4}$$

Solving equations (3) and (4) simultaneously for a_0 and a_1 gives $a_0 = 0.87$ and $a_1 = 0.80$, correct to two decimal places. Thus the regression line of extension on tensile force is of the form

$Y = 0.87 + 0.80\,X.$

Problem 2. For the data given in worked problem 1 above, determine the equation of the regression line of force on extension, assuming the regression line is linear.

In this case, the extension is taken as the independent variable and the force as the dependent variable. The equation of the regression line of X on Y is of the form $X = b_0 + b_1 Y$. The normal equations are:

$$\Sigma X = b_0 N + b_1 \Sigma Y \tag{5}$$
$$\text{and } \Sigma XY = b_0 \Sigma Y + b_1 \Sigma Y^2 \tag{6}$$

Using the data calculated in Table 2, and substituting in equations (5) and (6) gives:

$$93.5 = 6 b_0 + 80.0 b_1 \qquad (7)$$
$$1\,526.7 = 80.0 b_0 + 1\,295.66 b_1 \qquad (8)$$

Solving equations (7) and (8) simultaneously gives $b_0 = -0.72$ and $b_1 = 1.22$, correct to two decimal places. Hence the regression line of force on extension is

$$X = -0.72 + 1.22\, Y.$$

Problem 3. Plot the scatter diagram for the co-ordinates given in Problem 1. Draw the regression lines obtained in Problems 1 and 2 and estimate

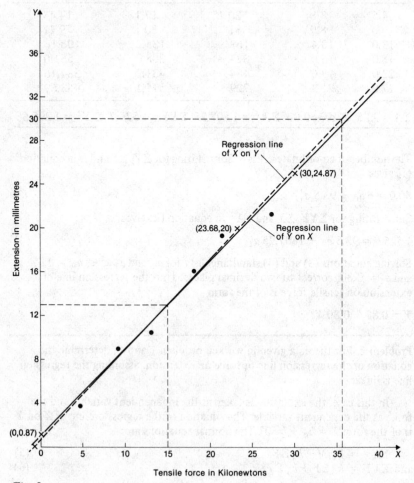

Fig. 3

(a) the value of extension when the force is 15 kN, and
(b) the value of force when the extension is 30 mm, assuming linearity extends beyond the range of values given.

By selecting any two values for X, say $X = 0$ and $X = 30$, the corresponding values of Y are calculated for the regression line of problem 1 which is $Y = 0.87 + 0.80 X$. This gives co-ordinates $(0, 0.87)$ and $(30, 24.87)$. The straight line passing through these two co-ordinate values gives the regression line of Y on X shown by the continuous line in Fig. 3. Similarly, by selecting any two values for Y, say $Y = 0$ and $Y = 20$, the corresponding values of X are calculated for the regression line of problem 2 which is $X = -0.72 + 1.22 Y$. This gives co-ordinates $(-0.72, 0)$ and $(23.68, 20)$. The straight line passing through these two co-ordinate values gives the regression line of X on Y shown as a broken line in Fig. 3.

(a) The value of extension when the force is 15 kN can either be determined from the graph in Fig. 3, using the regression line of Y on X, or by calculation, using the equation of the regression line of Y on X. From the graph, when the force is 15 kN, the extension is approximately 13 mm. Using the equation of the regression line of Y on X, when $X = 15$ kN,
$Y = 0.87 + 0.80 \times 15 = \textbf{12.87 mm}$.

(b) The value of force when the extension is 30 mm can either be determined from the graph in Fig. 3, using the regression line of X on Y or by calculation, using the equation of the regression line of X on Y derived in Problem 2 above. From the graph, when the extension is 30 mm, the force is approximately **35.6 kN**. Using the equation of the regression line of X on Y, when
$Y = 30$ mm,
$X = -0.72 + 1.22 \times 30 = \textbf{35.88 kN}$.

Further problems on regression lines may be found in the following Section (3) (Problems 1 to 10).

3 Further problems

Regression lines

1. The fuel consumption in cubic decimetres per hour, X, for an engine varies with the speed of the vehicle, Y, in kilometres per hour. Determine the regression coefficients for the regression line of Y on X for the values given below, correct to three significant figures.

X	10	30	40	60	80	90	110	140
Y	10	20	40	40	50	70	80	90

$$[a_0 = 5.45, a_1 = 0.636]$$

2. Determine the equation of the regression line of X on Y for the values given in Problem 1 above, expressing the regression coefficients correct to two decimal places. $\qquad [X = -5.00 + 1.50 Y]$

3. Draw a graph showing the regression lines of Y on X and of X on Y for the values given in Problem 1. Estimate the value of (a) Y when $X = 120$, and (b) X when $Y = 30$. (a) $[Y = 82]$ (b) $[X = 40]$

4. The data given below is the relationship between the heights and masses of ten people.

Height, X cm	175	180	193	165	187	171	198	168	184	177
Mass, Y kg	82	78	86	72	91	80	95	72	89	74

Determine the equation of the regression line of mass on height, expressing the regression coefficients correct to two decimal places.

$$[Y = -36.83 + 0.66 X]$$

5. Determine the regression coefficients, correct to two decimal places, for the regression line of height on mass for the data given in Problem 4 above. $[b_0 = 85.75, b_1 = 1.15]$

6. Draw a graph showing the regression lines of mass on height and of height on mass for the data given in Problem 4 above. Estimate the value of mass when the height is 175 cm and the height when the mass is 95 kg.

$$[79 \text{ kg}, 195 \text{ cm}]$$

7. The relationship between hardening temperature (X) and the percentage of carbon in steel, (Y), is as shown.

Carbon (%)	0.25	0.35	0.45	0.60	0.75	0.90	1.05
Hardening temperature (°C)	880	870	850	820	800	780	760

Calculate the regression equation for the percentage of carbon on hardening temperature, correct to two decimal places. $[Y = 918.64 - 154.14 X]$

8. The temperature at which a cadmium–bismuth alloy changes from a completely uniform liquid to solidified cadmium plus liquid varies with the percentage of cadmium in the alloy. The following results are obtained when determining part of the thermal equilibrium diagram relating the state of the alloy.

Cadmium (%) (X)	100	90	80	70	60	50	40
Temperature (°C) (Y)	321	295	265	230	205	160	140

Determine the equation of the regression line of percentage cadmium on temperature, correct to two decimal places.

$$[X = -3.69 + 0.32 Y]$$

9. The depth of case hardening varies with time of treatment at constant temperature. When 'nitriding' standard nitralloy steel at 500°C, the following results are obtained.

Time of treatment (Hours) (X)	10	20	30	40	50	60
Depth of case (mm) (Y)	0.14	0.26	0.34	0.45	0.55	0.62

Determine the values of the regression coefficients for the regression line of depth of case on time, correct to two significant figures.

$$[a_0 = 0.055, a_1 = 0.009 7]$$

10. The power needed to drive a lathe increases as the cutting angle of the tool increases when cutting at constant speed and depth of cut. The relationship for mild steel is:

Cutting angle (degrees) (X)	50	55	60	65	70	75	80	85	90
Power (kW) (Y)	6.2	6.8	7.6	8.2	8.1	8.8	9.7	10.0	10.4

Determine: (a) the equation of the regression line of power on cutting angle, and (b) the equation of the regression line of cutting angle on power, expressing the regression coefficients correct to three significant figures in each case.

(a) $[Y = 1.14 + 0.104\,X]$ (b) $[X = -9.27 + 9.41\,Y]$

Chapter 29

Chi-square tests for fitting theoretical distributions (EE 16)

1 Chi-square values

The significance tests introduced in Chapter 26 rely very largely on the normal distribution. For large sample numbers where z-values are used, the mean of the samples and the standard error of the means of the samples are assumed to be normally distributed (central limit theorem). For small sample numbers where t-values are used, the population from which samples are taken should be approximately normally distributed for the t-values to be meaningful. Chi-square tests, (pronounced KY and denoted by the Greek letter χ), which are introduced in this chapter, do not rely on the population or a sampling statistic such as the mean or standard error of the means being normally distributed. Significance tests based on z- and t-values are concerned with the parameters of a distribution, such as the mean and the standard deviation, whereas chi-square tests are concerned with the individual members of a set and are associated with **non-parametric tests**.

Observed and expected frequencies

The results obtained from trials are rarely exactly the same as the results predicted by statistical theories. For example, if a coin is tossed 100 times, it is unlikely that the result will be exactly 50 heads and 50 tails. Let us assume that, say, 5 people each toss a coin 100 times and note the number of, say, heads obtained. Let the results obtained be as shown below.

Person	A	B	C	D	E
Observed frequency	43	54	60	48	57
Expected frequency	50	50	50	50	50

A measure of the discrepancy existing between the observed frequencies shown in row 2 and the expected frequencies shown in row 3 can be determined by calculating the chi-square value. The chi-square value is defined as follows:

$$\chi^2 = \Sigma \left\{ \frac{(o-e)^2}{e} \right\},$$

where o and e are the observed and expected frequencies respectively. The χ^2-value for the data given above may be calculated by using a tabular approach as shown below.

Person	Observed frequency, o	Expected frequency, e	$o-e$	$(o-e)^2$	$\dfrac{(o-e)^2}{e}$
A	43	50	−7	49	0.98
B	54	50	4	16	0.32
C	60	50	10	100	2.00
D	48	50	−2	4	0.08
E	57	50	7	49	0.98
				$\chi^2 = \left\{ \dfrac{(o-e)^2}{e} \right\} =$	4.36

If the value of χ^2 is zero, then the observed and expected frequencies agree exactly. The greater the difference between the χ^2-value and zero, the greater the discrepancy between the observed and expected frequencies. χ^2-values are used in Section 2 following and the significance of the value obtained is also discussed in that section.

Problems on determining chi-square values may be found in Section 3 (Problems 1 to 3), page 409.

2 Fitting data to theoretical distributions

For theoretical distributions such as the binomial, Poisson and normal distributions, expected frequencies can be calculated. For example, from the theory of the binomial distribution, the probability of having $0, 1, 2, \ldots, n$ defective items in a sample of n items can be determined from the successive terms of $(q+p)^n$, where p is the defect rate and $q = 1 - p$. These probabilities can be used to determine the expected frequencies of having $0, 1, 2, \ldots, n$ defective items. As a result of counting the number of defective items when sampling, the observed frequencies are obtained. The expected and observed frequencies can be compared by means of a chi-square test and predictions can be made as to whether the differences are due to random errors, due to some fault in the method of sampling, or due to the assumptions made.

Table 1 Percentile values (χ_p^2) for the Chi-square distribution with v degrees of freedom.

v	$\chi_{0.995}^2$	$\chi_{0.99}^2$	$\chi_{0.975}^2$	$\chi_{0.95}^2$	$\chi_{0.90}^2$	$\chi_{0.75}^2$	$\chi_{0.50}^2$	$\chi_{0.25}^2$
1	7.88	6.63	5.02	3.84	2.71	1.32	0.455	0.102
2	10.6	9.21	7.38	5.99	4.61	2.77	1.39	0.575
3	12.8	11.3	9.35	7.81	6.25	4.11	2.37	1.21
4	14.9	13.3	11.1	9.49	7.78	5.39	3.36	1.92
5	16.7	15.1	12.8	11.1	9.24	6.63	4.35	2.67
6	18.5	16.8	14.4	12.6	10.6	7.84	5.35	3.45
7	20.3	18.5	16.0	14.1	12.0	9.04	6.35	4.25
8	22.0	20.1	17.5	15.5	13.4	10.2	7.34	5.07
9	23.6	21.7	19.0	16.9	14.7	11.4	8.34	5.90
10	25.2	23.2	20.5	18.3	16.0	12.5	9.34	6.74
11	26.8	24.7	21.9	19.7	17.3	13.7	10.3	7.58
12	28.3	26.2	23.3	21.0	18.5	14.8	11.3	8.44
13	29.8	27.7	24.7	22.4	19.8	16.0	12.3	9.30
14	31.3	29.1	26.1	23.7	21.1	17.1	13.3	10.2
15	32.8	30.6	27.5	25.0	22.3	18.2	14.3	11.0
16	34.3	32.0	28.8	26.3	23.5	19.4	15.3	11.9
17	35.7	33.4	30.2	27.6	24.8	20.5	16.3	12.8
18	37.2	34.8	31.5	28.9	26.0	21.6	17.3	13.7
19	38.6	36.2	32.9	30.1	27.2	22.7	18.3	14.6
20	40.0	37.6	34.4	31.4	28.4	23.8	19.3	15.5
21	41.4	38.9	35.5	32.7	29.6	24.9	20.3	16.3
22	42.8	40.3	36.8	33.9	30.8	26.0	21.3	17.2
23	44.2	41.6	38.1	35.2	32.0	27.1	22.3	18.1
24	45.6	43.0	39.4	36.4	33.2	28.2	23.3	19.0
25	46.9	44.3	40.6	37.7	34.4	29.3	24.3	19.9
26	48.3	45.9	41.9	38.9	35.6	30.4	25.3	20.8
27	49.6	47.0	43.2	40.1	36.7	31.5	26.3	21.7
28	51.0	48.3	44.5	41.3	37.9	32.6	27.3	22.7
29	52.3	49.6	45.7	42.6	39.1	33.7	28.3	23.6
30	53.7	50.9	47.7	43.8	40.3	34.8	29.3	24.5
40	66.8	63.7	59.3	55.8	51.8	45.6	39.3	33.7
50	79.5	76.2	71.4	67.5	63.2	56.3	49.3	42.9
60	92.0	88.4	83.3	79.1	74.4	67.0	59.3	52.3
70	104.2	100.4	95.0	90.5	85.5	77.6	69.3	61.7
80	116.3	112.3	106.6	101.9	96.6	88.1	79.3	71.1
90	128.3	124.1	118.1	113.1	107.6	98.6	89.3	80.6
100	140.2	135.8	129.6	124.3	118.5	109.1	99.3	90.1

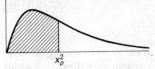

$\chi^2_{0.10}$	$\chi^2_{0.05}$	$\chi^2_{0.025}$	$\chi^2_{0.01}$	$\chi^2_{0.005}$
0.0158	0.0039	0.0010	0.0002	0.0000
0.211	0.103	0.0506	0.0201	0.0100
0.584	0.352	0.216	0.115	0.072
1.06	0.711	0.484	0.297	0.207
1.61	1.15	0.831	0.554	0.412
2.20	1.64	1.24	0.872	0.676
2.83	2.17	1.69	1.24	0.989
3.49	2.73	2.18	1.65	1.34
4.17	3.33	2.70	2.09	1.73
4.87	3.94	3.25	2.56	2.16
5.58	4.57	3.82	3.05	2.60
6.30	5.23	4.40	3.57	3.07
7.04	5.89	5.01	4.11	3.57
7.79	6.57	5.63	4.66	4.07
8.55	7.26	6.26	5.23	4.60
9.31	7.96	6.91	5.81	5.14
10.1	8.67	7.56	6.41	5.70
10.9	9.39	8.23	7.01	6.26
11.7	10.1	8.91	7.63	6.84
12.4	10.9	9.59	8.26	7.43
13.2	11.6	10.3	8.90	8.03
14.0	12.3	11.0	9.54	8.64
14.8	13.1	11.7	10.2	9.26
15.7	13.8	12.4	10.9	9.89
16.5	14.6	13.1	11.5	10.5
17.3	15.4	13.8	12.2	11.2
18.1	16.2	14.6	12.9	11.8
18.9	16.9	15.3	13.6	12.5
19.8	17.7	16.0	14.3	13.1
20.6	18.5	16.8	15.0	13.8
29.1	26.5	24.4	22.2	20.7
37.7	34.8	32.4	29.7	28.0
46.5	43.2	40.5	37.5	35.5
55.3	51.7	48.8	45.4	43.3
64.3	60.4	57.2	53.5	51.2
73.3	69.1	65.6	61.8	59.2
82.4	77.9	74.2	70.1	67.3

As for normal and t-distributions, a table is available for relating various calculated values of χ^2 to those likely because of random variations, at various levels of confidence. Such a table is shown in Table 1. In table 1, the column on the left denotes the number of degrees of freedom, v, and when the χ^2-values refer to fitting data to theoretical distributions, the number of degrees of freedom are usually $(N - 1)$, where N is the number of rows in the table from which χ^2 is calculated. However, when the population parameters such as the mean and standard deviation are based on sample data, the number of degrees of freedom are given by $v = N - 1 - M$, where M is the number of estimated population parameters. An application of this is shown in Worked Problem 3.

The columns on the far left of the table headed $\chi^2_{.995}$, $\chi^2_{.99}$, ... give the percentile of χ^2-values corresponding to levels of confidence of 99.5 per cent, 99 per cent, ... (i.e. levels of significance of 0.005, 0.01, ...). On the far right of the table, the columns headed ..., $\chi^2_{0.01}$, $\chi^2_{0.005}$ also correspond to levels of confidence of ... 99 per cent, 99.5 per cent and are used to predict the 'too good to be true' type results, where the fit obtained is so good that the method of sampling must be suspect.

The method in which χ^2-values are used to test the goodness of fit of data to probability distributions is shown in the worked problems following.

Worked problems on fitting data to theoretical distributions

Problem 1. As a result of a survey carried out of 200 families, each with five children, the distribution shown below was produced. Test the null hypothesis that the observed frequencies are consistent with male and female births being equally probable, assuming a binomial distribution, a level of significance of 0.05 and a 'too good to be true' fit at a confidence level of 95 per cent.

Number of boys (B) and girls (G)	5B, 0G	4B, 1G	3B, 2G	2B, 3G	1B, 4G	0B, 5G
Number of families	11	35	69	55	25	5

To determine the expected frequencies

Using the usual binomial distribution symbols, let p be the probability of a male birth and $q = 1 - p$ be the probability of a female birth. The probabilities of having 5 boys, 4 boys, ... 0 boys are given by the successive terms of the expansion of $(q + p)^n$. Since there are 5 children in each family, $n = 5$, and

$$(q + p)^5 = q^5 + 5q^4 p + 10q^3 p^2 + 10q^2 p^3 + 5qp^4 + p^5.$$

When $q = p = 0.5$, the probabilities of 5 boys, 4 boys, ... 0 boys are

0.031 25, 0.156 25, 0.312 5, 0.312 5, 0.156 25 and 0.031 25.

For 200 families, the expected frequencies, rounded off to the nearest whole number are.

6, 31, 63, 63, 31 and 6 respectively.

To determine the χ^2 value

Using a tabular approach, the χ^2-value is calculated using

$$\chi^2 = \Sigma \left\{ \frac{(o - e)^2}{e} \right\}$$

Number of boys (B) and girls (G)	Observed frequency, o	Expected frequency, e	$o - e$	$(o - e)^2$	$\dfrac{(o - e)^2}{e}$
5B, 0G	11	6	5	25	4.167
4B, 1G	35	31	4	16	0.516
3B, 2G	69	63	6	36	0.571
2B, 3G	55	63	-8	64	1.016
1B, 4G	25	31	-6	36	1.161
0B, 5G	5	6	-1	1	0.167

$$\chi^2 = \Sigma \left\{ \frac{(o - e)^2}{e} \right\} = 7.598$$

To test the significance of the χ^2-value

The number of degrees of freedom are given by $v = N - 1$ where N is the number of rows in the table above, thus $v = 5$. For a level of significance of 0.05, the confidence level is 95 per cent, i.e. 0.95 per unit. From Table 1, for $\chi^2_{.95}$ $v = 5$ the percentile value χ^2_p is 11.1. Because the calculated value of χ^2 is less than χ^2_p, **the null hypothesis that the observed frequencies are consistent with male and female births being equally probable is accepted.**

For a confidence level of 95 per cent, the $\chi^2_{.05}$ $v = 5$ value from Table 1 is 1.15 and because the calculated value of χ^2 is greater than this value of χ^2, **the fit is not so good as to be unbelievable.**

Problem 2. The deposition of grit particles from the atmosphere is measured by counting the number of particles on 200 prepared cards in a specified time. The following distribution was obtained.

Number of particles	0	1	2	3	4	5	6
Number of cards	41	69	44	27	12	6	1

Test the null hypothesis that the deposition of grit particles is according to a Poisson distribution at a level of significance of 0.01 and determine if the data is 'too good to be true' at a confidence level of 99 per cent.

To determine the expected frequency

The expectation or average occurrence is λ

$$= \frac{\text{total number of particles deposited}}{\text{total number of cards}}$$

$$= \frac{69 + 88 + 81 + 48 + 30 + 6}{200}$$

$$= 1.61.$$

The expected frequencies are calculated using a Poisson distribution, where the probabilities of there being $0, 1, 2, \ldots, 6$ particles deposited are given by the successive terms of the

$$e^{-\lambda} \left(1 + \lambda + \frac{\lambda^2}{2!} + \frac{\lambda^3}{3!} + \ldots \right),$$

taken from left to right, i.e. $e^{-\lambda}$, $\lambda e^{-\lambda}$, $\dfrac{\lambda^2 e^{-\lambda}}{2!}$, $\dfrac{\lambda^3 e^{-\lambda}}{3!}$, $\ldots$

Calculating these terms for a value λ of 1.61 gives:

Number of particles deposited	0	1	2	3	4	5	6
Probability	0.199 9	0.321 8	0.259 1	0.139 0	0.056 0	0.018 0	0.004 8
Expected frequency	40	64	52	28	11	4	1

To determine the χ^2-value

The χ^2-value is calculated, using a tabular method as shown below.

Number of grit particles	Observed frequency, o	Expected frequency, e	$o - e$	$(o - e)^2$	$\dfrac{(o - e)^2}{e}$
0	41	40	1	1	0.025 0
1	69	64	5	25	0.390 6
2	44	52	−8	64	1.230 8
3	27	28	−1	1	0.035 7
4	12	11	1	1	0.090 9
5	6	4	2	4	1.000 0
6	1	1	0	0	0.000 0

$$\chi^2 = \Sigma \left\{ \frac{(o - e)^2}{e} \right\} = 2.773$$

To test the significance of the χ^2-value

The number of degrees of freedom are $v = N - 1$, where N is the number of rows in the table above, giving $v = 6$. The percentile value of χ^2 is determined from Table 1, for $(\chi^2{}_{0.99}, v = 6)$, and is 16.8. Because the calculated value of χ^2 is smaller than the percentile value, **the hypothesis that the grit deposition is according to a Poisson distribution is accepted.**

For a confidence level of 99 per cent, the $(\chi^2{}_{0.01}, v = 6)$ value is obtained from Table 1, and is 0.872. Because the calculated value of χ^2 is greater than this value, **the fit is not 'too good to be true'.**

Problem 3. The diameter of a sample of 500 rivets produced by an automatic process have the following size distributions.

Diameter (mm)	4.011	4.015	4.019	4.023	4.027	4.031	4.035
Frequency	12	47	86	123	107	97	28

Test the null hypothesis that the diameters of the rivets are normally distributed at a level of significance of 0.05 and also determine if the distribution gives a 'too good' fit at a level of confidence of 90%.

To determine the expected frequencies

In order to determine the expected frequencies, the mean and standard deviation of the distribution is required. These population parameters, μ and σ, are based on sample data, $\bar{x}$ and s, and an allowance is made in the number of degrees of freedom used for estimating the population parameters from sample data.

The sample mean, $\bar{x} = \dfrac{12 \times 4.011 + 47 \times 4.015 + \ldots + 28 \times 4.035}{500}$

$$= \frac{2\,012.176}{500} = 4.024$$

The sample standard deviation, s is given by:

$$s = \sqrt{\left[\frac{12(4.011-4.024)^2 + 47(4.015-4.024)^2 + \ldots + 28(4.035-4.024)^2}{500} \right]}$$

$$= \sqrt{\frac{0.017\,212}{500}} = 0.005\,87$$

The class boundaries for the diameters are 4.009 to 4.013, 4.013 to 4.017, and so on, and are shown in column 2 of table 2. Using the theory of the normal probability distribution, the probability for each class and hence the expected frequency is calculated as shown in table 2.

In column 3, the z-values corresponding to the class boundaries are determined using $z = \dfrac{x - \bar{x}}{s}$, which in this case is $z = \dfrac{x - 4.024}{0.005\,87}$. The area between a z-value in column 3 and the mean of the distribution at $z = 0$ is determined using the table of partial areas under the standardised normal distribution curve given on page 350, and is shown in column 4. By subtracting the area between the mean and the z-value of the lower class boundary from that of the upper class boundary, the area, and hence the probability of a particular class is obtained, and is shown in column 5. There is one exception in column 5, corresponding to class boundaries of 4.021 and 4.025, where the areas are added to give the probability of the 4.023 class. This is because these areas lie immediately to the left and right of the mean value. Column 6 is obtained by multiplying the probabilities in column 5 by the sample number, 500. The sum of column 6 is not equal to 500 because the area under the standardised normal curve for z-values of less than -2.56 and more than 2.21 are neglected. The error introduced by doing this is 10 in 500, i.e. 2 per cent, and is acceptable in most problems of this type. If it is not acceptable, each expected frequency can be increased by the percentage error.

Table 2

1 Class mid-point	2 Class boundaries, x	3 z-value for class boundary	4 Area from 0 to z	5 Area for class	6 Expected frequency
	4.009	−2.56	0.494 8		
4.011				0.025 5	13
	4.013	−1.87	0.469 3		
4.015				0.086 3	43
	4.017	−1.19	0.383 0		
4.019				0.188 0	94
	4.021	−0.51	0.195 0		
4.023			Add	0.262 8	131
	4.025	0.17	0.067 8		
4.027				0.234 5	117
	4.029	0.85	0.302 3		
4.031				0.134 7	67
	4.033	1.53	0.437 0		
4.035				0.049 4	25
	4.037	2.21	0.486 4		
				Total:	490

To determine the χ^2-value

The χ^2-value is calculated, using a tabular method as shown below.

Diameter of rivets	Observed frequency, o	Expected frequency, e	$o-e$	$(o-e)^2$	$\dfrac{(o-e)^2}{e}$
4.011	12	13	−1	1	0.076 9
4.015	47	43	4	16	0.372 1
4.019	86	94	−8	64	0.680 9
4.023	123	131	−8	64	0.488 5
4.027	107	117	−10	100	0.854 7
4.031	97	67	30	900	13.432 8
4.035	28	25	3	9	0.360 0

$$\chi^2 = \Sigma \left\{ \frac{(o-e)^2}{e} \right\} = 16.265\ 9$$

To test the significance of the χ^2-value

The number of degrees of freedom is given by $N-1-M$, where M is the number of estimated parameters in the population. Both the mean and the standard deviation of the population are based on the sample value, $M = 2$, hence $v = 7 - 1 - 2 = 4$. From Table 1, the $\chi^2{}_p$-value corresponding to $\chi^2{}_{0.95}$ and v_4 is 9.49. **Hence the null hypothesis that the diameters of the**

rivets are normally distributed is rejected. For $\chi^2_{0.10}$, ν_4, the χ^2_p-value is 1.06, **hence the fit is not 'too good'.** Because the null hypothesis is rejected, the second significance test need not be carried out.

Further problems on fitting data to theoretical distributions may be found in Section (3) following (Problems 4 to 10).

3 Further problems

Chi-square values

1. Determine the χ^2-value for the data given below.

Observed frequency	5	23	32	27	12	3	1	
Expected frequency	6	20	30	25	14	5	1	[2.00]

2. A dice is rolled 240 times and the observed and expected frequencies are as shown.

Face	1	2	3	4	5	6
Observed frequency	49	35	32	46	49	29
Expected frequency	40	40	40	40	40	40

Determine the χ^2-value for this distribution.   [10.2]

3. The number of telephone calls received by the switchboard of a company in 200 five-minute intervals are shown in the distribution below.

Number of calls	0	1	2	3	4	5	6	7
Observed frequency	11	44	53	46	24	12	7	3
Expected frequency	16	42	52	42	26	14	6	2

Calculate the χ^2-value for this data.   [3.16]

Fitting data to theoretical distributions.

4. Test the null hypothesis that the observed data given below fits a binomial distribution of the form $250(0.6 + 0.4)^7$ at a level of significance of 0.05.

Observed frequency: 8   27   62   79   45   24   5   0

Is the fit of the data 'too good' at a level of confidence of 90 per cent?

[Expected frequencies: 7, 33, 65, 73, 48, 19, 4, 0; χ^2-value $= 3.62$; $\chi^2_{0.95}$, $\nu_7 = 14.1$, hence hypothesis accepted. $\chi^2_{0.10}$, $\nu_7 = 2.83$, hence data is not 'too good']

5. The data given below refers to the number of people injured in a city by accidents for weekly periods throughout a year. It is believed that the data fits a Poisson distribution. Test the goodness of fit at a level of significance of 0.05.

Number of people injured in the week	0	1	2	3	4	5	6
Number of weeks	5	12	13	9	7	4	2

$[\lambda = 2.788\,5$; expected frequencies: $5, 11, 14, 11, 7, 3, 1$. χ^2-value $= 1.86$; $\chi^2_{0.95}, v_6 = 12.6$, hence the data fits a Poisson distribution at a level of significance of 0.05]

6. The resistance of a sample of carbon resistors are as shown below.

Resistance (MΩ)	1.28	1.29	1.30	1.31	1.32	1.33	1.34	1.35	1.36
Frequency	7	19	41	50	73	52	28	17	9

Test the null hypothesis that this data corresponds to a normal distribution at a level of significance of 0.05.

$[\bar{x} = 1.32, s = 0.018\,0$; expected frequencies, $6, 17, 36, 55, 65, 55, 36, 17, 6$; χ^2-value $= 5.98$; $\chi^2_{0.95}, v_6 = 12.6$, hence the null hypothesis is accepted, i.e. the data does correspond to a normal distribution]

7. Four coins are tossed 200 times and the observed number on which 0, 1, 2, 3 and 4 heads are obtained are 16, 34, 102, 36 and 12 respectively. Test the null hypothesis that the four coins are unbiased at a level of significance of 0.05.

[If coins are unbiased, the data is a binomial distribution with $q = p = 0.5$. Expected frequencies 13, 50, 75, 50, 13; $\chi^2 = 19.5$; $\chi^2_{0.95}, v_4 = 9.49$, hence null hypothesis rejected, i.e. the coins are biased]

8. The quality assurance department of a firm selects 250 capacitors at random from a large quantity of them and carries out various tests on them. The results obtained are as follows:

Number of tests failed	0	1	2	3	4	5	6 and over
Number of capacitors	113	77	39	16	4	1	0

Test the goodness of fit of this distribution to a Poisson distribution at a level of significance of 0.05.

$[\lambda = 0.896$; expected frequencies are $102, 91, 41, 12, 3, 1, 0$; χ^2-value $= 5.10$. $\chi^2_{0.95}, v_6 = 12.6$, hence this data fits a Poisson distribution at a level of significance of 0.05]

9. Test the hypothesis that the maximum load before breaking supported by certain cables produced by a company follows a normal distribution at a level of significance of 0.05, based on the experimental data given below. Also test to see if the data is 'too good' at a level of significance of 0.05.

Maximum load (MN)	8.5	9.0	9.5	10.0	10.5	11.0	11.5	12.0
Number of cables	2	5	12	17	14	6	3	1

[$\bar{x} = 10.09$ MN; $\sigma = 0.733$ MN; expected frequencies, 2, 5, 12, 16, 14, 8, 3, 1; χ^2-value $= 0.563$; $\chi^2_{0.95}$, $v_5 = 11.1$. Hence hypothesis accepted. $\chi^2_{0.05}$, $v_5 = 1.15$, hence the results are 'too good to be true']

10. A sample of 100 electric lamps are tested and their lifetimes to failures are as shown below:

Lifetime (hours)	100–290	300–490	500–690	700–890	900–1090	1100–1290	1300–1490
Number of lamps	3	13	17	32	23	7	5

Test the null hypothesis that this distribution is a random sample from a normal population at a level of significance of 0.05.

[$\bar{x} = 795$, $s = 277.13$; expected frequencies 3, 10, 22, 28, 22, 10, 3; χ^2-value $= 4.89$; $\chi^2_{0.95}$, $v_4 = 9.49$, hence the hypothesis that the distribution fits a normal population is accepted]

Index